Recovery for Performance in Sport

**Institut National du Sport,
de l'Expertise et de la Performance
(INSEP)**

Christophe Hausswirth, PhD
INSEP

Iñigo Mujika, PhD
University of the Basque Country

Editors

Human Kinetics

Library of Congress Cataloging-in-Publication Data

Recovery for performance in sport / Christophe Hausswirth, Iñigo Mujika, editors; The National Institute of Sport for Expertise and Performance (INSEP).
    p. ; cm.
  Includes bibliographical references and index.
  I. Hausswirth, Christophe, 1968- II. Mujika, Iñigo. III. Institut national du sport et de l'éducation physique (France)
  [DNLM: 1. Athletic Performance--physiology. 2. Athletic Injuries--prevention & control. 3. Recovery of Function. QT 260]
  617'.03--dc23

2012037852

ISBN: 978-1-4504-3434-8

This book is a revised edition of *Récupération et Performance en Sport,* published in 2010 by INSEP/MAPI/Publications.

Reference herein to any specific commercial products, process, or service by trade name, trademark, manufacturer, or otherwise, does not necessarily constitute or imply its endorsement, recommendation, or favoring by Human Kinetics.

The web addresses cited in this text were current as of October 24, 2012, unless otherwise noted.

Acquisitions Editors: Karalynn Thomson and Amy N. Tocco; **Developmental Editor:** Judy Park; **Assistant Editors:** Brendan Shea, PhD, Kali Cox, Casey A. Gentis, Derek Campbell, and Erin Cler; **Copyeditor:** Joy Wotherspoon; **Indexer:** Bobbi Swanson; **Permissions Manager:** Dalene Reeder; **Graphic Designer:** Joe Buck; **Graphic Artist:** Tara Welsch; **Cover Designer:** Keith Blomberg; **Photograph (cover):** Courtesy of Jean-Romain Gautier; **Photo Asset Manager:** Laura Fitch; **Visual Production Assistant:** Joyce Brumfield; **Photo Production Manager:** Jason Allen; **Art Manager:** Kelly Hendren; **Associate Art Manager:** Alan L. Wilborn; **Illustrations:** © Human Kinetics, unless otherwise noted; **Printer:** Total Printing Systems

Printed in the United States of America   10 9 8 7 6 5 4 3 2

The paper in this book is certified under a sustainable forestry program.

Human Kinetics
Website: www.HumanKinetics.com

United States: Human Kinetics
P.O. Box 5076
Champaign, IL 61825-5076
800-747-4457
e-mail: info@hkusa.com

Canada: Human Kinetics
475 Devonshire Road Unit 100
Windsor, ON N8Y 2L5
800-465-7301 (in Canada only)
e-mail: info@hkcanada.com

Europe: Human Kinetics
107 Bradford Road
Stanningley
Leeds LS28 6AT, United Kingdom
+44 (0) 113 255 5665
e-mail: hk@hkeurope.com

For more information about Human Kinetics' coverage in other areas of the world,
please visit our website: www.HumanKinetics.com

E5770

WD500 REC

Contents

Contributors

François Bieuzen, PhD
National Institute of Sport, Expertise
and Performance (INSEP)
Research Department
Paris, France

Antoine Couturier, PhD
National Institute of Sport, Expertise
and Performance (INSEP)
Research Department
Paris, France

Anne Delextrat, PhD
Department of Sport and Health Sciences
Oxford Brookes University
Oxford, UK

Kevin De Pauw, MSc
University of Brussels
Department of Human Physiology & Sports
Medicine
Brussels, Belgium

Sylvain Dorel, PhD
National Institute of Sport, Expertise
and Performance (INSEP)
Research Department
Paris, France

Rob Duffield, PhD
School of Human Movement Studies
Charles Sturt University
Bathurst, Australia

Jean Fournier, PhD
National Institute of Sport, Expertise
and Performance (INSEP)
Research Department
Paris, France

Charles-Yannick Guézennec, PhD
Laboratory of Performance, Health and Altitude
University of Perpignan Via Domitia
Font-Romeu, France

Christophe Hausswirth, PhD
National Institute of Sport, Expertise
and Performance (INSEP)
Research Department
Paris, France

Michael Lambert, PhD
MRC/UCT Research Unit for Exercise Science
and Sports Medicine
Sports Science Institute of South Africa
Newlands, South Africa

Yann Le Meur, PhD
National Institute of Sport, Expertise
and Performance (INSEP)
Research Department
Paris, France

Nicolas Lemyre, PhD
Department of Coaching and Psychology
Norwegian School of Sport Sciences
Oslo, Norway

Frank E. Marino, PhD
School of Human Movement Studies
Charles Sturt University
Bathurst, Australia

Romain Meeusen, PhD
University of Brussels
Department of Human Physiology & Sports
Medicine
Brussels, Belgium

Iñigo Mujika, PhD
Department of Physiology
Faculty of Medicine and Odontology
University of the Basque Country
Leioa, Spain

Giuseppe Rabita, PhD
National Institute of Sport, Expertise
and Performance (INSEP)
Research Department
Paris, France

Melissa Skein, PhD
School of Human Movement Studies
Charles Sturt University
Bathurst, Australia

Preface

Physical training regularly exposes the body to workloads greater than those usually experienced in order to induce a significant modification of the functions involved in executing the task. However, for high-level athletes who do intensive training, severe workloads do not necessarily improve their performance. Indeed, a complex combination of central and peripheral responses can induce marked decreases in performance following intense training sessions. This performance decline will only be reversed when adequate recovery is provided. If recovery periods are insufficient or are not programmed in an appropriate way, the state of fatigue may become so severe that complete rest will be the only efficient remedy.

A state of acute or chronic fatigue needs to be recognized; its causes must be analyzed and appropriate periods of recovery must be strategically planned in combination with practical recovery modalities that are proportional to the training load and the level of fatigue induced. Signs of fatigue must therefore be heard and perceived both consistently and early on to adequately establish the athlete's recovery needs. This is an absolute necessity to induce training adaptation, yield an improved performance, and allow the athlete to repeat this performance multiple times.

How This Book Will Help You Practice Your Sport

Most athletes, coaches, and sport scientists are aware of the key role of recovery in improving performance in both training and competition, but very few are certain about the most suitable recovery strategies for their individual needs. This element is often managed by coaches, physiotherapists (physical therapists), and the athletes themselves. It is usually not programmed every day in an efficient or optimal way. When should recovery start? How long should it last? Will the athletes recover every day in the same way and at the same time? Because there are no simple answers to these questions, the recovery modalities are often planned and designed following experiences based on a trial-and-error approach. As a consequence, athletes and their support teams use a range of recovery strategies to optimize and enhance sport performance and to avoid overtraining. However, the most efficient methods to prepare athletes for elite-level competition are those based on proven scientific principles.

This book compiles for the first time the available scientific data on recovery. It presents its physiological and psychological effects, as well as how these effects relate to athletic performance and its circumstances. It also presents the experience-based practical knowledge of some of the world's most successful experts and the recovery methods currently used. A unique feature of this book is that it addresses the concerns of sport scientists and students, physicians and physical therapists, elite athletes and national coaches, and all those athletes and coaches in between who want to improve performance, prevent injury, and avoid overreaching. We have done this by providing a scientific base of information within a 17-chapter text. To this highly technical text, we have added several specific and practical elements to enable those who are not interested in the scientific details to understand the basis of what really happens on the field. These include case studies written by coaches or sport scientists involved in elite sports.

Parts I and II of the book describe the physiology of optimal training, how to prevent overtraining, and how to peak for optimal performance. Part III focuses on the best way to optimize recovery by using several recovery strategies, and lists their practical applications in each chapter. Part IV presents special situations for consideration. Each part includes relevant scientific investigations in thorough detail, both in the text and in figures and tables. All studies mentioned are fully referenced, so readers can explore the original studies for more details. However, the book also contains many features that are fully accessible to nonscientists, such as the case studies presented.

How This Book Is Organized

Recovery and Performance in Sport is divided into four major parts, which consistently describe scientific, practical applications and case studies according to recent literature.

Part I, Fundamentals of Fatigue and Recovery, contains two chapters that deal with the most relevant scientific aspects regarding the balance between training and recovery for optimal adaptation to training. The introduction sets the scene with an overview to the key concepts of fatigue and recovery and a brief historical review of the available research on this topic. Chapter 1 focuses on the physiology of training overload and adaptation, key principles of the training process, and the outcome of an adequate balance between training and recovery. Chapter 2 covers overtraining syndrome—the continuum from optimal training to fatigue and overtraining. Relevant information is provided regarding the balance between training and recovery, functional and nonfunctional overreaching, and the prevention and detection of overtraining syndrome.

Part II, Periodization and Managing Recovery, includes three chapters. Chapter 3 focuses on preventing excessive fatigue and overtraining by adequately planning an athlete's training program. To this aim, training periodization and its different models are discussed and recommendations are made on how to monitor athletes' exercise tolerance, fitness, and fatigue. This chapter thoroughly covers the topic of overtraining, helping the reader to gain comprehensive understanding on the etiology and the diagnosis of overtraining syndrome. The end of the chapter provides information on practical considerations for coaches and physicians, as well as a checklist for the diagnosis of this unwanted potential outcome of the training process.

Chapter 4 gathers recent data on the benefits of active recovery after intense or prolonged exercise. This type of recovery is described as used after an exercise session or practice, or within a training session. A table summarizing the effectiveness of this recovery method for all modes of locomotion and specific graphs from keystone studies are of particular interest. Finally, the psychological aspects of recovery are presented in chapter 5, which discusses various questionnaires developed on this theme and offers different tools for psychological evaluation.

Part III, Strategies for Optimizing Recovery, contains nine chapters gathering practical methods for optimizing and enhancing the recovery processes. The reader will find all well- and long-known recovery methods, which do not necessarily refer to specific techniques, but that are inherent to the athlete's everyday routine. Chapter 6 describes stretching techniques, including their direct or indirect relationship to performance. Different stretching methods are presented through the most recent scientific results in this field, and a specific section is dedicated to the issue of stretching after eccentric exercise. Chapters 7 and 8 cover adequate hydration and nutrition during recovery, emphasizing how to efficiently rehydrate according to fluid losses, and the undeniable necessity of consuming protein and carbohydrate after training. All effects of different nutritional and hydration strategies are described according to the latest findings. Timing of postexercise feeding is stressed as a key factor for recovery in sports training. Chapter 9 revisits the subject of sleep, reviewing its different stages, and explains how particular factors can improve or perturb sleep. One section is dedicated to the benefits of napping, an important part of the high-level athlete's daily schedule.

Chapter 10 describes massage techniques used in sports and their effect on subsequent performance. It details other techniques, like pressotherapy, electrostimulation, and luminotherapy, and presents the considerable advantages they can bring to performance. Chapter 11 reviews the effectiveness of compression garments, describing their effects during exercise and especially in the postexercise period. Chapter 12 provides information on the relative influence of local thermal applications for recovery from muscle trauma or altered muscle function following exercise. It also describes cooling jackets, a tool increasingly used for postexercise cooling. Thermal variations are presented in chapter 13, including the influence of sauna or steam baths and more contemporary techniques, such as whole-body cryotherapy, in enhancing the athlete's recovery. Methods of immersion in water at various temperatures make up chapter 14.

Part IV, Unique Considerations for Recovery, covers unique variables to consider for recovery techniques, such as gender, climate, and altitude. Chapter 15 specifies differences between men and women in postexercise recovery in great detail. Chapter 16 provides a contemporary understanding of the thermoregulatory responses and adaptations to exercise and heat stress. It also considers the interventions used to alleviate thermal strain and examines the limitations of various recovery strategies following exercise performance in the heat. Chapter 17 deals with altitude, another environmental stressor. It describes the physiological responses to altitude exposure and its effect on performance and various factors related to recovery, and provides practical recommendations to facilitate altitude adaptation and recovery.

The main purpose of this book is to summarize the scientific evidence and experience-based knowledge from case studies on the physiological, psychological, and performance consequences of numerous and varied recovery strategies. It aims to first help athletes, followed by coaches, sport scientists, and physical therapists, to adequately address this important issue, which should be a key ingredient of any sound training program. It could lead coaches and athletes to become more confident and secure in how to benefit from training sessions and avoid excessive fatigue when the training load increases. Even though *Recovery for Performance in Sport* does not intend to answer all questions regarding recovery and exercise, we strongly hope that the information provided will solidify readers' own knowledge and guide them in their practical implementation of recovery periods and strategies.

eBook
available at
HumanKinetics.com

Introduction

Christophe Hausswirth, PhD, and Iñigo Mujika, PhD

An efficient training program that improves physical capacities and performance must include intense, fatigue-inducing sessions. Fatigue is a state resulting from physiological and psychological constraints leading to a reduction in physical or mental performance. For a long time, fatigue was only recognized through its consequences, such as lowered work output. Athletes are mainly concerned with acute fatigue, a type that is perceived as normal, since it affects healthy people and has identifiable origins. Characterized by a rapid onset, acute muscular fatigue plays a protective role: By forcing athletes to reduce work output or to stop exercising and rest, it prevents their biological constants from getting too far removed from homeostasis. The disturbances to the body's biological constants brought on by exercise then trigger adaptive reactions that counter the metabolic changes and repair the structural damage caused by the training session.

The recovery period during which these adaptive anabolic reactions occur must therefore be adequate, both in duration and quality, to allow for the full repair of the various forms of damage incurred. The majority of research on exercise training and performance has focused solely on training methods, although most of the sought-after adaptations to training actually take place during the recovery period. Recovery is one of the least understood and most underresearched constituents of the exercise-adaptation cycle, even though dedicated athletes spend much more time in recovery than in active training. From a practical perspective, we define recovery as the whole set of processes that result in an athlete's renewed ability to meet or exceed a previous performance. Further, the recovery period is also defined as the time necessary for various physiological parameters, which were modified by exercise, to return to resting values.

The importance of the recovery period cannot be overstated. Failing to respect an athlete's recovery needs may lead to an inappropriate accumulation of fatigue, resulting not only in reduced workload tolerance and hence decreased performances, but also in an increased risk for injuries and cognitive and mood disturbances (irritability, difficulty concentrating, poor sleep), which may lead to an overtrained state. Coaches and the athletes themselves must, imperatively, pay close attention to the onset of signs of excessive physical and psychological fatigue in order to avoid reaching this state. Overtrained athletes require long periods of complete rest in order to fully recover and return to training, losing valuable periods of training and competing.

The different physiological constraints linked to muscular exercise lead to adaptation of the body, which contributes to improved aptitude, and thus, to better performance levels. Physical training therefore consists in exposing the body to higher workloads than usual in order to improve the functions necessary for completing these tasks. In high-performance athletes, who have particularly intense training regimes, the link between performance and training quantity is not so clear. The capacities of an individual athlete to adapt to training loads can, effectively, have limits. In addition, the level of solicitation can be such that, in confirmed athletes, a complex combination of local and central effects can cause marked reductions in performance levels over the days or weeks following an intense training period. The accumulation of excessive workloads with inadequate recovery periods can lead to a state of persistent fatigue that can only be improved by several weeks of rest.

Nutritional intake and hydration constitute primordial elements of the recovery process. The impressive training volumes executed by high-level athletes result in a considerable increase in nutrition and hydration needs. Meeting the specific caloric, macronutrient, and water requirements of each athlete in a timely manner is a crucial, yet sometimes tricky, aspect of recovery. Particular attention must be paid to athletes who are training or competing multiple times per day, since managing to ingest enough nutrients between sessions or performances can be challenging and requires a thorough organization and knowledge of the athlete's needs on the part of support staff. Proper restitution of water and electrolyte balance and timely replenishment of glycogen stores (aided by the ingestion of protein) will kick-start the recovery processes and bolster the athlete's capacity to tolerate the next performance.

Proper recovery is also necessary to maintain the athlete's psychological well-being, another crucial element in the ability to pursue intense physical training and to deliver satisfying performances. Athletes must have sufficient downtime to recover psychologically from the rigors of training and competition. Changes in mood and attitudes, such as increased anxiety and irritability, difficulty concentrating, a generally negative mood, and overall lack of energy and motivation are some of the most obvious signs indicating that the recovery time allotted (or its quality) is not adequate for that athlete. To avoid the drastic consequences of burnout or overtraining, these signs must be recognized early, and training and recovery periods must be adjusted accordingly.

When and how to use each type of therapy, how much rest and sleep are needed, and what, when, and how much to eat and drink are questions that gain complexity as training and competition schedules become busier. Athletes and coaches may be unsure of how to optimize the quality of recovery periods given the imposed constraints. Although some aspects must be customized to each athlete, it is important to first understand the rationale supporting each component of recovery and each modality of therapy. Next, coaches must know which strategies to adopt and when to apply them. Ultimately, improved quality of the recovery periods will help athletes further raise the bar of high-level performance. Since coaches already focus greatly on planning and customizing training schedules and workloads for each of their athletes, those who also incorporate realistic yet optimized recovery sessions into these schedules will likely be rewarded with stronger and, likely, better athletes with fewer injuries.

From a physiological standpoint, training improves and optimizes an athlete's physical abilities. Proper adaptation to a training load demands an initial phase of disturbance, during which the athlete's biological equilibrium is challenged.

Indeed, the homeostasis in which skeletal muscle cells develop must be perturbed incessantly by imposing a stressor followed by a refractory recovery period. During this time, the processes of adaptation and regeneration occur. For athletes, these processes are synonymous with reconstruction. Physical training is therefore inseparable from the state of fatigue, a normal physiological state inherent to daily life. However, this state must be recognized, analyzed, and then associated with appropriate periods of recovery, as well as coupled with practical recovery modalities that are adequate to the training load. Signs of fatigue must therefore be heard and perceived in order to establish a routine of recovery periods early on. Athletes absolutely need this rest in order to improve performance, as well as to repeat this performance multiple times.

Ever since the contributions of Claude Bernard, the field of exercise physiology has taught us that recovery within sport is concomitant to the entire restitution of athletes' biological constants, which are themselves associated with the integrity of their abilities. Functional activity may therefore always be accompanied by processes of organic destruction and recovery. If high-level sport is centered on the scheduling and management of training loads, then the recovery process truly represents adaptation for different training modalities. When considering these training cycles, athletes must imperatively brace themselves for their state of fatigue. In this manner, the brain imposes a state of fatigue before serious disturbances or disorders can develop.

This book also elaborates on the topic of recovery at the muscular level. The majority of recovery methods used in sport appears to act directly on muscle. These processes often involve nutritional strategies for recovery (and their effect on the repletion of glycogen stores), attempts to improve venous return through electrostimulation, and immersion procedures (with their proven benefits on markers of muscular damage).

The justification of these procedures for aiding muscular recovery takes on greater significance when we consider that the muscle secretes substances that inform the brain via hormonal or nervous paths about its metabolic state, thereby influencing the athlete's behavior.

To understand recovery in sport, we must first understand why it is an inevitable aspect of training: We believe that we must first improve the recovery process in order to then improve performance by means of recovery.

This recovery must lead to at least four major goals:

1. Easier adaptation to training loads
2. Decreased risk of overload
3. Reduced risk of injury
4. Improvement in the repeatability of performances: Winning the European championships is great, but winning the World or Olympic title is better!

Ever since the scientific community turned its attention toward the constraints arising from the practice of exercise, curiosity quickly grew about the physiological aspects of recovery. This scientific interest emerged in full force as soon as clinical signs of intolerance to exercise or training were identified. In this manner, the potential risk of overreaching or overtraining, which every high-level athlete must handle on a daily basis, has paved the way, albeit sporadically, for the studies that have attempted to characterize the physiological or physiopathological mechanisms of recovery in high-level athletes who are subjected to repeated training sessions.

Accurately gauging and quantifying each workload will allow us all to obtain a better grasp on adaptive systems. From there, we can program not the training sessions, which are left up to the coaches, but rather the recovery. This double programming of training and recovering can help us lead athletes to a higher level of performance.

The recovery process in sport is strategic; its structured, organized, and efficient application is applied to meet the goals, needs, and constraints encountered at the highest level of performance. The credibility of scientific findings within this field, associated with their practical, daily application, must facilitate the athlete's adaptation. Moreover, even though the significant variability that exists among individual athletes must be considered, we attempt to move beyond it in order to obtain standards that are applicable to the majority. As we approach excellence, the aspect of athletic recovery gains increasing importance.

This text, titled *Recovery and Performance in Sport,* covers a multitude of strategies for improving this recovery process to benefit athletes and to optimize their training time. This resource gathers current scientific knowledge and practices linked to athletic recovery. Over the course of a year, researchers, physical therapists, physicians, and coaches have compiled reference articles in order to extract exhaustive information that would facilitate comprehension of the plurality of recovery modes available to athletes of international rank. The transversal nature of this work undeniably yields a broad perspective on the recovery process by covering each one of its modalities in depth. While many modalities throughout the world help improve athletes' performances through optimized recovery, this recovery in itself must also be improved by means of better planned recovery periods following training sessions. The coordination of this book allows for a more thorough comprehension of just how paramount this concept of recovery has become within the sport world.

All the contributors have brought pertinent and sincere answers and reflection to this book. Once again, our goal is vested with capital importance: We must understand, as well as convey this understanding, that an adequate, individualized, and justified recovery is inherent to the process leading to improved athletic performance.

This book aims to reach a wide audience within the world of sport and exercise. By pulling together the most pertinent research findings on the numerous aspects of recovery and on the different techniques available to bolster it, this work will provide researchers and exercise physiologists with detailed and relevant knowledge on the scientific bases of physiological, metabolic, and psychological aspects of recovery. For physical education or kinesiology students, this book represents an approachable yet thorough learning resource whose content may complement classic resources focusing on physical training methods and theories. Finally, team physicians, therapists, coaches, and athletes will also appreciate the practical aspects of the information conveyed here. The ultimate goal of this book is to promote and facilitate the implementation of scientifically sound, efficient recovery methods into athletes' busy routines, thereby improving their ability to consistently perform at the highest level.

Fundamentals of Fatigue and Recovery

CHAPTER 1

Physiology of Exercise Training

Michael I. Lambert, PhD, and Iñigo Mujika, PhD

The concept that exercise training improves physical performance is not new. Since ancient times, trainers and coaches have guided athletes. Initially, the principles behind exercise training were based largely on intuition and folklore. Successful coaches often had a feel for the sport. They were able to manipulate training loads to induce changes, which were accompanied by an improvement in the athletes' performance.

However, in the early 1900s, physiologists started applying their skills to study the physiology and biochemistry associated with exercise and adaptations to exercise training. Gradually, the knowledge base increased, explaining the mechanisms behind muscle contraction, breakdown and regeneration, fatigue, and motor coordination. As this knowledge became more accessible, it started to be applied in various aspects of society. Consider, for example, how the productivity and well-being of miners increased once their heat tolerance and acclimatization to hot environments became better understood (Wyndham 1974). Many modern ideas about exercising in the heat can be attributed to work done in the laboratories attached to mining companies. Another example is that of scientists working with astronauts. They learned that astronauts needed to engage in physical activity that simulated weight-bearing activities when they were orbiting in space to protect themselves from muscle and bone loss in the gravity-free environment (White, Berry, and Hessberg 1972). Clinicians and health professionals have also learned that regular physical activity induces adaptations that protect against various noncommunicable diseases, such as hypertension and diabetes. Indeed, lack of physical activity was recognized as a risk factor for coronary artery disease (ACSM 1998). The global Exercise Is Medicine program was launched in the United States in 2007 to promote the benefits of exercise (www.exerciseismedicine.org). However, it was only in the late 1970s that sport physiology was applied to sport training, with the goal of improving performance (Tipton 1997). This launched the concept of a scientific approach to training.

As the scientific approach became more mainstream, coaches started to realize the importance of understanding and applying the principles of biology, particularly the principle of dose and response (i.e., the dose of exercise is the stimulus of a training session, and the response is the outcome from that training session). It became clear that athletes whose coaches understood and applied these principles had an advantage over their competitors. The scientific approach to training initiated the principles such as interval training, circuit training, endurance training, and the periodization of training. Other strategies to improve performance, such as the application of principles of nutrition, biomechanics, and psychology, also started to influence preparation for peak performances.

Overload Physiology

Perhaps the most important principle of biology that has applications for training and athletic performance is that of overload. This principle was first studied systematically more than 50 years ago (Hellenbrandt and Houtz 1956). Knowledge of this principle has continued to grow since then, forming the basis of all training programs. A more precise application of the principle of overload in training programs has coincided with a better understanding of the physiological and molecular mechanisms of adaptation after repeated exposure to a physical stress.

Understanding this principle is fundamental for coaches and trainers. In particular, applying it ensures that athletes' peak performances are more likely to coincide with their competitive events. Although it is not necessary for coaches or trainers to have a detailed knowledge of the biological steps associated with overload and adaptations, understanding the broad concepts will certainly help them to do their work in a more evidence-based manner.

Acute Stimulus

A training session can be explained in terms of physiological stress. As with all forms of physiological stress, homeostasis is disturbed in the cells of the muscles. Homeostatic adjustments occur to maintain the constancy of the body's internal milieu during exercise. Examples of these acute adjustments that occur during exercise include elevated heart rate and ventilation rate, redistribution of blood flow, increases in body temperature, and altered metabolic flux.

After the exercise session, the acute adaptations revert to the form they had before the start of the exercise. However, each physiological change has a different time course during the recovery period. This depends on the duration, intensity, and modality of exercise. For example, heart rate, blood lactate, and body temperature may take minutes to return to their pre-exercise levels, whereas oxygen consumption and cognitive function may take hours. If the bout of exercise was prolonged, muscle glycogen would also take days to restore to pre-exercise levels. If it caused muscle damage, circulating creatine kinase and muscle soreness would take days to diminish before they were comparable to the pre-exercise values. Muscle function and neuromuscular coordination may take weeks to recover fully, and muscle fibers may take months to regenerate completely (figure 1.1).

Training Stimulus

When the body is exposed to repetitive bouts of physiological stress, it will react to that stress by incurring adaptations. Long-term training adaptations are due to the cumulative effects of each short-term exercise (Coffey and Hawley 2007). The specific nature of the exercise-induced adaptations depends on the type of exercise stimulus. It manifests as changes in morphology, metabolism, and neuromuscular function (figure 1.2). These adaptations, also known as training-induced adaptations, enable the body to cope better with

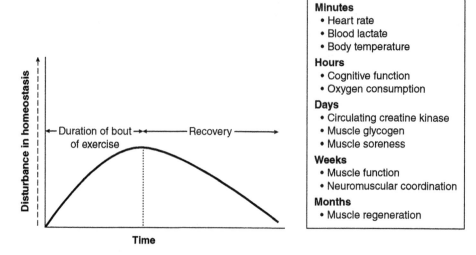

Figure 1.1   A bout of exercise disturbs homeostasis. During the recovery period, the different physiological systems take anywhere from minutes to months to recover, depending on the duration, intensity, and modality of exercise.

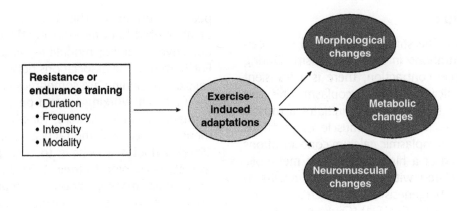

Figure 1.2   The specific exercise-induced adaptations depend on the stimulus, and are influenced by the duration, frequency, intensity, and modality of the bout of exercise. The changes can be classified as morphological, metabolic, or neuromuscular.

the next exposure to physiological stress or the next bout of exercise.

From a biological perspective, completing a bout of exercise after athletic training disturbs homeostasis less than performing that same task while in an untrained state.

The mechanisms of adaptation induced by resistance and endurance-based training are distinct, with each mode of exercise either activating or repressing specific subsets of genes and cellular-signaling pathways (Hawley 2009). The primary training stimulus described previously seems, in part, to be associated with an increase in intramuscular temperature, since muscle treated with cryotherapy after every training session does not incur training-induced adaptations at the same rate as muscle that is not cooled after each session (Yamane et al. 2006).

The physical side of training is interpreted as a stressful stimulus, as described previously. However, nontraining stressors, such as sleep deprivation, emotional stress, anxiety, and poor health, must also be accommodated by homeostatic mechanisms. Athletes often choose to minimize extraneous stress before a competition, tending just to eat, sleep, and train. Regular monitoring of athletes during the different phases of training can track various stressors and identify which athletes are vulnerable to excessive stress or over-reaching (Lambert and Borresen 2006; Meeusen et al. 2006).

Although, at the practical level, the training stimulus can be altered by the duration, frequency, and intensity of the training session, at the cellular level, the primary stimulus is a combination of the

load on the muscle, metabolic stress, and calcium flux (Baar 2009).

Load on Muscle

The type of muscle contraction (i.e., shortening or lengthening under tension) affects the training load. Muscle hypertrophy occurs after the load is increased. This occurs through the activation of the pathways of protein synthesis and a suppression of the protein degradation pathways. This is a relatively slow process that takes several weeks because protein synthesis must exceed protein degradation for an extended period before there is a net gain in myofibrillar protein (Baar 2009).

Metabolic Stress

The metabolic stress stimulus depends on a combination of the type of exercise, its intensity, and the prevailing nutritional state of the muscle cells (i.e., glycogen concentration in the muscle). Metabolic stress is associated with a high rate of ATP utilization, and is influenced by the substrate availability in the muscle cell (Baar 2009). It follows that the training stimulus for a muscle cell that has low levels of glycogen will be different from the stimulus of a muscle cell that has adequate levels of glycogen, since the substrate availability and endocrine response under these conditions will be different. Repeated stimuli during a high rate of ATP utilization cause the mitochondrial content and oxidative enzyme activity to increase, resulting in reduced glycogen utilization and enhanced capacity to oxidize fat during submaximal exercise, thereby increasing the capacity to resist fatigue and enhancing endurance performance.

Calcium Flux

Calcium flux is the stimulus associated with calcium concentrations in the sarcoplasm. During a single muscle contraction, there is a transient increase in calcium in the sarcoplasm. This calcium binds to troponin C and initiates a muscle contraction. As the rate of muscle contractions increase, the sarcoplasmic calcium concentrations are maintained at a higher level. This metabolic state is associated with a signal that results in mitochondrial biogenesis and an increase in the glucose transporter (GLUT 4) (Ojuka et al. 2003).

Exercise-Specific Responses

Exercise can be divided broadly into endurance and weight training types. The phenotypic consequences of each type of exercise are different. The morphological, metabolic, and neuromuscular adaptations are associated with improved performance. Depending on the type of exercise, this may be increased power output, delayed fatigue, or refined motor recruitment.

During endurance training, the load is low and the metabolic stress is high, and the calcium flux consists of short intermittent bursts for long periods. This combination results in a stimulus that increases mitochondrial mass and oxidative enzyme activity. In contrast, during resistance training, the load is high, the metabolic stress is moderate, and the calcium flux is high. This increases protein synthesis, resulting in muscle-fiber hypertrophy (Baar 2009).

Neuromuscular adaptations occur after resistance training, particularly during the first few weeks. During this phase, the gains in strength are rapid, despite the lack of muscle hypertrophy. Examples of neural adaptations that increase strength include changes in the firing rate and synchronization of motor units. Sensory receptors, such as the Golgi tendon organs, may also change, resulting in disinhibition during a muscle contraction and an increase in force production. Another example of neuromuscular adaptations is reduced antagonistic coactivation during muscle contractions (Gabriel, Kamen, and Frost 2006).

Overload Variables

Although coaches and athletes universally agree that physical performance improves with training (Foster et al. 1996), there is not as much consensus about the specific guidelines on how to achieve peak performance. This lack of agreement can be attributed to differences in athletes' goals and objectives, training modalities, and adaptability. Furthermore, the associations between the physical side of training are sometimes confounded by the fact that working toward peak sporting performance includes training for physical development and technical and tactical exercises (Bompa 1999). Athletes also have to train psychological aspects pertaining to their sport. In team sports, they must develop good communication and understanding to ensure harmony within the team structure. The combined success of the different types of training is difficult to assess; therefore, it is difficult to pinpoint the problem when athletes underperform in competition.

An overload can be imposed in several different ways. The most common is to adjust the frequency, duration, and intensity of the training session, as well as the recovery period between training sessions. These parameters are most frequently used to describe an exercise session, particularly at the practical level of coaching and exercise prescription.

Frequency

The training frequency describes the number of training sessions in a defined period. For example, at the elite level, training frequency may vary between 5 and 14 sessions per week, depending on the sport and stage of training cycle (Smith 2003).

Duration

This refers to the time of the exercise session. Athletes competing at the international level usually train about 1000 h per year (Bompa 1999).

Intensity

Exercise intensity, which is related to power output, defines how difficult an exercise session is. It can be monitored by measuring submaximal oxygen consumption (Daniels 1985), heart rate (Lambert, Mbambo, and St Clair Gibson 1998), blood lactate (Swart and Jennings 2004), the weight lifted during the exercise (Sweet et al. 2004), or the perception of effort (Foster et al. 2001). The exercise intensity lies somewhere on a continuum between rest and supramaximal exercise. Rest coincides with basal metabolic rate. Supramaximal exercise is defined by short-duration, high-intensity work done at a load exceeding that associated with maximal oxygen consumption. Athletes should only incorporate high-intensity training into their programs after they have developed a sufficient base (Laursen and Jenkins 2002). Incorporating too much high-intensity training into the program too soon increases the risk of fatigue associated with overreaching and overtraining (Meeusen et al. 2006).

For endurance training, the duration and intensity of the session are the main aspects that can be altered to manipulate the overload stimulus. Some ultra-endurance athletes also create a unique training stress by starting some training sessions with low concentrations of muscle glycogen. The goal of this is to mimic the metabolic signal that occurs at the end of a long race, when fatigued muscles have low glycogen levels. The scope of options for manipulating the workload for resistance training is wider.

A TRAINING STIMULUS CAN BE ALTERED TO IMPOSE TRAINING OVERLOAD IN THE FOLLOWING WAYS:

Endurance training	Resistance training
Duration of training session	Number of sets and reps
Intensity of training session	Weight used
Rest period between training sessions	Number of sets per exercise
Rest periods between interval sessions	Type of exercise
Altered nutrition status	Order of exercises
	Rest period between sets
	Rest period between training sessions

Overload must be carefully applied and increased to improve performance in a systematic way. The concept of varying training volume at various stages of the season is explained by the principle of periodization.

If the training stimulus is kept constant, the athlete soon reaches a plateau. Performance levels stagnate, and the athlete may show signs of staleness. This situation can be overcome by applying an overload training stimulus. This altered physiological stress induces further adaptations (figure 1.3).

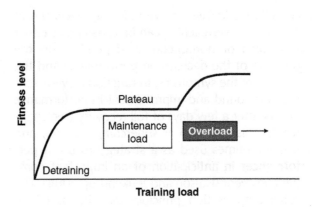

Figure 1.3 Plateau occurs if the maintenance load is not altered by an overload.

Taper Period

Reduced training and tapering are a key part of the preparation for peak performances. Just as the frequency, duration, and intensity of training are manipulated to impose a training stimulus, they can also be manipulated in the taper period to induce desired changes. The ultimate goal for the athlete during the taper period is to maintain the training-induced adaptations without experiencing the fatigue that typically accompanies a period of sustained training. If the tapering period is too long, the training-induced adaptations start to regress. This has a negative effect on performance. The performance gain associated with a properly implemented tapering program varies from about 0% to 6% in trained athletes (Mujika and Padilla 2003). The key features of the taper are its duration, the magnitude of reduction in training, and the interaction of the taper with the preceding phase of training (Pyne, Mujika, and Reilly 2009).

A general guideline for a taper is as follows (Pyne, Mujika, and Reilly 2009):

- Incorporate a progressive nonlinear reduction (40–60%) in training volume over 2 or 3 weeks.
- Maintain training intensity.
- Reduce training frequency a modest amount (approximately 20%), if at all.

Training and Recovery

When the relationship between training and recovery becomes imbalanced, symptoms of fatigue develop, followed by decreased performance. This condition is known as overreaching (Meeusen et al. 2006). Overreaching can be classified as either functional or nonfunctional, depending on the duration of the decrease in performance and the severity of the symptoms. In functional overreaching, a rebound and improvement in performance occurs after a few days of full recovery (see chapter 2). This is known as *supercompensation,* and it is sometimes used as a strategy to boost performances in anticipation of an important competitive event (Halson and Jeukendrup 2004). The supercompensation principle may be described as a planned breakdown as a result of sustained training, followed by recovery and enhanced performances (Kentta and Hassmen 1998). Nonfunctional overreaching is a more severe condition, with decreased performance lasting for weeks or months (Meeusen et al. 2006).

As shown in figure 1.1, the disturbance in homeostasis is reduced after athletes stop the exercise. Various biological systems require different time periods for recovery. This seemingly passive period is important for the subsequent adaptations. Consider, for example, that the mRNA expression of several oxidative enzymes is upregulated 24 h after the bout of exercise (Leick et al. 2010). This suggests that an important action associated with training-induced adaptations occurs long after the acute effects of the bout of exercise have subsided. Inadequate recovery between training sessions will result in maladaptations, with the accompanying symptoms of fatigue and impaired muscle function.

Insufficient recovery prevents athletes from training at the required intensity or completing the required load at the next training session. To enhance the recovery process, athletes often undertake proactive recovery strategies after training, such as massage, cryotherapy, immersion in water of contrasting temperatures, compression, and stretching (Barnett 2006). These sessions shift the stress–recovery balance in favor of the recovery processes. A purported consequence of this is that athletes can tolerate higher training volumes. The positive effects of the training load may also be enhanced (see part III).

PRACTICAL APPLICATION

A successful training program should progressively increase the training load by adequately manipulating training frequency, duration, intensity, and recovery, both within and between sessions. This overload principle applies to both endurance and strength training. Adequate recovery should be integrated into the training program to ensure a balance between the training stimulus and recovery and to avoid nonfunctional overreaching (excessive fatigue). To achieve supercompensation and an optimal fitness peak, coaches and athletes should consider tapering their training for 2 or 3 weeks before major competitions, reducing their training volume during this phase by 40% to 60% while maintaining high intensity and frequency.

Summary

Training is a process that exposes athletes to repetitive stimuli with the goal of inducing adaptations. These changes are matched to a desirable function, such as delaying fatigue, increasing power output, refining motor coordination, or reducing the risk of injury. The outcome of the training depends on the type of stimulus. Coaches and trainers who understand this cause-and-effect relationship between training dose and response can prescribe exercise training accordingly.

CHAPTER 2

Overtraining Syndrome

Romain Meeusen, PhD, and Kevin De Pauw, MSc

The goal in working with competitive athletes is to provide training loads that improve their performance. Athletes may go through several stages within a competitive season of periodized training. These phases of training range from undertraining, during the period between competitive seasons or during active rest and taper, to overreaching (OR) and overtraining (OT), which include maladaptations and diminished competitive performance (Meeusen et al. 2013a, b). When prolonged, excessive training is combined with other stressors and insufficient recovery time, decreased performance can result in chronic maladaptations, leading to overtraining syndrome (OTS).

The importance of optimal recovery must be highlighted. Coaches and researchers suggest that enhanced recovery allows athletes to train more, thus improving their overall fitness, technique, and efficiency (Kellmann 2010). In order to help athletes recover faster and perform better, coaches and researchers investigated different approaches, such as active recovery, compression, massage, cryotherapy, contrast water therapy, and combined recovery interventions.

Confusion exists in research literature about the definition and usage of the word *overtraining*. Therefore, this chapter not only presents the current state of knowledge on OTS, but also highlights the prevalence and recognition of the syndrome.

Although in recent years, knowledge of the central pathomechanisms of overtraining syndrome has significantly increased, there is still a strong demand for relevant tools for the early diagnosis of the condition. OTS is characterized by a sport-specific decrease in performance, combined with persistent fatigue and mood disturbances (Urhausen and Kindermann 2002; Meeusen et al. 2013a, b; Armstrong and Van Heest 2002; Halson and Jeukendrup 2004). This underperformance persists, despite a prolonged recovery period, which may last several weeks or months. Since no tool exists that identifies OTS, a diagnosis can only be made by excluding all other possible influences on changes in an athlete's performance and mood. Therefore, if no explanation for the observed changes can be found, the athlete is diagnosed with the syndrome. Early and unequivocal recognition of OTS is virtually impossible because the only certain sign of this condition is a decrease in performance during competition or training. The definitive diagnosis of OTS always requires ruling out an organic disease (e.g., endocrinological disorders of the thyroid or adrenal gland, or diabetes), iron deficiency with anemia, or infectious diseases (Meeusen et al. 2013a, b). Other major disorders or eating disorders, such as anorexia nervosa and bulimia, should also be excluded. However, many endocrinological and clinical findings due to nonfunctional overreaching (NFOR) and OTS can mimic other diseases. The line between under- and overdiagnosis is very difficult to judge (Meeusen et al. 2013a, b).

Continuum of Fatigue

Successful training must involve a specific exercise intensity or overload in order to achieve supercompensation, which is a positive adaptation or improvement in performance. It must also avoid the combination of excessive training and inadequate recovery. If the exercise intensity is too low, athletes cannot expect to improve their performance (de Graaf-Roelfsema et al. 2007). Athletes commonly intensify their training to enhance their performance. As a result of an intense training period (or even a single training session that is particularly taxing), they may experience acute feelings of fatigue and decreased performance. If adequate rest follows the training session, supercompensation may result. This process is the basis of effective training programs (figure 2.1). However, when the athlete's capacity for adaptation becomes oversolicited through insufficient recovery, a maladaptive training response may occur.

Overtraining syndrome is very difficult to diagnose, both because it is a complex condition and because research definitions of the different training statuses that are recognized as precursors have been very inconsistent. This complicates the detection of a reliable marker for the early diagnosis of OTS.

Several authors consider overtraining as an inevitable result of normal training, beginning with overreaching (OR) and finally ending in OTS. Since the states of OR and OTS have different defining characteristics, this concept of an overtraining continuum is probably an oversimplification. Overtraining syndrome results from more than the presence of training errors, coinciding particularly with other stressors (Meeusen et al. 2013a, b). However, as the joint consensus statement of the European College of Sport Science and the American College of Sports Medicine states (Meeusen et al. 2006, 2013a, b), these definitions indicate that the difference between overtraining and overreaching is the amount of time needed for performance restoration, not the type or duration of training stress or the degree of impairment. These definitions imply that an absence of psychological signs may be associated with the conditions. Since it is possible to recover from OR within 2 weeks (Lehmann, Foster, et al. 1999; Halson et al. 2002; Jeukendrup et al. 1992; Kreider 1998; Steinacker et al. 2000), it may be argued that this condition is a relatively normal and harmless stage of the training process. However, athletes who are suffering from OTS may need months or even years to completely recover.

Intensified training can result in a decline in performance. However, when appropriate periods of recovery are provided, athletes may exhibit enhanced performance compared to baseline levels. This supercompensation often occurs when athletes go to a short training camp in order to achieve functional overreaching. Studies typically not only follow athletes during the increased volume or intensity of training, but also register recovery from this training status. Usually athletes show temporary performance decrements that disappear after a taper period. In this situation, the physiological responses compensate for the stress related to training (Steinacker et al. 2004).

If this intensified training continues, athletes can progress to a state of extreme overreaching, or nonfunctional overreaching, that leads to a stagnation or decrease in performance. They will not be able to resume their previous levels for several weeks or months.

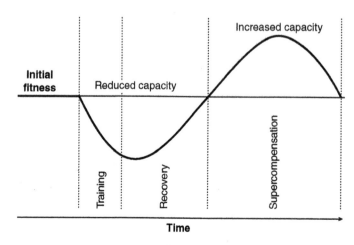

Figure 2.1   Physical response to training, recovery, and supercompensation.

Overreached athletes in both the functional and nonfunctional categories can fully recover after sufficient rest. In nonfunctional overreaching, progress along the overtraining continuum is determined both quantitatively (i.e., by the increase in training volume) and qualitatively (e.g., signs and symptoms of psychological or endocrine distress occur). This is in line with recent neuroendocrine findings through a double-exercise test (Urhausen, Gabriel, and Kindermann 1998; Urhausen et al. 1998; Meeusen et al. 2004, 2010).

Although it is possible to recover from short-term (functional) overreaching within 2 weeks, the recovery time for nonfunctional overreaching is less clear. This is probably because not many studies have tried to define the subtle difference between extreme overreaching, which requires several weeks or even months for recovery (Meeusen et al. 2013a, b), and overtraining syndrome, which may take months or even years to heal. Even top athletes who suffer from OTS frequently need to end their sports careers due to long recovery times. Another consideration is the use of the expression *syndrome,* which emphasizes multiple factors and acknowledges that exercise (training) is not necessarily the sole cause of the condition.

The definition used in this chapter is based on the joint consensus developed by the European College of Sport Science and the American College of Sports Medicine (Meeusen et al. 2006, 2013a, b). Table 2.1 illustrates the different training stages that lead to OTS.

Prevalence of Overtraining

Reports on athletes suffering from OTS are mostly case descriptions. This is because it is not only unethical, but probably also impossible, to train an athlete with a high training load while simultaneously including other stressors, especially since the symptoms of OTS differ for each person. The most-cited study on athletes suffering from OTS is the paper by Barron and colleagues (1985). The authors report clear hormonal disturbances in four long-distance runners. An insulin-induced hypoglycemic challenge was administered to assess hypothalamic–pituitary function in the overtrained athletes and control participants. In this study, performance was not measured. The authors declared that the athletes recovered hormonal function after 4 weeks. This might indicate that the athletes were in a state of nonfunctional overreach, rather than suffering from OTS. Rowbottom and colleagues (1995) report on differences in glutamine in overtrained athletes. Again, this study gives no clear indication how performance decrements were registered.

Hedelin and associates (2000) reported on a cross-country skier who, after several months of intensive training (up to 20 h/week), showed increased fatigue, reduced performance, and disturbances on psychometric tests. After ruling out other illnesses, this athlete was diagnosed with OTS. In order to investigate changes in the autonomic nervous system, heart-rate variability was recorded in the athlete. They registered a shift toward increased heart-rate variability, particularly in the high-frequency range, along with a reduced resting heart rate. This indicates an autonomic imbalance, with extensive parasympathetic modulation. Although it is not clear how performance decrements were measured, the authors report that the athlete needed 8 weeks to recover.

Meeusen and associates (2004) report on differences in normal training status, functional overreaching (after a training camp), and compare the endocrinological results to a double-exercise test with an OTS athlete. Athletes were

Table 2.1   Different Stages of Training Leading to Overreaching (OR) and Overtraining Syndrome (OTS)

Process	Training (overload)	Intensified training		
Outcome	Acute fatigue	Functional (short-term) OR	Nonfunctional (extreme) OR	Overtraining syndrome (OTS)
Recovery	Days	Days–weeks	Weeks–months	Months–years
Performance	Increase	Temporary performance decrement (e.g., training camp)	Stagnation decrease	Decrease

Based on Meeusen et al. 2006, Meeusen et al. 2013, "Prevention, diagnosis and treatment of the overtraining syndrome: Joint consensus statement of the European College of Sport Science and the American College of Sports Medicine," *Medicine and Science in Sports and Exercise.*

tested in a double-exercise protocol (two exercise tests separated by 4 hrs) in order to register their recovery capacity. Performance was measured as the time required to reach voluntary exhaustion. They compared the first and the second exercise tests in order to verify if the athletes were able to maintain the same performance. The training camp reduced exercise capacity in the athletes. Performance decreased 3% between the first and second tests. In the functional overreaching category, there was a 6% performance decrease. The OTS subject showed an 11% decrease in time to exhaustion. The OTS athletes also showed clear psychological and endocrinological disturbances (figure 2.2).

Uusitalo and colleagues (2004) report on an overtrained athlete who showed an abnormal reuptake of serotonin. This case presentation was well documented; however, the authors give no indication of the time the athlete needed to recover from OTS.

The border between optimal performance and performance impairment due to OTS is subtle. This applies especially to physiological and biochemical factors. The apparent vagueness surrounding OTS is further complicated by the fact that the clinical features are varied from one athlete to another. Factors are nonspecific, anecdotal, and numerous.

Probably because of the difference in the definition used, prevalent data on overtrained athletes are dispersed. Studies have reported that up to 60% of distance runners show signs of overtraining during their careers, while data on overtrained swimmers vary between 3% and 30% (Morgan et al. 1987; Lehmann, Foster, and Keul 1993; Hooper, Mackinnon, and Hanrahan 1997; O'Connor et al. 1989; Raglin and Morgan 1994). If the definition of OTS as stated previously is used, the incidence figures will probably be lower. We therefore suggest that a distinction be made between NFOR and OTS. Athletes should only be defined as suffering from OTS when a clinical exclusion diagnosis establishes the condition.

Since OTS is difficult to diagnose, it is therefore important to prevent it (Foster et al. 1988; Kuipers 1996; Uusitalo 2001). Moreover, because OTS is mainly due to an imbalance in the training–recovery ratio (too much training and too many competitions; too little recovery time), athletes must frequently record their training load using retrospective questionnaires, training diaries, physiological screening, and the direct observational method (Hopkins 1991).

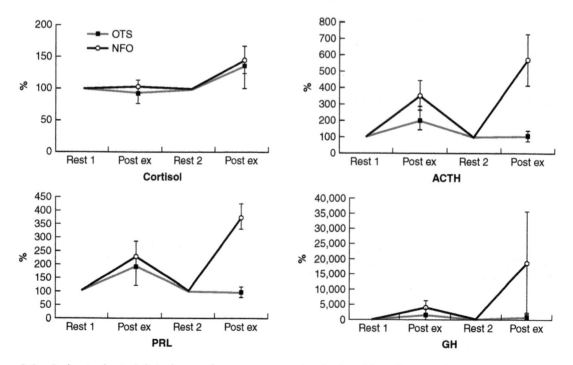

Figure 2.2 Endocrinological disturbances for an overtrained (NFOR) athlete during a double-exercise protocol.

Overtraining Syndrome Diagnosis Research

One approach to understanding the etiology of OTS involves the exclusion of organic diseases or infections and factors such as dietary caloric restriction (negative energy balance), insufficient carbohydrate or protein intake, iron deficiency, magnesium deficiency, and allergies, along with identification of initiating events or triggers (Meeusen et al. 2013a, b). One of the most certain triggers is a training error that results in an imbalance between load and recovery. Other possible triggers might be the monotony of training (Foster et al. 1996; Foster 1998), too many competitions, personal and emotional (psychological) problems, and emotional demands of occupation. Less commonly cited possibilities are altitude exposure and exercise-heat stress. Scientific evidence is not strong for most of these potential triggers. Many triggers, such as glycogen deficiency (Snyder et al. 1993) or infections (Rowbottom et al. 1995; Gabriel and Kindermann 1997), may contribute to NFOR or OTS, but they might not be present at the time the athlete sees a physician.

Performance Testing

Hallmark features of OTS are the inability to sustain intense exercise and decreased sport-specific performance capacity when the training load is maintained or even increased (Urhausen, Gabriel, and Kindermann 1995; Meeusen et al. 2004, 2010). Athletes suffering from OTS are usually able to start a normal training sequence or a race at their normal pace, but they are not able to complete the training load they are given or to race as usual. The key indicator of OTS can be considered an unexplainable decrease in performance (figure 2.3). Therefore, an exercise or performance test is considered essential for the diagnosis of OTS (Budgett et al. 2000; Lehmann, Foster, et al. 1999; Urhausen, Gabriel, and Kindermann 1995).

It appears that both the type of performance test employed and the duration of the test are important in determining the changes in performance associated with OTS. Debate exists as to which performance test is the most appropriate when attempting to diagnose OR and OTS. In general, time-to-fatigue tests will most likely

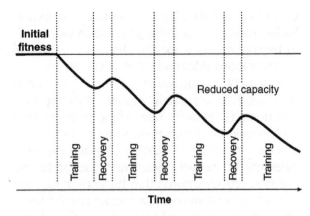

Figure 2.3 Imbalance between training and recovery resulting in reduced performance.

show greater changes in exercise capacity as a result of OR and OTS than incremental exercise tests (Halson and Jeukendrup 2004). Additionally, they allow the assessment of substrate kinetics, hormonal responses, and submaximal measures to be made at a fixed intensity and duration. In order to detect subtle performance decrements, it might be better to use sport-specific performance tests.

Urhausen, Gabriel, and Kindermann (1998) and Meeusen and colleagues (2004, 2010) have shown that multiple tests that are carried out on different days (Urhausen, Gabriel, and Kindermann 1998) or that use two maximal-incremental exercise tests (separated by 4 h) can be valuable tools for assessing the performance decrements usually seen in OTS athletes (figure 2.4). A decrease in exercise

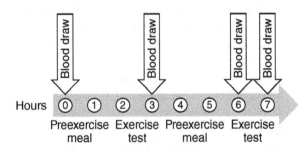

Figure 2.4 Schematic overview of the protocol (Meeusen et al. 2010).

time of at least 10% is necessary to be significant. Furthermore, this decrease in performance needs to be confirmed by specific changes in hormone concentrations (Meeusen et al. 2004, 2010).

A variety of measurement tools have been developed to accurately monitor training load and quantify recovery rate. These measurement tools should be noninvasive, sensitive to change, and easy to administer. One such tool is the measurement of recovery heart rate. Lamberts and Lambert (2009) showed that the chosen exercise intensity of a submaximal test should elicit a heart rate between 85% and 90% of maximum heart rate because the lowest day-to-day variations were found at this intensity. From test to test, a change in heart rate recovery (HRR) of more than 6 beats per minute (bpm), or the change in submaximal heart rate of more than 3 bpm, can be regarded as a meaningful change under controlled conditions. These changes can be caused by an improved or decreased training status or by the accumulation of fatigue as a result of functional overreaching (Lamberts et al. 2010).

Regular peak performance testing disrupts an athlete's training program. A variety of other measurement tools exist, such as the resting heart rate, the profile of mood states questionnaire, the daily analysis of life demands for athletes questionnaire, and the total quality of recovery scale. These have been used to monitor and predict changes in training status and performance, but none has been able to consistently monitor and predict changes in these parameters. Therefore, Lamberts and colleagues (2010) developed a submaximal test, the Lamberts and Lambert submaximal cycle test (LSCT), with the purpose of monitoring and predicting changes in cycling performance. This submaximal test does not interfere with normal training habits and has the potential to detect subtle changes in performance as a result of training-induced fatigue. In addition, monitoring the cumulative fatigue could contribute to detecting the status of functional or nonfunctional overreaching. The LSCT could indicate the development of fatigue by increased levels in rating of perceived exertion (RPE), a sudden increase in mean power, and a change in HRR.

Ratings of perceived exertion (RPE) offer a valuable determinant of impending fatigue during exercise testing. They are widely accepted as an effective tool for prescribing aerobic exercise intensity. Foster and associates (1996) introduced the concept of session RPE (sRPE), in which an athlete subjectively estimates the overall difficulty of an entire workout after its completion. Green and colleagues (2007) noted that sRPE may prove useful in evaluating recovery across exercise sessions, but more work is needed to determine the precision of this measure and to further elaborate on both the physiological and psychological factors influencing sRPE.

Biological Indications

Most of the blood parameters (e.g., blood count, c-reactive protein [CRP], erythrocyte sedimentation rate [SR], creatine kinase [CK], urea, creatinine, liver enzymes, glucose, ferritin, sodium, potassium) are not capable of detecting OR or OTS, but they are helpful in providing information on the actual health status of the athlete, and are therefore useful in the exclusion diagnosis (Meeusen et al. 2013a, b).

It is clear that the immune system is extremely sensitive to stress, both physiological and psychological. Thus, potentially immune variables could be used as an index of stress in relation to exercise training. Unresolved viral infections are not routinely assessed in elite athletes, but they may be worth investigating in athletes experiencing fatigue and underperformance in training and competition.

Overtraining syndrome can be understood partly within the context of the general adaptation syndrome (GAS) of Selye (1936). Concomitant to this stress disturbance, the endocrine system is called on to counteract the stress situation. The primary hormone products (adrenaline, noradrenaline, and cortisol) all serve to redistribute metabolic fuels, maintain blood glucose, and enhance the responsiveness of the cardiovascular system. Repeated exposure to stress may lead to different levels of responsiveness to subsequent stressful experiences, depending on the stressor as well as on the stimulus paired with the stressor. This leads to unchanged, increased, or decreased function of neurotransmitters and receptors (Meeusen 1999).

Behavioral adaptation (neurotransmitter release, receptor sensitivity, receptor binding) in higher brain centers certainly influences hypo-thalamic output (Lachuer et al. 1994). Lehmann and colleagues (1993; Lehmann, Gastmann, et al. 1999) introduced the concept that hypothalamic function reflects the state of OR or OTS because the hypothalamus integrates many of the stress-ors. It has been shown that acute stress not only increases release of hypothalamic monoamine, but consequently increases secretion of cortico-trophin-releasing hormone (CRH) and adrenocor-ticotropic hormone (ACTH) (Shintani et al. 1995). Chronic stress and the subsequent chronically ele-vated adrenal glucocorticoid secretion could play an important role in the desensitization of higher brain center response to acute stressors (Duclos et al. 1997, 1998, 1999; Duclos, Gouarne, and Bonnemaison 2003), since it has been shown that during acute and chronic stress, the responsive-ness of hypothalamic CRH neurons rapidly falls (Barron et al. 1985; Lehmann, Foster, and Keul 1993; Cizza et al. 1993; Urhausen, Gabriel, and Kindermann 1998).

However, simply knowing the effect of a hor-mone is not sufficient to understand its actual role in metabolic control. Each hormone has a predefined exercise-induced pattern. When investigating hormonal markers of training adap-tation, it is therefore important to target specific hormones for their informational potential and to synchronize their sampling in accordance with their response patterns. Recent research findings (Barron et al. 1985; Meeusen et al. 2004, 2010) support that athletes experiencing maladaptive training and performance adaptation problems seem to suffer from a dysfunctional HPA-axis (hypothalamic-pituitary-adrenal axis) response to exercise, resulting in an altered hormonal response to intense training and competition. When inves-tigating elite athletes, the HPA axis is believed to offer valuable information about an athlete's state of adaptation (Steinacker and Lehmann 2002).

Meeusen and colleagues (2004, 2010) pub-lished a test protocol with two consecutive maximal-exercise tests separated by 4 h. With this protocol, they found that in order to detect signs of OTS and to distinguish them from normal training responses (or FOR), this method may be a good indicator not only of the recovery capac-ity of the athlete, but also of the ability to nor-mally perform the second bout of exercise. The use of two bouts of maximal exercise to study neuroendocrine variations showed an adapted exercise-induced increase of ACTH, prolactin (PRL), and growth hormone (GH) to a two-exer-cise bout (Meeusen et al. 2004, 2010). The test could therefore be used as an indirect measure of hypothalamic-pituitary capacity. In a FOR stage, a less pronounced neuroendocrine response to a second bout of exercise on the same day is found (De Schutter et al. 2004; Meeusen et al. 2004), while in a NFOR stage, the hormonal response to a two-bout exercise protocol shows an extreme increased release after the second exercise trigger (Meeusen et al. 2010). With the same protocol, it has been shown that athletes suffering from OTS have an extremely large increase in hormonal release in the first exercise bout, followed by a complete suppression in the second exercise bout (Meeusen et al. 2004, 2010). This could indicate a hypersensitivity of the pituitary followed by an insensitivity or bout of exhaustion afterward. Pre-vious reports that used a single exercise protocol found similar effects (Meeusen et al. 2004, 2010). It appears that the use of two exercise bouts is more useful in detecting OR for preventing OTS.

Early detection of OR may be very important in the prevention of OTS. However, testing of central hypothalamic-pituitary regulation requires functional tests, which are considered invasive, and require diagnostic experience. These tests are time consuming and expensive.

Psychological Parameters

General agreement exists that OTS is character-ized by psychological disturbances and nega-tive affective states. It has been suggested that although the psychological processes underpin-ning OTS are important, the phenomenon occurs only when these psychological processes are com-bined with a negative training adaptation (Silva 1990). Sustained failure to adapt to training gen-erates excessive fatigue. Training when the body's adaptivity is lost leads to OTS (Silva 1990; Urhau-sen, Gabriel, and Kindermann 1995; Foster and Lehmann 1997). Athletes then suffer from a neu-roendocrine imbalance, and typically experience a noticeable drop in performance (Lemyre 2005).

When athletes suffer from OTS, they typically experience chronic fatigue, poor sleep patterns, a drop in motivation, and episodes of depression and helplessness (Lemyre 2005). Not surprisingly, their performance is considerably impaired. Full recovery from OTS represents a complex process that may necessitate many months, or even years, of rest and removal from sport (Kellmann 2002; Kentta and Hassmen 1998).

Several questionnaires, such as the profile of mood states (POMS) (Morgan et al. 1988; Raglin, Morgan, and O'Connor 1991; O'Connor 1997; O'Connor et al. 1989; Rietjens et al. 2005), recovery-stress questionnaire (RestQ-Sport) (Kellmann 2002), daily analysis of life demands of athletes (DALDA) (Halson et al. 2002), and the self-condition scale (Urhausen et al. 1998), have been used to monitor psychological parameters in athletes. Other tests, such as attention tests (finger pre-cueing tasks) (Rietjens et al. 2005) or neurocognitive tests (Kubesch et al. 2003), also serve as promising tools to detect subtle neurocognitive disturbances registered in OR or OTS athletes. It is important to register the current state of stress and recovery and to prospectively follow the evolution for each athlete individually (Morgan et al. 1988; Kellmann 2002). The great advantage of psychometric instruments is the quick availability of information (Kellmann 2002), especially since psychological disturbances coincide with physiological and performance changes, and are generally the precursors of neuroendocrine disturbances. In OTS, the depressive component is more expressed than in OR (Armstrong and Van Heest 2002). Changes in mood state may be a useful indicator of OR and OTS. However, it is necessary to combine mood disturbances with measures of performance.

Training Logs

Hooper and colleagues (1995) used daily training logs during an entire season in swimmers to detect staleness (OTS). The distances swum, the dryland work time, and subjective self-assessment of training intensity were recorded. In addition to these training details, the swimmers also recorded subjective ratings of quality of sleep, fatigue, stress and muscle soreness, body mass, early morning heart rate, occurrence of illness, menstruation, and causes of stress. Swimmers were classified as having OTS if their profile met five criteria. Three of these criteria were determined by items of the daily training logs: fatigue ratings in the logs of more than 5 (scale 1–7) lasting longer than 7

days, comments in the page provided in each log that athletes felt they responded poorly to training, and a negative response to a question regarding the presence of illness in the swimmer's log, together with a normal blood leukocyte count.

Foster and associates (1996; Foster 1998) have described training load as the product of the subjective intensity of a training session, using session RPE and the total duration of the training session expressed in minutes. If these parameters are summated on a weekly basis, it is called the total training load of an athlete. The session RPE has been shown to be related to the average percent heart rate reserve during an exercise session, and to the percentage of a training session during which the heart rate is within blood lactate–derived heart-rate training zones. With this method of monitoring training, they have demonstrated the utility of evaluating experimental alterations in training and have successfully related training load to performance (Foster et al. 1996). However, training load is clearly not the only training-related variable contributing to the genesis of OTS. In addition to weekly training load, daily mean training load and the standard deviation of training load were calculated during each week. The daily mean divided by the standard deviation was defined as the monotony. The product of the weekly training load and monotony was calculated as strain. The incidence of simple illness and injury was noted and plotted together with the indices of training load, monotony, and strain. They noted the correspondence between spikes in the indices of training and subsequent illness or injury. Thresholds that allowed for optimal explanation of illnesses were also computed (Foster 1998).

Assessment of Overtraining

The four methods most frequently used to monitor training and prevent overtraining are retrospective questionnaires, training diaries, physiological screening, and the direct observational method (Hopkins 1991). Also, the psychological screening of athletes (Berglund and Safstrom 1994; Hooper et al. 1995; Hooper and Mackinnon 1995; Raglin, Morgan, and O'Connor 1991; Urhausen et al. 1998; Morgan et al. 1988; Kellmann 2002; Steinacker and Lehmann 2002) and the ratings of perceived exertion (RPE) (Acevedo, Rinehardt, and Kraemer 1994; Callister et al. 1990; Foster et al.

1996; Foster 1998; Hooper et al. 1995; Hooper and Mackinnon 1995; Kentta and Hassmen 1998; Snyder et al. 1993) currently receive increasing attention.

The need for definitive diagnostic criteria for OTS is reflected in much of the research on over-reaching and overtraining by a lack of consistent findings. A reliable marker for the onset of OTS must fulfill several criteria: The marker should be sensitive to the training load and should ideally be unaffected by other factors (e.g., diet). Changes in the marker should occur prior to the establishment of OTS and changes in response to acute exercise should be distinguishable from chronic changes. Ideally, the marker should be relatively easy to measure and not too expensive (form 2.1). However, none of the currently available or suggested markers meets all of these criteria (Meeusen et al. 2013a, b). When choosing several markers that might give an indication of the training or overtraining status of the athlete, one needs to take into account several possible problems that might influence decision making.

When testing the athlete's performance, the intensity and reproducibility of the test should be sufficient to detect differences (max test, time trial, 2 max tests). Baseline measures are often not available; therefore, the degree of performance limitation may not be exactly determined. Many of the performance tests are not sport specific. Heart rate variability seems a promising tool in theory, but it needs to be standardized when tested. At present, it does not provide consistent results. A performance decrease of more than 10% on two tests separated by 4 h can be indicative of OTS if other signs and symptoms are present. Biochemical markers, such as lactate or urea, as well as immunological markers, do not have consistent enough reports in the literature to be considered as absolute indicators for OTS. Many factors affect blood-hormone concentrations. These include factors linked to sampling conditions and conservation of the sampling, such as stress of the sampling, and intra- and interassay coefficient of variability. Others, such as food intake (nutrient composition or pre- vs. postmeal sampling), can significantly modify either the basal concentration of some hormones (cortisol, DHEA-S, total testosterone) or their concentration change in response to exercise (cortisol, GH). Diurnal and seasonal variations of the hormones are important fac-

tors that need to be considered. In female athletes, the hormonal response will depend on the phase of the menstrual cycle. Hormone concentrations at rest and following stimulation (exercise = acute stimulus) respond differently. Stress-induced measures (exercise, prohormones) need to be compared with baseline measures from the same athlete. Poor reproducibility and feasibility of some techniques used to measure certain hormones can make the comparison of results difficult. Therefore, the use of two maximal performance (or time trial) tests separated by 4 h could help in comparing the individual results.

Psychometric data always need to be compared with the baseline status of the athlete. The lack of success induced by a long-term decrement of performance could be explained by the depression in OTS. The differences between self-assessment and the questionnaires, given by an independent experimenter and the timing of the mood-state assessment, are important. Questionnaires should be used in standardized conditions. Other psychological parameters different from mood state (attention span, anxiety) might also be influenced.

One of the disadvantages of the traditional paper-and-pencil method is that data collection can be complicated. Immediate feedback is not always possible. Another problem is that when athletes are at an international training camp or competition, immediate data computing is not possible. It might therefore be useful to have an online training log (Cumps, Pockelé, and Meeusen 2004; Pockelé et al. 2004) with specific features for detecting not only slight differences in training load, but also the subjective parameters (muscle soreness, mental and physical well-being) that have been proven to be important in the detection of OTS.

Athletes and the field of sports medicine in general would benefit greatly if a specific, sensitive, simple, diagnostic test existed to identify OTS. At present, no such test meets these criteria, but there certainly is a need for a combination of diagnostic aids to pinpoint possible markers for OTS. In particular, there is a need for a detection mechanism for early triggering factors.

Therefore, a flowchart, as presented in the consensus statement of the European College of Sports Science, could help to establish the exclusion diagnosis for the detection of OTS (see form 2.1).

Form 2.1 Diagnosis of OTS Checklist

Performance and Fatigue

Is the athlete suffering from any of the following?

- ❏ Unexplainable underperformance
- ❏ Persistent fatigue
- ❏ Increased sense of effort in training
- ❏ Sleep disorders

Exclusion Criteria

Are there confounding diseases?

- ❏ Anemia
- ❏ Epstein Barr virus
- ❏ Other infectious diseases
- ❏ Muscle damage (high CK)
- ❏ Lyme disease
- ❏ Endocrinological diseases (diabetes, thyroid, adrenal gland)
- ❏ Major disorders connected with eating behavior
- ❏ Biological abnormalities (increased SR, increased CRP, creatinine, ferritin, increased liver enzymes)
- ❏ Injury (musculoskeletal system)
- ❏ Cardiological symptoms
- ❏ Adult-onset asthma
- ❏ Allergies

Are there training errors?

- ❏ Training volume increased (>5%) (h/week, km/week)
- ❏ Training intensity increased significantly
- ❏ Training monotony present
- ❏ High number of competitions
- ❏ In endurance athletes : Decreased performance at anaerobic threshold
- ❏ Exposure to environmental stressors (altitude, heat, cold)

Other confounding factors:

- ❏ Psychological signs and symptoms (disturbed POMS, RestQ-sport, RPE)
- ❏ Social factors (family, relationships, financial, work, coach, team)
- ❏ Recent or multiple time-zone travel

Exercise Test

- ❏ Are there baseline values to compare with (performance, heart rate, hormonal, lactate)?
- ❏ Maximal exercise test performance
- ❏ Submaximal or sport-specific test performance
- ❏ Multiple performance tests

Reprinted, by permission, from R. Meeusen et al., 2006, "Consensus statement ECSS: Prevention, diagnosis and treatment of the overtraining syndrome," *European Journal of Sport Science* 6(1): 1-14.

PRACTICAL APPLICATION

Until a definitive diagnostic tool for OTS is present, coaches and physicians need to rely on performance decrements as verification that the syndrome exists. However, if sophisticated laboratory techniques are not available, the following considerations may be useful:

- Maintain accurate records of performance during training and competition. Be willing to adjust daily training intensity or volume, or allow a day of complete rest, when performance declines or the athlete complains of excessive fatigue.
- Avoid excessive monotony of training.
- Always individualize the intensity of training.
- Encourage and regularly reinforce optimal nutrition, hydration status, and sleep.
- Be aware that multiple stressors such as sleep loss or sleep disturbance (e.g., jet lag), exposure to environmental stressors, occupational pressures, change of residence, and interpersonal or family difficulties may add to the stress of physical training.
- Treat OTS with rest! Reduced training may be sufficient for recovery in some cases of overreaching.
- Resumption of training should be individualized on the basis of the signs and symptoms, since there is no definitive indicator of recovery.

- Communication with the athletes (maybe through an online training diary) about their physical, mental, and emotional concerns is important.
- Include regular psychological questionnaires to evaluate the emotional and psychological state of the athlete.
- Maintain confidentiality regarding each athlete's condition (physical, clinical, and mental).
- Importance of regular health checks performed by a multidisciplinary team (physician, nutritionist, psychologist).
- Allow the athlete time to recover after illness or injury.
- Note the occurrence of upper respiratory tract infections (URTI) and other infectious episodes. The athlete should be encouraged to suspend training or reduce the training intensity when suffering from an infection.
- Always rule out an organic disease in cases of performance decrement.
- Unresolved viral infections are not routinely assessed in elite athletes, but it may be worth investigating this in those experiencing fatigue and underperformance in training and competition.

Summary

Training is a process of overload that is used to disturb homeostasis. It results in acute fatigue that leads to an improvement in performance. When training continues or when athletes deliberately use a short-term period (e.g., training camp) to increase their training load, they can experience short-term performance decrements without severe psychological, or other negative and prolonged, symptoms. This functional overreaching (FOR) eventually leads to an improvement in performance after recovery.

However, when athletes do not sufficiently respect the balance between training and recovery, nonfunctional overreaching (NFOR) can occur. At this stage, the first signs and symptoms of prolonged training distress, such as performance decrements, psychological disturbance (decreased vigor, increased fatigue), and hormonal disturbances will occur. The athletes will need weeks or months to recover. Several confounding factors, such as inadequate nutrition (energy or carbohydrate intake), illness (most commonly URTIs), psychosocial stressors (related to work, team, coaches, or family) and sleep disorders may be present. At this stage, the distinction between NFOR and OTS is very difficult, and will depend on the clinical outcome and exclusion diagnosis. Athletes often show the same clinical and hormonal signs and symptoms. Therefore, the diagnosis of OTS can often only be made retrospectively when the time course can be overseen. A keyword in the recognition of OTS might be *prolonged maladaptation,* not only of the athlete, but also of several biological, neurochemical, and hormonal regulation mechanisms.

The physical demands of intensified training are not the only elements in the development of OTS. It seems that a complex set of psychological factors are important in the development of the syndrome, including excessive expectations from a coach or family members, competitive stress, personality structure, social environment, relationships with family and friends, monotony in training, personal or emotional problems, and school- or work-related demands. While no single marker can be taken as an indicator of impending OTS, the regular monitoring of a combination of performance, physiological, biochemical, immunological, and psychological variables would seem to be the best strategy to identify athletes who are failing to cope with the stress of training.

Much more research is needed to get a clear-cut answer about the origin and detection of OTS. We therefore encourage researchers and clinicians to report as much as possible on individual cases of athletes who are underperforming and, by following the exclusion diagnosis, to find that they are possibly suffering from OTS.

Periodization
and Managing Recovery

CHAPTER 3

Overtraining Prevention

Michael I. Lambert, PhD, and Iñigo Mujika, PhD

One of the biggest challenges for high performance and recreational athletes alike is the ability to regulate their training programs to induce optimal training adaptations. Failure to do so results in either undertraining or overtraining. In response to this requirement, the concept of periodization of training has emerged. A general interpretation of a periodized training program includes a structured approach and a long-term plan for training. However, beyond the general description, a periodized approach to training means different things to different people. Most coaches and athletes agree that a periodized training program has important outcomes, such as reduced risks of injury and of developing symptoms of overtraining, as well as a better chance of attaining peak performances during competition. While not much debate exists about the outcomes, the process of periodized training still divides opinions.

The lack of consensus can perhaps be attributed to the jargon that has tracked the concept and development of periodized training. Furthermore, the terms associated with the discipline of periodized training have sometimes been misinterpreted during their translation from other languages into English. This confusion over nomenclature has made the concept of periodized training more complicated than it needs to be. As a result, it has alienated some coaches. Another aspect to consider is that most of the knowledge about periodization comes from observational evidence, anecdotal data, and inferences from related studies, such as research on overtraining

(Stone et al. 1999). Skeptics are quick to point out that the concept is not completely supported by science. That may be the case, but explanations for the lack of good scientific evidence exist. These will be discussed later.

Although the origin of periodization is often attributed to Russian Professor Matvejev, he really just used the term to describe the planning phases of an athlete's training. A much earlier account of periodization can be found in the writings of the ancient Greek scientist Philostratus from the second century AD. He described the pre-Olympic preparation of athletes, which consisted of a compulsory 10-month period of purposeful training followed by 1 month of centralized preparation prior to the Olympic Games (Issurin 2010). According to Professor Michael Stone of East Tennessee State University in the United States, the term *periodization* was first used around the turn of the 20th century in reference to seasons, particularly seasonal periods of the sun. Coaches noted that athletes were able to train and perform better during the summer months, when days were longer and warmer. It was also believed that athletes could perform better during this period of warmer months because more fresh vegetables and produce were available. During the 1920s and 1930s, the term started to be applied to training methods. Up to this point, the concept was mainly used in Russia and East Europe. It was only in the 1960s and 1970s that Americans began to seriously consider periodization as a training strategy (Haff 2004).

Despite the debate about nomenclature and nuances of the different phases of a periodized training program, the fundamentals can be reduced to a few concepts. For example, periodization is the process of systematic planning of short- and long-term training programs by varying training loads and incorporating adequate rest and recovery (Lambert et al. 2008). Therefore, it provides a structure for controlling the stress and recovery associated with training, serving as a template for coaches and athletes and providing variation in the training load. A similar training load applied over several days or weeks causes training monotony, which is related to negative adaptations to training (Foster 1998). Varying the training load in a planned way reduces the risk of monotony.

The basic goals of a periodized program are to apply a training stimulus at optimal times, manage fatigue, and prevent stagnation or overtraining (Plisk and Stone 2003). The main aim is to induce the desired physiological adaptations and the technical and tactical skill associated with peak performance. Despite an apparent lack of scientific rigor and evidence to support some of the principles, periodization is widely practiced. It is recommended as the most appropriate training method (Turner 2011).

Periodization Research

As mentioned before, there is a mismatch between the practice of periodization and the evidence arising from scientific experimentation. Although several hundred studies have attempted to gain a deeper understanding about the application of periodized training, most have failed to add much value to what coaches have already been doing. The following factors make it difficult to conduct meaningful studies on different aspects of periodization (Cissik, Hedrick, and Barnes 2008).

LIMITATIONS TO THE RESEARCH ON PERIODIZATION:

- Research is primarily applicable to strength/power sports, not team sports.
- Research rarely uses athletes as subjects. It is more convenient to use students, who are often moderately trained or untrained. Their rate of adaptation and ability to recover from a training load differs from that of elite athletes.

- Use of volume and intensity (the variables manipulated in most studies) is contentious.
- Research studies are rarely long enough to elicit long-term training effects. It is very difficult to have good compliance with studies that last several months or years.

A meta-analysis of studies that used periodized training (all studies between 1962 and 2000 that met certain criteria) for strength and power revealed that the approach is more effective than nonperiodized training (Rhea and Alderman 2004). This meta-analysis showed that the overall effect size (ES) for increasing strength and power (periodized versus nonperiodized) was ES = 0.84 for all groups. This statistic can be interpreted as a moderate effect. The effect was greater in untrained participants than for trained athletes. However, it must be pointed out that most of the studies included in the meta-analysis were limited by at least one of the points shown in the preceding list.

Models of Periodization

Several different models for periodizing training exist (Bompa 1999). These models differ depending on the sport, but they all share a common principle in having phases of general preparation, specific preparation, competition preparation, competition, and transition or active rest. The models also differ for athletes competing in "cgs" sports (i.e., sports in which performance is measured in centimeters, grams, or seconds) (Moesch et al. 2011), where the main focus is developing superior physical capabilities. Team sports have a different set of demands for success. The terminology for dividing the cycles is as follows:

- *Macrocycles:* long plan, usually about a year in duration
- *Mesocycles:* shorter plan from about 2 weeks to several months
- *Microcycles:* short plan of about 7 days (Stone et al. 1999)

The time period around these training blocks varies depending on the sport. Athletes who choose to peak every 4 years coinciding with the Olympics have a much longer macrocycle. The annual plan (macrocycle) is normally designed around the number of competitive phases in a year. For sports that have one competitive phase a year, the year is divided into a preparatory, competitive, and transition or recovery phases. Each phase elicits a specific goal. For example, the preparatory phase is usually in the off or pre-season. It may focus on developing aerobic capacity, muscle hypertrophy and strength endurance, basic strength, and strength/power. The recovery phase can be a period of complete rest, active rest, or cross-training, with the goal of recovery and regeneration for the next cycle of training.

For sports that have winter and summer competitive schedules (e.g., track and field and swimming), the year is divided into two cycles. The planning process for team sports, which is more complicated, is discussed in more detail in the next section.

Nonlinear Periodization

Other terminology used to describe periodization is the undulating or nonlinear model, where training loads are varied acutely on a daily and weekly basis (Brown 2001). The variation factor in training is particularly emphasized in the nonlinear model (Issurin 2010). The linear model is the more classical method, in which training volume is decreased while training intensity is increased as the competitive period approaches (Brown 2001). This model is subdivided into different phases. Each phase has a specific training focus (e.g., hypertrophy, strength, power), and usually starts with high-volume, low-intensity workouts and then progresses to low-volume, high-intensity workouts. The main aim is to maximize the development of strength and muscle power (Haff 2004).

Block Periodization

As the number of competitions in sports has increased, athletes have had more difficulty following a highly regulated plan, so coaches have started using a new approach called block periodization. This model has concentrated training loads of about 2 to 6 weeks that focus on the consecutive development of physical and technical abilities (García-Pallarés et al. 2010). Arguably the best study on periodization was one that circumvented the problems outlined in the research section. This study, done on world-class kayakers, compared traditional periodization to block periodization (García-Pallarés et al. 2010). It showed that the kayakers halved their training volume when they changed from the traditional periodization program to a block periodization model

(i.e., short 5-week training phases). However, despite the lower training volume with the block periodization, kayakers experienced similar gains in peak oxygen consumption and larger gains in paddling speed, peak power output, and stroke rate at peak oxygen consumption, compared to the more traditional periodized training plan.

Team Sports

In team sports, the demands for planning training are quite different, presenting coaches and strength and conditioning trainers with unique challenges. In most sports, it is possible to include the phases of general preparation, specific preparation, precompetition, and competition (which lasts the duration of the season). However, a common problem is to determine the appropriate training loads during the competition phase of the season (Kelly and Coutts 2007). Kelly and Coutts summarized the points that make periodization difficult with team sports:

- Extended season of competition
- Multiple training goals
- Interaction of strength training and conditioning
- Time constraints imposed by concurrent technical and tactical training
- Effect of physical stresses from games

To elaborate on these points, consider the game of rugby union (the same logic can be applied to other sports). First, international-level players are involved in tournaments or tours that last for 10 months. That means they have 2 months for a combined transition from competition to preseason. Given that the career of an international player is 8 to 10 years, failure to incorporate general strength training during the season results in a gradual loss of the physical characteristics that are expected at that level.

The players are required to be strong and powerful, to resist fatigue associated with repetitive bouts of short-duration, high-intensity exercise, and to have fast sprinting speed while maintaining their ball skills and cognitive function. Maintaining each of these characteristics for the duration of the season imposes unique challenges for the trainer, who is always competing with the technical coach for time with the players. It is clear that a traditional periodization model cannot be applied in this case.

When trainers plan the periodization of training loads between matches, they have to consider factors such as the quality of the opposition, the number of training days between matches, and any travel associated with playing games away from home (Kelly and Coutts 2007). They also have to consider the stage of season, the importance of the match, and which players need to be rotated in the squad. These demands, coupled with the expectations of the fans and the sponsors of the team, make the task of periodizing training a continual challenge!

Monitoring Exercise Tolerance

Monitoring how athletes respond to training is imperative for the periodized plan to work effectively. High-level human performance requires years of training and careful planning to achieve a peak. Athletes can get to a certain level on talent alone, but careful, well-constructed training is required to move to the next level.

The capacity to absorb a training load varies significantly among athletes. Even for an individual athlete, the relationship between training load and performance varies over a career and according to the different phases of training. A study of Olympic swimmers showed that their response to a given training volume varied, not only between seasons, but even between training sessions (Hellard et al. 2005). This study showed that at the elite level, training variables accounted for only 30% of the variation in performance (Hellard et al. 2005). This supports the concept that training programs need to be highly individualized for elite athletes. Some athletes who are prone to injury when training load is increased respond well to low-volume training, whereas other athletes need to sustain a high training volume to reach their potential. Indeed, there was a fourfold difference in training volume between runners of similar ability in the Two Oceans marathon (56 km) (Lambert and Keytel 2000).

It is clear that the ideal training load for each athlete needs to be carefully individualized. This can only be achieved by manipulating the training load in response to symptoms of how the athlete is responding to the training program. It is clear that without frequent monitoring, athletes may conceivably over- or undertrain. To monitor the training status effectively so that the training load can be adjusted accordingly, a few questions should be answered after every training session (Lambert and Borresen 2006):

1. How hard did the athlete find the session?
2. How hard was the session?
3. How did the athlete recover from the session?
4. How is the athlete coping with the cumulative stress of training?

Answers to these questions are imperative for the coach or fitness trainer to make informed decisions about whether or not the planned training needs to be modified. The saying "You cannot manage what you don't measure" is particularly relevant in this context.

While it is important to have a training plan, the day-to-day implementation of the plan should not be rigid. Rather, it should be modifiable based on the symptoms of the athlete. Certain tools can be used to gather this information, and the whole process of gathering the information takes only a matter of minutes (Lambert and Borresen 2006).

Fitness and Fatigue Relationship

A review study by Smith (2003) set out a comprehensive framework of the components of performance and training that should be addressed when developing short- and long-term training plans for elite performance. This review also summarized the most common errors associated with training for high performance (Smith 2003).

These are summarized as follows:

- Recovery is often neglected, and mistakes are made in the micro- and macrocycle sequences. General exercise sessions are not adequately used for recovery.
- Demands on athletes are made too quickly relative to capacity, compromising the adaptive process.
- After a break in training due to illness or injury, the training load is increased too rapidly.
- High volumes of both maximal and submaximal intensity training are used.

- The overall volume of intense training is too high when an athlete is primarily engaged in an endurance sport.

- Excessive attention and time are spent in complex technical or mental aspects, without allowing adequate recovery phases or downtime.

- Excessive amount of competitions with maximum physical and psychological demands are combined with frequent disturbance of the daily routine and insufficient training.

- Training methodology is biased, with insufficient balance.

- The athlete lacks trust in the coach due to high expectations or goal setting, leading to frequent performance failure.

All these points are related to the concept of periodization, particularly a deviation from the general rules that govern the periodization model. These points can also be described using the theory of the fitness–fatigue relationship (Banister 1991; Chiu and Barnes 2003). According to this theory, an athlete's readiness for performance is defined by the difference between the two consequences of a training session: fitness and fatigue (Plisk and Stone 2003; Morton 2001). Following each training session, a state of fatigue occurs, which is at its greatest immediately after the session, but then decreases with time. Another property, referred to loosely as *fitness,* also occurs after each training session. Fitness, which encompasses the benefits of the session, is associated with the acute and more long-term adaptations following the bout of exercise. Fitness benefits decay with time.

The readiness for performance gradually increases after the training session, since fatigue dissipates at a faster rate than fitness decays. A window of opportunity will appear, during which readiness for performance is heightened (figure 3.1). It is obviously counterintuitive to expose athletes to a hard training session when their levels of fatigue are high (unless the plan is to intentionally induce an early stage of overreaching) (Meeusen et al. 2006). Any strategies that minimize fatigue or accelerate recovery after training are theoretically beneficial to athletes, since they will accelerate and increase the duration of the readiness for further training and competitive performance (Barnett 2006).

Regular monitoring ensures that coaches are aware of athletes' fitness–fatigue state (figure 3.1) at any stage of training so that they can adjust the planned training if necessary.

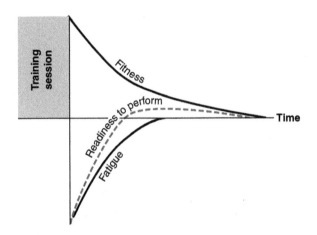

Figure 3.1   The fitness–fatigue relationship

Plisk and Stone 2003; Turner 2011; Banister 1991; Chiu and Barnes 2003.

PRACTICAL APPLICATION

Athletes and coaches are advised to use a periodized training approach when designing their training plans. This implies systematically planning short- and long-term training, varying training loads, providing adequate rest and recovery, and dividing training periods into macrocycles, mesocycles, and microcycles. Younger and less experienced athletes may benefit from a traditional periodization model, whereas older and more experienced athletes may opt for a block periodization approach to training. Whatever the model used, monitoring how the athlete responds to training and making adjustments as needed are imperative for the periodized plan to work effectively. Strategies to accelerate recovery after training increase athletes' readiness for further training and performance in competition.

Summary

Symptoms of overtraining can be avoided if a periodized training program is followed. Periodization is not an exact science in which specific training programs can be designed for all athletes months in advance and adhered to without any flexibility. The basic features of periodized training should include a structured plan that introduces variation into training by modifying training volume, intensity, and type of exercises, according to the specific needs of each individual athlete.

Furthermore, the plan should include structured recovery and restoration (Haff 2004). A common theme for all periodization protocols is the need to manipulate training loads to avoid monotony, to progress from general to sport-specific training, and to focus on eliminating fatigue (Turner 2011). Daily monitoring of an athlete's response to training is an integral part of a periodized training program. Based on this information, coaches should be able to adjust the planned training accordingly.

CHAPTER 4

Active Recovery

Yann Le Meur, PhD, and Christophe Hausswirth, PhD

Active recovery consists of maintaining sub-maximal work after fatiguing exercise with the aim of preserving performance level between events. This is supposed to enhance the recovery mechanisms at the energetic, muscular, and psychological levels. Active recovery can be planned at different times, either as part of a training session or during the cool-down phase. In this case, it generally precedes other recovery methods, such as stretching or massage. It is also possible to include active recovery during the days following an intense training session or competition. This chapter addresses active recovery through this temporal prism, examining its efficacy and how it is applied in relation to the available scientific data.

Recovery Methods and Interval Work

Performance in many athletic activities depends on the athlete's capacity to reconstitute the energy required for the organic functions used during the exercise, in the form of a molecule known as adenosine triphosphate (ATP) (Knicker et al. 2011). Depending on the duration of the exercise, its intensity, and its continuous or intermittent nature, performance depends on the contribution of the athlete's anaerobic and aerobic pathways (Gastin 2001). For example, brief, high-intensity activities, such as sprinting, mainly involve anaerobic metabolism, while long-distance races rely more on aerobic contribution. Depending on the pace variations imposed by the activity, many disciplines involve mixed energy sources, which activate both the anaerobic and aerobic resynthesis pathways to significant levels (Gastin 2001). An example is in team sports where decisive, repeated sprints are separated by periods of less-intense activity (Carling et al. 2008; Gray and Jenkins 2010).

Improvements in anaerobic and aerobic resynthesis capacities are closely linked to the development of some physical athletic qualities, such as speed, endurance, or the capacity to repeat high-intensity efforts. Many methods are used to do this, often relying on the overload principle. According to this principle, the athlete must repeatedly perform at almost maximal level to induce the appropriate physiological adaptations that promote improved physical performance levels (Issurin 2010; Laursen 2010). Coaches and trainers often use interval work, during which periods of high-intensity effort are alternated with periods of low-intensity effort, to prolong the duration of work at supramaximal intensities and to maximize the biochemical and genetic adaptations that favor performance. Because of this, recovery, in terms of duration and the method used (active or passive), must be managed carefully, given that it can directly influence the metabolic response to the exercise and the chronic physiological adaptations induced by the training.

In energy terms, anaerobic performance appears to rely on the muscle's capacity to break down phosphocreatine (PCr) (Hirvonen et al. 1987) and glycogen stocks as rapidly (or for as long) as possible, depending on the intensity and duration of the effort to be produced (Gastin 2001; Glaister 2008; Ward-Smith and Radford 2000). These metabolic adaptations are linked to increases in the activity of key enzymes involved in energy-resynthesis mechanisms, increases in endogenous stores of intramuscular substrates, and increases in the muscle's ability to prevent the accumulation of metabolites linked to fatigue (H^+, Pi, ADP) (Ross and Leveritt 2001). The training methods used to induce these adaptive mechanisms are heterogeneous, but they generally consist of repeating short or long sprints while manipulating the type and duration of recovery between repetitions (Ross and Leveritt 2001). This type of protocol has been used successfully to improve performance levels and to induce positive adaptations of anaerobic metabolism (Cadefau et al. 1990; Dawson et al. 1998; Harridge et al. 1998; Linossier et al. 1993; McKenna et al. 1993; Nevill et al. 1989; Ortenblad et al. 2000), although some studies found no significant effect (Allemeier et al. 1994; Jacobs et al. 1987).

Short-Sprint Repetition

This method involves repeating short-duration sprints (2 to 6 s), between which passive or active recovery periods are placed. These can vary depending on the effect sought (ranging from 10 s to 2 or 3 min). During this type of exercise, ATP resynthesis is mainly due to anaerobic metabolism. The ATP reserves, the alactic and lactic anaerobic resynthesis pathways, and the aerobic pathway are responsible for 10%, 55%, 32%, and 3%, respectively, of the total energy supplied during a 3 s maximal performance (Spencer et al. 2005), and for 6%, 45%, 41%, and 8% during a 6 s sprint (Gaitanos et al. 1993).

In metabolic terms, several studies showed no effect of an increase in intramuscular PCr store (by oral creatine supplementation) on sprint performance (Dawson et al. 1995; Snow et al. 1998). This suggests that progress during maximal exercises of very short duration does not depend on this factor. In contrast, other studies suggested that progress in sprint performance is at least partly related to an improved capacity to deplete PCr stores within a given time (Abernethy, Thayer, and Taylor 1990). Hirvonen and colleagues (1987) showed that highly trained sprinters (maximal racing speed: 10.07 ± 0.13 m/s) depleted their PCr stocks to a greater extent after maximal races over 80 m and 100 m than a group of less-trained athletes (maximal racing speed: 9.75 ± 0.10 m/s). Repeated performance of short exercises (<6 s), separated by complete recovery periods, is likely to improve this PCr-depletion capacity by allowing total resynthesis of the PCr store and repeated almost-maximal speeds of depletion (Dawson et al. 1998). This type of exercise should increase the athlete's maximal racing speed and acceleration capacity by repeated solicitations close to maximum level within each repetition.

To achieve this effect, the duration and the recovery method (active or passive) must be carefully managed to ensure almost total replenishment of the PCr store, since its depletion follows a rapid exponential curve during maximal exercise. While the intramuscular PCr store is about 75 to 85 mmol/kg of fresh muscle, ATP turnover can reach up to 9 mmol ATP $\cdot$ kg^{-1} $\cdot$ s^{-1} (Hultman and Sjoholm 1983). This can easily deplete the store after only 10 s of exercise (Walter et al. 1997). A maximal sprint over 10 to 12.5 s results in the depletion of approximately 40% to 70% of the initial store (Hirvonen et al. 1992; Jones et al. 1985), while the drop is lower for a maximal exercise lasting only 6 s (35–55%) (Dawson et al. 1997; Gaitanos et al. 1993). From this point of view, optimization of the characteristics of intersprint recovery is mainly affected by the level of depletion induced by the exercise and the kinetics of PCr resynthesis.

Post-exercise PCr resynthesis takes place in two phases, which are characterized by monoexponential kinetics, revealing an initial, faster phase, followed by a slower phase. Thus, 50% of PCr reserves can be resynthesized in less than 25 s, while it takes 5 to 8 min to fully restore pre-exercise levels (Harris et al. 1976) (figure 4.1).

Several studies have shown that a number of factors influence the kinetics of PCr resynthesis. Interestingly, their results converge to indicate that this resynthesis involves exclusively aerobic metabolism (McMahon and Jenkins 2002). Thus, PCr resynthesis depends on oxygen availability (providing oxygen to the muscles significantly increases the speed of PCr resynthesis) (Harris et al. 1976; Haseler, Hogan, and Richardson 1999), the subject's aerobic capacity (endurance-trained athletes resynthesize the PCr store more rapidly) (Buchheit and Ufland 2011; da Silva, Guglielmo, and Bishop 2010; Yoshida 2002), and on the type of recovery (passive recovery allows faster resynthesis) (Spencer et al. 2006; Spencer et al. 2008;

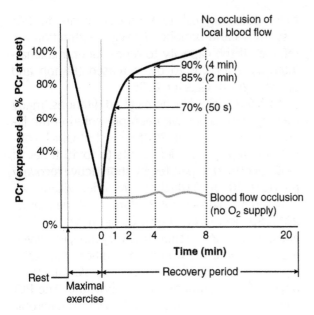

Figure 4.1 PCr resynthesis in the quadriceps femoris during recovery from exhaustive dynamic exercise, with and without occlusion of local blood flow.

Adapted from Harris et al. 1976.

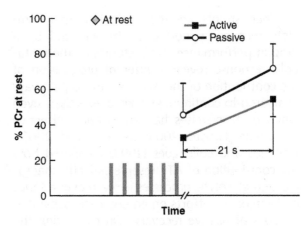

Figure 4.2 Changes in muscle phosphocreatine (PCr) and during and after a 6 × 4 s repeated-sprint test (sprints are represented by vertical bars), with either active or passive recovery in nine moderately trained male subjects. Values are expressed as percentage of resting value ± SD. Note that with active recovery, the percent of resting [PCr] was lower, and it approached significance post-test (32.6 ± 10.6% vs. 45.3 ± 18.6%; p = 0.06; effect size (ES) = 0.8) and after 21 s of recovery (54.6 ± 9.6% vs. 71.7 ± 14.1%; p = 0.06; ES = 1.2).

Reprinted, by permission, from M. Spencer et al., 2006, "Metabolism and performance in repeated cycle sprints: Active versus passive recovery," *Medicine and Science in Sports and Exercise* 38: 1492-1499.

Yoshida, Watari, and Tagawa 1996). For example, after six 4 s sprints on an ergocycle separated by 21 s recovery, Spencer and colleagues (2006) showed that the intramuscular PCr concentration in the vastus lateralis was lower when recovery was based on active (~32% $\dot{V}O_2$max), rather than passive, methods (figure 4.2). They and others hypothesized that active recovery reduces the availability of oxygen for PCr resynthesis in the active muscles compared to passive recovery (McAinch et al. 2004; Spencer et al. 2006; Spencer et al. 2008), since a significant fraction of the oxygen delivered to the muscles would be used to ensure the resynthesis of ATP necessary for active recovery by aerobic metabolism. Results from Dupont and colleagues (2004) support this hypothesis by showing that muscle oxygenation is lower when active recovery (rather than passive) is used.

These results suggest that, when seeking to improve the athlete's capacity to produce short-term (<6 s) efforts at close to their maximum capacity, it is preferable to favor full recovery by a passive method, with adequate time between efforts. In trained subjects, passive recovery for 30 s is enough to maintain a maximal level of performance over 40 repeated 15 m sprints (Balsom et al. 1992b). In contrast, during repeated 40 m

sprints, recovery for at least 2 min is required to maintain the performance level (Balsom et al. 1992a). For the coach, performance progression during the session constitutes an indicator of a well-calibrated duration of recovery. Deterioration in performance during sprints can indicate incomplete recovery and significant modifications to the metabolic response during a series of repeated sprints (Balsom et al. 1992a).

In many sports, such as team or racket sports, performance can be assessed through repeated maximal efforts, rather than based on a single performance (Carling et al. 2008; Gray and Jenkins 2010). In field sports, such as hockey, rugby, and football, studies have shown that the duration of high-intensity efforts varies between 4 and 7 s (Bangsbo, Norregaard, and Thorso 1991; Carling et al. 2008). Thus, even though the ratio between the durations of high- and low-intensity efforts remains in favor of the recovery time, a reduction in the capacity to repeat maximal performance levels has been noted after this type of event. These results indicate the importance of developing the capacity to reproduce maximal efforts (Spencer et al. 2005).

When sprints are repeated after an incomplete recovery period (i.e., characterized by a drop in performance level), studies of the metabolic response reveal a different progression of the contribution of the ATP resynthesis pathways compared to continuous maximal exercises, even when both exercises had an identical effective work time (Glaister 2005; Spencer et al. 2005). Gaitanos and colleagues (1993) compared how the contribution of ATP reserves, alactic anaerobic, and glycolytic processes evolved over a series of 10 maximal efforts on an ergocycle separated by 30 s of passive recovery. On comparing the first to the tenth sprint, they observed that the glycolytic contribution dropped from 44% to 16%, even reaching a null value during the last sprint in 4 out of 7 subjects tested (figure 4.3). At the same time, the relative contribution of the alactic anaerobic pathway compared to the global anaerobic contribution increased from 50% to 80%. These results, which were confirmed by subsequent studies, indicated that during repeated short sprints separated by brief recovery periods, the evolving metabolic environment induced by the exercise led to a gradual inhibition of glycolysis (Gaitanos et al. 1993; Parolin et al. 1999; Putman et al. 1995). While the reasons explaining this phenomenon remain to be determined (e.g., reduction in glycogen store, drop in phosphofructokinase activity linked to the lower pH or to accumulation of citrate in the cytosol),

these data suggest that improvements to PCr resynthesis and peripheral oxygen extraction are required if the capacity to repeat brief, maximal-intensity efforts is to be increased (Bishop and Spencer 2004; Glaister 2005).

To favor these biochemical adaptations, many training protocols have been presented in the literature (Spencer et al. 2005), most of which consisted in repeating brief sprints (5 to 15 times of 3–6 s of effort), separated by short active recovery periods (~10–30 s) (Buchheit, Mendez-Villanueva, et al. 2010; Glaister 2008; Hunter, O'Brien, et al. 2011; Spencer et al. 2005). Using these protocols and considering the short recovery time, PCr levels are incompletely restored. In addition, maintaining submaximal work reduces the availability of oxygen for resynthesis of this substrate. The PCr store thus reduces progressively, while the aerobic contribution to ATP production increases during exercise. This training method therefore ensures solicitation of mixed ATP resynthesis pathways, which are all extensively activated during this type of exercise (Gaitanos et al. 1993).

Long-Sprint Repetition

While the results illustrated in figure 4.3 show a significant contribution of the alactic-anaerobic ATP-resynthesis pathway for short-burst efforts (<6 s), they also show that glycolysis is only activated for a very short time. Gastin's meta-analysis (2001) also shows that ATP resynthesis is mainly

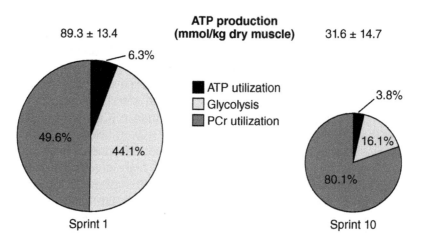

Figure 4.3 Anaerobic adenosine triphosphate (ATP) production (excluding energy provision related to lactate efflux) during the first and tenth sprints of 10 × 6 s maximal sprints separated by 30 s recovery periods. PCr = phosphocreatine.

Reprinted, by permission, from G.C. Williams et al., 1993, "Human muscle metabolism during intermittent maximal exercise," *Journal of Applied Physiology* 75: 712-719.

due to lactic anaerobic metabolism during maximal exercises lasting between 10 s and 1 min. For example, studies have shown that the anaerobic and aerobic processes contribute to ATP resynthesis for 88% and 12%, respectively, during a maximal exercise for 15 s (Medbo, Gramvick, and Jebens 1999) and 73% and 27% when the duration is extended to 30 s (Medbo and Tabata 1993). Beyond 75 s of effort, oxidative mechanisms are the main energy sources, although glycolytic activity continues to contribute 15% of the energy for a 3000 m event in trained athletes (Duffield, Dawson, and Goodman 2004, 2005).

From this point of view, developing glycolytic qualities constitutes a priority training axis for many sporting disciplines. Scientific research has shown that the metabolic adaptations associated with the development of glycolytic capacities are mainly linked to an improved buffering capacity, which represents the muscle's capacity to limit the drop in pH despite the accumulation of H^+ ions induced by a supramaximal exercise (Juel 1998; Parkhouse and McKenzie 1984; Ross and Leveritt 2001). Other studies have also shown that sprinting over long distances leads to an increase in activity of key glycolytic enzymes (Cadefau et al. 1990; MacDougall et al. 1998; Parra et al. 2000; Roberts, Billeter, and Howald 1982), but how these adaptations explain the variations in performance levels remains controversial (Ross and Leveritt 2001). Similarly, the influence of sprint training on the increase in intramuscular glycogen store is debated, and there is a lack of scientific proof affirming that the level of glycolytic performance is linked to this adaptive mechanism (Ross and Leveritt 2001).

Given that lactic anaerobic activity is progressively reduced during repeated short sprints (Gaitanos et al. 1993), the most commonly adopted strategy to develop glycolytic capacity consists in repeating long sprints, lasting between 20 s and 2 min, separated by recovery periods (Dawson et al. 1998; MacDougall et al. 1998; Putman et al. 1995). In metabolic terms, the aim of this type of session is to improve the athlete's anaerobic capacities by improving the muscle's capacity to fight against the accumulation of metabolites (in particular H^+, Pi, ADP) and to tolerate these metabolic disturbances during prolonged supramaximal exercises. Inter-exercise recovery is generally active and extended if the duration of total high-intensity work is long (>2 min). This allows the athlete to reach and maintain an almost maximal intensity throughout the series. Active recovery allows an earlier return to homeostasis by accelerating metabolite clearance (see section on active recovery and repeated performances for more information). If the number of repetitions is moderate (total effective work time < 2 min), recovery can be induced, but it may lead to an earlier deterioration in performance levels (Dupont et al. 2007; Roberts, Billeter, and Howald 1982). During this type of session, active recovery accelerates the return to homeostasis and reduces the oxygen debt accumulated at the start of the exercise (Dupont et al. 2007). Therefore, it also solicits aerobic metabolism to a greater extent. Passive recovery improves PCr resynthesis and hemoglobin and myoglobin re-oxygenation, and also leads to a greater oxygen debt at the start of exercise. This promotes more extensive solicitation of anaerobic metabolism during sprints (Dupont et al. 2007).

Recovery Methods and Development of Aerobic Fitness

$\dot{V}O_2$max represents the highest level of oxygen consumption that someone can achieve when exercising at sea level. This parameter is identified as a determinant for performance in endurance activities. Thus, a high $\dot{V}O_2$max is required to reach a higher level of performance in aerobic sports (Joyner and Coyle 2008). Numerous scientific studies have investigated methods to promote and maintain a high fraction of $\dot{V}O_2$max (Billat 2001a, b; Laursen and Jenkins 2002). Their results indicate that intermittent exercises involving >90% $\dot{V}O_2$max improve aerobic performance levels in previously trained athletes.

Interval training consists of alternating high-intensity exercises with passive or active recovery periods. Introducing recovery periods between the phases of intense exercise allows athletes to maintain the intensity for longer than when exercise is performed continuously until exhaustion. Thus, the duration and method of recovery (active or passive) significantly affects the bioenergetic response during interval training sessions. Many trainers and researchers recommend active recovery between repeated brief, intense efforts to maintain a high $\dot{V}O_2$max, to promote the return to homeostasis (elimination of lactate ions and protons), and thus to maintain the exercise for longer.

This hypothesis was validated with a recovery period of at least 30 s (Dorado, Sanchis-Moysi, and Calbet 2004; Thevenet et al. 2007). For example, Dorado and colleagues (2004) compared the influence of the recovery method (pedaling at 20% $\dot{V}O_2$max, stretching, passive recovery) used for 5 min between four exercise series where 110% of maximal power was maintained for as long as possible during an incremented maximal test. They observed that with active recovery, compared to the other two recovery methods tested, the global maintenance time and aerobic contribution were increased by 3% to 4% and 6% to 8%, respectively. This difference was associated with faster $\dot{V}O_2$ kinetics and a higher peak $\dot{V}O_2$ level for active recovery, although no difference in aerobic energy supply was noted among the three conditions.

On the other hand, when recovery was of short duration (5–15 s), no difference between active and passive recovery in terms of duration of work at close to $\dot{V}O_2$max (>90% $\dot{V}O_2$max) was observed (Dupont and Berthoin 2004). Indeed, these authors showed that during a series of alternating 15 s periods of running at 120% of maximal aerobic speed (MAS) with 15 s recovery phases, with passive recovery, the lower proportion of time spent close to $\dot{V}O_2$max is compensated by a longer-limit exercise time (43% of the time at >90% $\dot{V}O_2$max over 745 s versus 64% over 445 s, for passive and active recovery methods, respectively). It therefore seems appropriate to favor passive recovery during intermittent "short–short" work targeting the development of aerobic qualities (such as 15 s–15 s). Indeed, the numerous speed changes associated with this type of work are linked to more significant variations in peripheral oxygen extraction, given that the relative exercise time during which the heart rate adjusts to the intensity of the workload is longer.

Active Recovery and Repeated Performances

In many sporting activities, because of how competitions are designed, athletes are forced to reproduce performances within a short time frame. In judo and swimming, for example, com-

petitors must often reproduce several maximum performances over the course of a single day, with recovery times that are sometimes less than 30 min. In other disciplines, the organization of tournaments or qualification legs takes place over a longer period, which may extend to several weeks, with competition days and recovery phases interspersed throughout. For example, the winner of a tennis Grand Slam must win 7 matches, which may be played as the best of 5 sets over 2 weeks, while during the football (soccer) World Cup, the best teams play approximately 2 matches per week over a month. Logically, recovery plays an essential role in these sports, all the more so since the most difficult events are generally played toward the end of these competitions. Although active recovery is often used by players after matches and the following day, the results in the scientific literature have not validated its advantages.

Return to Reference Performance Level

Many studies have compared the influence of the recovery method (active or passive) on mean capacity to repeat a maximal performance. The results have generally shown a positive effect of active recovery on performance-level maintenance, but a large number of the protocols used were not in line with the conditions of real sporting practice as part of competitions. Table 4.1 presents the results of recent studies that were correlated better with the real-life context of high-level performance.

As indicated in table 4.1, active recovery generally seems to have a positive effect on maintaining a performance level within a relatively short time frame (10–20 min) (Franchini et al. 2009; Franchini et al. 2003; Greenwood et al. 2008; Heyman et al. 2009; Thiriet et al. 1993), but it does not appear to provide any significant benefit when events are separated by several hours or days (King and Duffield 2009; Lane and Wenger 2004; Tessitore et al. 2007).

Interestingly, Heyman and colleagues (2009) showed that active recovery involving large muscle mass favors the maintenance of performance levels, including for tasks not involving the same muscle groups (e.g., pedaling with lower limbs for climbers, who are mainly faced with problems due to repeated isometric contractions, which lead to temporary ischemia in the forearm muscles). Green and colleagues (2002) showed

Table 4.1 Effects on Performance of Active and Passive Recovery

Sport	Tiring exercise	Time between tests	Active recovery method	POST-RECOVERY PERFORMANCE		Significance
				Passive recovery	Active recovery	
Swimming (Greenwood et al. 2008)	Maximal performance over 200 m freestyle	10 min	10 min at the lactic threshold	116 s	112.4 s	Significant difference, positive effect of active recovery
Artistic gymnastics (Jemni et al. 2003)	Olympic circuit (6 elements)	10 min between each element	5 min passive recovery followed by 5 min active recovery (self-managed)	38.39 pts	47.28 pts	Significant difference, positive effect of active recovery
Judo (Franchini et al. 2003)	5 min Randori	15 min	15 min running at 70% of the anaerobic threshold	Upper body Wingate test: 570 W	Upper body Wingate test: 571 W	No significant difference
Judo (Franchini et al. 2009)	5 min Randori	15 min	15 min at 70% of the anaerobic threshold	3/8 (victories/defeats)	9/4 (victories/defeats)	Significant difference on the outcome of the combat when only one judoka recovers actively (positive effect of active recovery). No difference during an upper body Wingate test, nor during a specific judo fitness test.
Climbing (Heyman et al. 2009)	Continuous climbing until completion of a 6-b-graded route (~8 min)	20 min	20 min pedaling with the legs at 30–40 W (increase in brachial arterial circulation noted)	Time limit on the route: −28% of pre-test value	Time limit on the route: −3% of pre-test value	Significant difference, positive effect
Cycling (Monedero and Donne 2000)	5 km time trial (~6 min)	20 min	50% $\dot{V}O_2$max	+9.9 s	(+2.3 s for the combination of active recovery + massage)	Significant difference only when active recovery was combined with massage, positive effect

(continued)

Table 4.1 *(continued)*

Sport	Tiring exercise	Time between tests	Active recovery method	POST-RECOVERY PERFORMANCE		Significance
				Passive recovery	Active recovery	
Cycling (Thiriet et al. 1993)	4 time limits (~2 min each)	20 min	Arm or leg pedaling	Significant decrease of performance between pre- and post-recovery test sessions	No significant decrease of performance between pre- and post-recovery test sessions	Significant difference (whether recovery was performed using the arms or the legs), positive effect
Running (Coffey, Leveritt, and Gill 2004)	T_{lim} at 130% MAS / 15 min passive recovery / T_{lim} at 100% MAS	4 h	15 minutes at 50% of MAS	+2.9 s over 400 m +4.8 s over 1000 m +16.7 s over 5000 m	+2.7 s over 400 m +2.2 s over 1000 m +1.4 s over 5000 m	No significant difference
Futsal (Tessitore et al. 2008)	Futsal match (1 h)	6 h	8 min jogging, 8 min alternating walking / side skipping, 4 min stretching (2 methods: normal and aqua jogging)	CMJ: −2.3% Bounce jumping height: −1.2% 10 m sprint time: +1.7%	CMJ: −2.0% Bounce jumping height: +0.0% 10 m sprint time: +1.7%	No significant difference during CMJ tests, bounce jumping, and 10 m sprint
Cycling (Lane and Wenger 2004)	18 min maximal performance	24 h	15 min at 30% $\dot{V}O_2$max	−2%	No degradation in performance level	Effect of recovery not significant
Netball (King and Duffield 2009)	Exercise simulating a netball match	24 h	15 min running at 40% $\dot{V}O_2$max	Percentage decrement in 5 × 20 m sprint time and vertical jump: −2.3% and +0.8%	Percentage decrement in 5 × 20 m sprint time and vertical jump: 0.0% and +1.6%	No significant difference
Soccer (Andersson et al. 2008)	Football match (90 min)	72 h	20 min pedaling at 45% of $\dot{V}O_2$peak, 30 min circuit training <50% 1RM, 10 min pedaling at 45% of $\dot{V}O_2$peak	+0.02 s over 20 m Vertical jump: −0.9 cm	+0.01 s over 20 m Vertical jump: −1.3 cm	No significant difference during CMJ tests, 20 m sprint, maximal force at 5 h, 21 h, 45 h, 51 h, 69 h after the match

This table summarizes the results from sample studies comparing the effects of active and passive recovery on performance. The studies listed were ranked according to the recovery time between the two performance tests, during which the recovery method was used. From all these studies, it appears that the benefits of active recovery mainly depend on the time lapse between repeat performances and the type of exercise involved.

T_{lim}: Limit-time to fatigue

MAS: Maximal aerobic speed

CMJ: Countermovement jump

that pedaling with the lower limbs also induces an increase in local blood flow to the upper limbs. This is particularly interesting for athletic activities involving prolonged holds, such as climbing, judo (through kumi-kata), and windsurfing.

Thus, the results reported in table 4.1 reveal a positive effect of active recovery on exercises involving a significant anaerobic contribution (Franchini et al. 2009; Franchini et al. 2003; Greenwood et al. 2008; Heyman et al. 2009; Thiriet et al. 1993), during which fatigue is associated with metabolite accumulation in the muscles and blood (Knicker et al. 2011). Active recovery used for a short time (10–20 min) between two exercises of this type accelerates the return to homeostasis.

Return to Homeostasis

Most studies investigating the effects of active recovery on the kinetics of return to homeostasis and the reduction in muscle fatigue have focused on the kinetics of lactate clearance. Quite unanimously, they demonstrated that active recovery allows a faster return to resting lactatemia levels than passive recovery (Ahmaidi et al. 1996; Belcastro and Bonen 1975; Choi et al. 1994; Gisolfi, Robinson, and Turrell 1966; Greenwood et al. 2008; Hermansen and Stensvold 1972; Stam-

ford et al. 1981; Taoutaou et al. 1996; Watts et al. 2000). Maintaining submaximal activity after exercise, leading to a significant increase in lactatemia, effectively favors lactate oxidation, mainly by the active muscle fibers (Bangsbo et al. 1994). This is what causes lactatemia to return more rapidly to resting values than with passive recovery.

Many studies have also attempted to identify the optimal exercise intensity for active recovery by studying the kinetics of postexercise lactatemia. However, these reported heterogeneous results indicate an interval between 25% and 63% of $\dot{V}O_2$max, depending on the study (Boileau et al. 1983; Bonen and Belcastro 1976; Dodd et al. 1984; Hermansen and Stensvold 1972). In these studies, the optimal intensity for recovery was expressed as a function of the athletes' maximal oxygen consumption capacity, whereas lactate production progresses in a nonlinear manner with exercise intensity. More recent work expressed the exercise intensity as a function of submaximal parameters. These studies reported more homogeneous results and identified the anaerobic threshold as the intensity for which lactate clearance is most effective (Baldari et al. 2004, 2005; Greenwood et al. 2008; Menzies et al. 2010) (figure 4.4). Interestingly, Menzies and colleagues (2010) also showed that trained athletes spontaneously choose

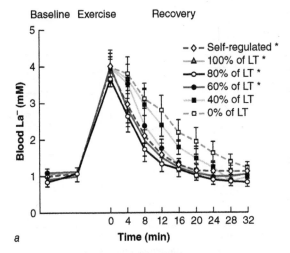
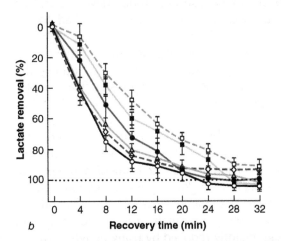

Figure 4.4   (a) Blood-lactate concentration ($[La^-]$) at baseline (warm-up), after a 5 min run at 90% $\dot{V}O_2$max (exercise, 0 min on x-axis), and during active and passive recovery at exercise intensities of 0–100% of the individual lactate threshold (LT) and a self-regulated exercise intensity for active recovery (79 ± 5% of lactate threshold). (b) Normalized blood-lactate clearance during active and passive recovery. The dashed line indicates baseline.

*Significantly different from passive recovery (0% of lactate threshold, $p < 0.01$) and 40% of lactate threshold ($p < 0.05$).

Reprinted, by permission, from P. Menzie et al., 2010, "Blood lactate clearance during active recovery after an intense running bout depends on the intensity of the active recovery," *Journal of Sports Sciences* 28: 975-982.

an intensity of exercise during recovery close to this target intensity.

Several studies extended these results by showing that active recovery also allows an earlier return to a resting pH value than passive recovery (Fairchild et al. 2003; Yoshida, Watari, and Tagawa 1996). Active recovery therefore appears to reduce the effects of exercise-induced acidosis in the short-term, both peripherally and centrally. This may help preserve the neuromuscular system's function during subsequent events. Takahashi and Miyamoto (1998) showed that this faster return to homeostasis might be linked to improvements in venous return induced by active recovery. Given that transport of lactate ions from the intramuscular medium toward the blood is based on cotransport with H^+ ions, maintenance of a local high rate of blood flow during active recovery would also allow a faster return to muscle homeostasis during the postexercise phase.

To our knowledge, only the study by Yoshida and colleagues (1996) investigated the influence of maintaining a submaximal exercise intensity after an anaerobic-dominant exercise on changes to the intracellular inorganic-phosphate (Pi) concentration. These changes can perturb the mechanisms of muscle contraction (Allen, Lamb, and Westerblad 2008). This study also investigated how peak values of muscle proton (H^+) and Pi concentrations were affected during exercise and recovery. Indeed, several studies have shown that two exercise-related Pi peaks are linked to pH kinetics (Laurent et al. 1992; Mizuno et al. 1994): one during the exercise phase and another during the recovery phase. The first peak, during the exercise phase, is due to a very significant depletion in phosphocreatine (PCr) stocks at the start of exercise. During passive recovery, a second peak occurs when the PCr stores are depleted, once again to resynthesize the ATP that the cell needs to ensure a return to homeostasis, because as the pH remains low in these conditions, glycolytic activity is limited. However, Yoshida and colleagues (1996) indicate that this phenomenon can be significantly reduced by using active recovery. Indeed, this ensures a better supply of oxygen to the muscles, which favors a faster return to resting pH levels. This would thus reactivate the glycolytic pathway and, in return, limit the accumulation of Pi during the recovery period. This hypothesis indicates that, by maintaining an adequate local blood flow, active recovery can help eliminate metabolites accumulated during exercise.

Muscle-Glycogen Resynthesis

Another important consideration, although not often taken into account when postexercise recovery is planned, involves the influence of the recovery method on muscle glycogen resynthesis. Several studies have shown that passive recovery allows faster resynthesis (Bonen et al. 1985; Choi et al. 1994; Fairchild et al. 2003) (figure 4.5), while others observed no significant difference between active and passive recovery (Bangsbo et al. 1994; McAinch et al. 2004; Peters Futre et al. 1987). However, two of these studies probably did not use an adequate recovery duration (10 min and 15 min for Bangsbo et al. 1994 and McAinch et al. 2004, respectively) to allow effective resynthesis. Bonen and colleagues (1985) also showed glycogen resynthesis to be lower, including when athletes use nutritional strategies to maximize the recovery mechanism (1.2 g/kg immediately after exercise, then $1.2 \text{ g} \cdot \text{kg}^{-1} \cdot \text{h}^{-1}$ over the following 4–6 h). In the other studies, the athletes were not provided with a source of carbohydrate.

In energy terms, active recovery may thus potentially constitute a disadvantage for athletes who train several times per day or for whom com-

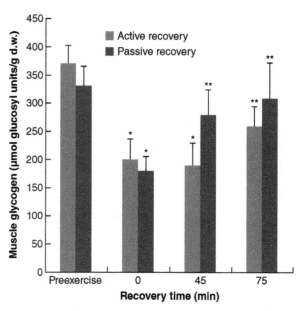

Figure 4.5 Muscle glycogen during passive and active recovery. Results are expressed as mean ± SEM (*n* = 8). Symbols * and ** indicate a significant difference from levels determined before exercise or at 0 min of recovery, respectively.

Reprinted, by permission, from T.J. Fairchild et al., 2003, "Glycogen synthesis in muscle fibers during active recovery from intense exercise," *Medicine and Science in Sport and Exercise* 35: 595-602.

petitions rely on repeating events over the same day. Repeated mobilization of glycogen stores, coupled with an increase in basal metabolism due to the stress of the competition is, in effect, likely to significantly deplete the store and to limit the athlete's performance level. Therefore, the best strategy appears to be to plan active recovery when the time between two events is less than 30 min and to use longer periods to favor good nutrition using a recovery drink (see chapters 7 and 8 for hydration and nutrition strategies).

Single-Performance Recovery

A final active recovery technique that is commonly used consists of maintaining low-intensity activity immediately after the main part of the training session or competition, or even over the following days. This recovery strategy is generally considered a gradual cool-down, during which the intensity of exercise is progressively lowered. It relies on using exercise (running, pedaling, swimming) at moderate intensity for 10 to 30 min. To our knowledge, very few studies have investigated the advantages of this, despite the fact that it is very commonly used by athletes and it is recommended by most trainers to help promote metabolic recovery (return to the resting state), accelerate muscle recovery (reduction in severity of DOMS), or promote psychological disengagement from the session (cool-down).

Energy Considerations

In metabolic terms, active recovery after a high-intensity exercise accelerates metabolite elimination and a return to homeostasis in the muscles and blood (Ahmaidi et al. 1996; Belcastro and Bonen 1975; Choi et al. 1994; Gisolfi, Robinson, and Turrell 1966; Greenwood et al. 2008; Hermansen and Stensvold 1972; Stamford et al. 1981; Taoutaou et al. 1996; Watts et al. 2000). Given that the biochemical perturbations induced by the exercise activate metabosensitive III and IV afferent nerves, and thus increase the difficulty of the exercise (Gandevia 2001), it appears logical that active recovery would favor a better mental state after a training session than rest alone. This hypothesis is supported by several studies revealing improved perceived recovery when active rather than passive recovery methods are used (Suzuki et al. 2004). In the longer term, however, the benefit appears questionable, given that

a return to basal blood- and muscle-metabolite concentrations is generally noted less than 1 h after discontinuing exercise, even with passive recovery (Baldari et al. 2004, 2005).

Recovery From Muscular Damage

Active recovery is also suggested to promote recovery from muscle damage. By favoring an increase in local blood flow, and thus promoting elimination of muscle-cell debris as well as nutrient transport to damaged tissues, it is suggested for accelerating the muscle-regeneration process (Ballantyne 2000; Hedrick 1999; Mitchell-Taverner 2005). The results from the scientific literature on this question do not, however, confirm this hypothesis. Some studies have reported a positive effect of active recovery on reducing muscle damage (Gill, Beaven, and Cook 2006), while others report no effect (Andersson et al. 2008; Martin et al. 2004), and still others indicate delayed muscle recovery (Sherman et al. 1984). Gill and colleagues (2006) thus showed that active recovery (i.e., 7 min pedaling at 150 W) immediately after a rugby match significantly reduced the creatine-kinase concentration measured in the forearm of professional players by transdermal sampling. However, this measurement technique has not yet been validated. A study by Suzuki and associates (2004) showed no significant effect of active recovery used immediately after a rugby match on muscle injury up to 48 h after the match. Martin and colleagues (2004) compared the effect of 30 min running at 50% $\dot{V}O_2$max to passive recovery over 4 days following intense eccentric work, leading to significant muscle damage. Their results showed no effect of the recovery method on maximal-voluntary or electrically induced force. Similarly, Andersson and associates (2008) observed no significant effect of active recovery performed 24 h and 48 h after a football match by international players, either in terms of perceived muscle pain or biochemical markers of muscle damage (creatine kinase, urea, uric acid). Recently, King and Duffield (2009) confirmed this result by showing perceived muscle pain to be similar in netball players 24 h after a match simulation, whether they had run for 20 min at low intensity (40% $\dot{V}O_2$max) afterward or not. On the other hand, values were higher than those measured with cold-water or contrasting-temperature immersion-based recovery.

Finally, a study by Sherman and associates (1984) showed that running at low intensity (50%

$\dot{V}O_2$max) for 20 to 40 min slowed the return to basal-work capacity of the lower limbs after a marathon (figure 4.6). Up to 1 week after the race, they thus reported a significant difference between the group of runners who continued their training after the competition and a control group that used passive recovery over the same time period.

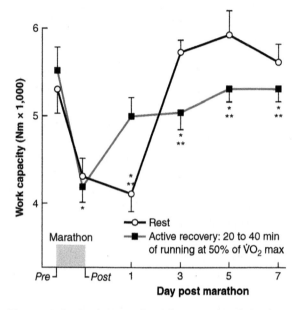

Figure 4.6 Evolution of work capacity of the knee extensors after a marathon: influence of active recovery.

*Significantly different from the pre-marathon value.

**Significantly different from passive recovery.

Reprinted, by permission, from W.M. Sherman et al., 1984, "Effect of a 42.2-km footrace and subsequent rest or exercise on muscular strength and work capacity," *Journal of Applied Physiology* 57: 1668-1673.

Based on all these results, it seems that active recovery does not reduce muscle injury induced by mechanical constraints or exercise-related oxidative stress (Barnett 2006). It even appears preferable to avoid running after exercises that induce significant muscle injury in the lower limbs, given that this activity is likely to increase the time before muscle regeneration can set in (Sherman et al. 1984). Activities such as cycling, swimming, and aqua jogging, where the weight is borne by an external element, are probably better adapted in these conditions.

Immune Response

Unexpectedly, some studies have shown a positive effect of active recovery after an intense training session on postexercise immune defenses (Wigernaes et al. 2000; Wigernaes et al. 2001). These authors showed that maintaining a low level of aerobic exercise at the end of a session (15 min at 50% $\dot{V}O_2$max) regulates the number of white blood cells, whereas these are usually seen to drop. This drop has been associated with a reduction in immune defenses after intense exercise. These results suggest that active recovery could potentially limit the risk of infection in athletes, particularly those with a high training workload, when the risk of infection is high (Walsh, Gleeson, Shephard, et al. 2011). Other studies will be required to confirm this hypothesis and to add to our understanding of the relationship between immunity and active recovery.

PRACTICAL APPLICATION

When performances must be repeated in a short period (<30 min), active recovery should be planned because it accelerates the return to homeostasis. For this reason, active recovery generally seems to have a positive effect on maintaining a performance level within a relatively short time frame, when it involves large muscle masses. In contrast, no clear benefit appears for maintaining submaximal exercise intensity when maximal exercises are interspersed by longer recovery periods. In this case, other strategies, including rest, massage, or cold-water immersion, are preferred for promoting recovery. Nutrition strategies should also be privileged immediately after exercise cessation in order to replenish glycogen stores.

Are We Talking About Recovery or Training?

From all these results, it emerges that the expected benefits of active recovery used at the end of a session or the next day are not confirmed by the scientific data (table 4.1). However, the strategy remains widely used by both trainers and athletes. We believe that the experimental data show that maintaining sub-maximal activity the day after a competition or a very intense training session could actually be a means to continue training without increasing the state of fatigue, rather than a means to truly accelerate the recovery process. Seiler and colleagues (2007) showed that in endurance-trained athletes, performing a low-intensity exercise (60% $\dot{V}O_2$max) does not affect the autonomous central nervous system, even when the exercise is maintained for several hours. Recent work suggests that the duration of low-intensity work is an adequate stimulus to favor the expression of an aerobic phenotype. By activating the appropriate signaling pathways (figure 4.7), the expression of muscle factors favoring endurance performances is induced (mitochondrial biogenesis, capillarization, increase in oxidative activity) (Coffey and Hawley 2007; Laursen 2010). Future studies should add to our understanding of these adaptive mechanisms and allow us to better describe the chronic physiological responses linked to low-intensity work, of which active recovery techniques are an integral part. So, is it really active recovery or continued training? The question remains.

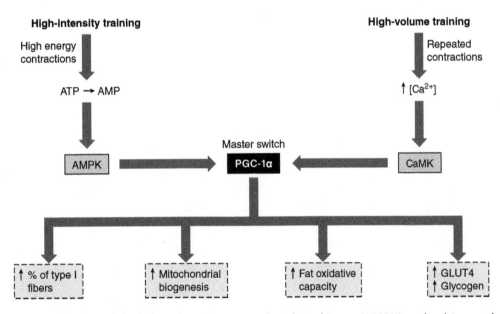

Figure 4.7 Simplified model of the adenosine monophosphate kinase (AMPK) and calcium–calmodulin kinase (CaMK) signaling pathways, as well as their similar downstream target (PGC-1α). This master switch is thought to be involved in promoting the development of the aerobic muscle phenotype. High-intensity training appears more likely to signal via the AMPK pathway, while high-volume training appears more likely to operate through the CaMK pathway. ATP: adenosine triphosphate; AMP: adenosine monophosphate; GLUT4: glucose transporter 4; [Ca²⁺]: intramuscular calcium concentration.

Reprinted, by permission, from P.B. Laursen, 2010, "Training for intense exercise performance: high-intensity or high-volume training?" *Scandinavian Journal of Medicine and Science in Sports* 20 Suppl 2: 1-10.

Summary

The recovery modality during interval training directly influences the metabolic response to the exercise as follows:

- Passive recovery between short maximal sprints (<6 s) increases phosphocreatine resynthesis (and performance maintenance), while active recovery reduces the availability of oxygen for resynthesis of this substrate.

- Active recovery between long sprints (>20 s) accelerates the return to homeostasis and reduces the oxygen debt accumulated at the start of exercise.

- During interval training aiming to develop $\dot{V}O_2$max using long intervals (≥30 s), active recovery increases aerobic contribution through faster $\dot{V}O_2$ kinetics and higher $\dot{V}O_2$ level during recovery. In contrast, during short interval training (5–15 s), passive and active modalities lead to similar accumulated time near $\dot{V}O_2$max.

- When performances must be repeated in a short period (<30 min), active recovery should be planned because it accelerates the return to homeostasis. No clear benefit appears for maintaining submaximal exercise intensity when maximal exercises are interspersed by longer recovery periods.

CHAPTER 5

Psychological Aspects of Recovery

Pierre-Nicolas Lemyre, PhD, and Jean Fournier, PhD

With: Christophe Hausswirth, PhD, and Jean-Francois Toussaint, PhD

In sport psychology, few studies deal directly with the benefits of recovery (Kellmann and Kallus 2001), a topic mainly linked to physiology. The term *recovery* is mainly used in sport psychology to indicate convalescence after injury (Brewer 2003), or by athletes themselves to indicate days or periods of rest (Gustafsson et al. 2007).

The sport psychology literature follows the general trends of psychology. Traditionally, it has primarily covered the negative aspects of practicing a sport. Thus, the study of psychological aspects, such as anxiety, stress, or concentration problems, is more common than an examination of the factors or characteristics defining good performance (e.g., the optimal performance state). The same can be said of recovery. This chapter is based on the existing scientific literature, which explains why it mainly deals with the negative aspects of intense training.

It is currently believed that, in recent years, elite athletes have been subjected to increasingly stringent requirements (Holt 2007). Sport psychology studies have shown that the pressure placed on athletes taking part in competitions leads to a reduced involvement with the sport and an increased rate of burnout (Gould and Dieffenbach 2003). *Burnout* is defined by Freudenberger (1980) as a syndrome characterized by progressive disillusionment linked to psychological and physiological symptoms that result in reduced self-esteem. It is accepted today that some athletes who train excessively or inappropriately, participating in too many competitions, are also often under pressure from parents, relatives, and coaches. The effects can be observed in terms of physical performance, health, and well-being (Schaal et al. 2011).

However, the psychological aspects of overtraining, intense engagement, and recovery have not been extensively studied (Goodger et al. 2007). Only two special issues of sport psychology journals have been devoted to burnout (the *Journal of Applied Sport Psychology*, 1990, and the *International Journal of Sport Psychology*, 2007). Due to this dearth of data on training, it is difficult to present what we currently know about the psychological aspects of recovery.

We therefore chose to deal with recovery by selecting three recent research trends that all use biological indicators and are all based on psychosocial aspects. The first—the only one of the three dealing directly with recovery—was introduced by Kellmann. It suggests that athletes suffer more from the effects of poor recovery than from stress (Kellmann 2002). Thus, rather than suffering from overtraining, they suffer from underrecovery. The second research trend highlights the importance of motivation in relation to the risk of burnout (Lemyre, Roberts, and Stray-Gundersen 2007). Athletes faced with growing difficulties because of an excessive workload or an excessive desire to win may lose motivation and disengage with their sport. The question of motivation (or absence of motivation, since we are dealing with poorly adapted motivation patterns) is therefore crucial, both to understand the situation and to adapt the recovery strategy. Finally, a third psychosocial trend focuses on describing athletes' engagement with their sport, rather than describing burnout itself. This approach is based on the problems that may lead to burnout, since an ill-adapted motivation pattern can lead to burnout with significant time and number of training sessions. The third series of studies focuses more on a positive psychology approach (Seligman and Csikszentmihalyi 2000) on burnout (Raedeke and Smith 2001). Interest in studies on athletes' commitment to their sport and how it is affected will probably grow in the future, largely because commitment is suggested as a conceptual opposite to burnout, making these complementary studies two sides of the same coin. We end the chapter with a description of the tools currently used at INSEP [National Institute of Sport, Expertise and Performance] in France as part of multidisciplinary research to evaluate some psychological aspects of recovery in sport.

Effects of Poor Recovery

The Kellmann model (Kellmann 2002) relies on the recovery-stress questionnaire for sport (RESTQ-Sport), which can identify athletes who underrecover. The RESTQ-Sport is part of psychological follow-up to training, where it can be used to track the progression of stress and recovery indicators for individual athletes. The recovery scales specific to sport (see items 16 to 19 in the following section) relate to perceived physical condition, relationship with team members, confidence, and motivation.

The questionnaire identifies athletes whose stress recovery is inadequate, given the intensity of the training-induced workload. The emphasis is placed, not on reducing the training level, but on increasing the quality and quantity of recovery. Kellmann's idea of recovery, as assessed through specific subscales, focuses on emotional and social aspects. Social relationships, outings, and time spent with friends compensate for the stress related to intense training sessions. Obviously, this type of psychological recovery can be beneficial on the condition that it does not hinder adequate physiological recovery (e.g., through excessive alcohol drinking and late nights). According to Kellmann, early intervention can help the athlete to manage training-related stress, contributing to the prevention of underrecovery or overtraining. In addition to social or distracting aspects, it is recommended to plan recovery times or less-intense activities (e.g., tennis matches, golf played for leisure, or cinema outings) over the year.

The RESTQ-Sport assesses the frequency of stressful and recovery events over the 3 days and nights prior to the tests. Low scores on the stress scales indicate a low level of stress, while high scores on the recovery scales indicate good recovery. The tool uses 76 questions linked to 19 scales, as shown in the following form.

Because the RESTQ-Sport reflects a state at a given time in the athlete's life (see figure 5.1 for an example), it can evolve within a few days. In addition, due to personal interpretation of each question, results are very individual and subjective. It therefore seems important not to try to establish rules for the different scales, but rather to compare results, either between the athlete and a reference group, or for the same athlete at two different times (e.g., before and after a rest period). However, Kellmann recently suggested that profiles indicating underrecovery combined with overtraining could be used as a reference to detect athletes who may be at risk (Kellmann 2009). This questionnaire relates to events that occurred over the last 3 days or nights. It is therefore useful in the aftermath of intense training periods or extended rest periods (e.g., balneotherapy, holidays). However, due to the length of the test (76 questions) and how the questions are asked (over the last 3 days and nights), it is not so good for studying daily variations. We consider it preferable to use this questionnaire for long-term assessments, for example, over a season.

The recently developed Kellmann approach is centered on stress and recovery. It has been used alongside work on burnout in sport that was pioneered by Dan Gould and colleagues on tennis players in the '90s.

RESTQ-Sport Scales

Seven general stress scales

1. *General stress.* Subjects with high scores on this scale describe themselves as stressed, depressed, unbalanced, and listless.

2. *Emotional stress.* High values on this scale are linked to feelings of anxiety, inhibition, irritability, and aggression.

3. *Social stress.* High values tend to indicate a highly strung state (they are likely to enter into conflict and are highly emotional), tendency toward irritation with others, and lack of humor.

4. *Conflicts/pressure.* High scores are found when, over the last 3 days, conflicts were left unresolved, unpleasant tasks were performed, goals were not reached, and some negative thoughts could not be dismissed.

5. *Fatigue.* This scale is characterized by pressure felt during work, training, or school, by feeling frequently disturbed during the performance of important tasks, and by significant fatigue.

6. *Lack of energy.* This scale can highlight inefficiency at work, combined with problems in concentration, decision making, and reaction time.

7. *Physical complaints.* High values indicate physical problems and frequent somatic complaints.

Five general recovery scales

8. *Success.* A scale to assess current success, pleasure derived from work and training, and creativity.

9. *Social recovery.* Athletes who have recently had pleasant or amusing social contacts and exchanges have high scores on this scale.

10. *Physical recovery.* This scale assesses physical recovery, physical well-being, and fitness.

11. *Well-being.* Athletes with high scores on this scale are in good humor, have a general sense of well-being, and are relaxed and content.

12. *Sleep quality.* High scores indicate an absence of sleep disorders and adequate sleep-related recovery.

Three scales for sport-related stress

13. *Disrupted breaks.* High scores indicate frequent interruptions of rest periods and difficulty recovering during and outside training.

14. *Emotional exhaustion.* Athletes who feel overexerted and exhausted, or who have felt like abandoning their sport over the last 3 days, will have a high score on this scale.

15. *Physical fitness/injuries.* High scores indicate injuries or vulnerability to injury with reduced physical fitness.

Four scales for sport-related recovery

16. *Physical condition/fitness.* Athletes with high scores describe themselves as physically effective, fit, and reactive.

17. *Exhaustion/personal accomplishment.* High values indicate athletes who feel integrated in their team, communicate effectively with teammates, and derive pleasure from their sport.

18. *Self-efficacy.* This scale reflects how athletes rate their preparation and the quality of their training over the last 3 days.

19. *Self-regulation.* This scale assesses how well athletes use their mental capacity to prepare their performance (motivation techniques and defining goals).

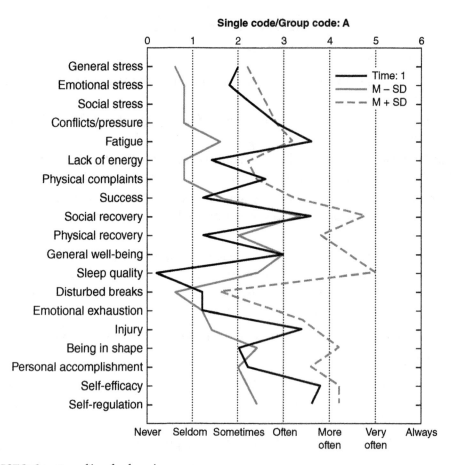

Single code/Group code: A

Figure 5.1   RESTQ-Sport profile of a female rower.

Reprinted, by permission, from M. Kellmann and W. Kallus, 2002, *Recovery stress questionnaire for athletes: User manual* (Champaign, IL: Human Kinetics), 14; Adapted from Springer and the original publisher *Recovery for Performance in Sport,* 2007, "Mood, recovery-stress state, and regeneration," edited by M. Kellmann and K.W. Kallus, pg. 112. With kind permission from Springer Science and Business Media.

Motivation and Burnout

Tentatively describing and explaining the phenomenon of athlete burnout, the models of Smith (1986), Coakley (1992), and Silva (1990) all preceded the Gould studies (1996; Gould et al. 1997). These studies rely on the hypothesis that burnout results from a complex interaction between personal and situational factors, leading to high perceived stress with subsequent repercussion on the athletes' motivation. It is assumed that the athletes' initial engagement with the sport is the result of strong motivation. When experiencing success, athletes experience great pleasure, which leads to further engagement. Athletes' excessive investment in their sport puts them at risk for burnout by narrowing their identity. This changes motivation, causing athletes to switch from a strong desire for success combined with strong engagement, to physical and psychological

disengagement. Gould (1996) calls this inversion of intent "motivation gone awry."

Following up on these studies linking motivational processes to burnout, Hall and colleagues (1997) believe that it is not so much the intensity of the motivation that leads to burnout, but the fact that the reasons for participating in sports at the start of the athletic career were adaptive to begin with, but became maladaptive in the face of important challenges. Burnout is the inevitable consequence of this type of maladaptive motivational profile, which is difficult to assess in some athletes experiencing initial normative progression and early success.

Theoretical analysis of motivation, which gives meaning to thoughts, emotions, and behavior in athletes, can be used to understand why the wrong motivation would lead an athlete to burn out. Using this perspective as an initial starting point, Lemyre and associates (2006) suggest that any motivation patterns leading athletes to engage with their sport may be adaptive to begin with, when experiencing the positive outcomes associated with initial success. However, poor adaptation has been observed in some athletes when facing increasing difficulties, severe challenges, and normative failure. This is what generally happens to high-performance athletes competing at a higher level than previously, combined with the necessarily increased number of training sessions. When these athletes aim only for success, without protecting their self-esteem against the consequences of repeated failure and permanent difficulties, they will rapidly become disenchanted with their sport, and will tend to withdraw. Motivation is therefore crucial for all the aspects depending on personal and environmental attitudes (the athlete's aims, the context or climate of training) and those affected by linked variables, such as perfectionism. For more information, consult more general works dealing with the aims of accomplishment (Nicholls 1989; Roberts 2001), the theory of self-determination (Deci and Ryan 2002), and the motivational climate (Ames 1992). The following summary is based on the French versions of these theories (Cury and Sarrazin 2001), but it cannot replace a fuller understanding of these major psychological concepts. A few basic concepts will be set out here to facilitate understanding for those who are not familiar with this literature. We explain three of these important concepts to facilitate the understanding of the role of motivation in maximizing recovery in athletes and minimizing the risks for burnout.

A distinction must be made between the athlete's psychological state (which can vary rapidly, depending, for example, on the athletic situation) and personality dispositions (which are more stable). According to achievement goal theory within the sport context (Roberts 2001), the motivational state depends on both the athlete's personal dispositions and the environmental and situational cues. Athletes are typically motivated by goals involving mastery (e.g., demonstrating high effort, learning new skills and techniques) or ego (e.g., demonstrating normative sporting success). This is a result of a combination of their dispositional tendencies and the motivational climate in which they practice sports. The motivational climate, which is typically influenced by motivational cues from coaches, parents, and peers, cor-

responds to the environment the athlete is training and performing in. In a mastering climate, athletes must learn, master abilities, and perform at their best. In a performance climate, the trainer and the system value comparisons, results, and winning.

In self-determination theory, motivation serves to satisfy primary needs. Deci and Ryan (2002) suggest that that all humans, including athletes, seek autonomy, competence, and relationships in order to achieve psychological adjustment and personal growth. In their theory, athletes adopt self-determined styles to meet these primary needs in a competitive context, and their perceived level of satisfaction influences their level of motivation. Athletes who show little self-determination or autonomy generally feel anxious and stressed, and are prone to self-criticism. Those who show greater self-determination generally feel happier and have higher self-esteem. In the context of competition and sport accomplishments, several studies have linked autonomy to higher levels of performance, perseverance, and well-being. Results from Lemyre's studies add to the concept, suggesting that athletes who cannot satisfy their need for autonomy have lower self-determined motivation. This lack of personal ownership over reasons for training and competing in sports is believed to hamper the process of recovery and to increase the probability of experiencing burnout (Lemyre, Roberts, and Stray-Gundersen 2007).

People get involved in activities such as sports for a variety of reasons that can be placed on a continuum reflecting their degree of motivation, ranging from the most autonomous form of motivation (i.e., intrinsic) to the absence of motivation, or *amotivation*. Intrinsic motivation is linked to learning (e.g., "I train to learn new skills"), accomplishment (e.g., "My sport allows me to master new techniques."), and emotions (e.g., "I practice my sport because I like the feelings it creates."). However, some behaviors may also be motivated by external sources. Extrinsic motivation includes several dimensions that are quite distinct from intrinsic motivation, ranging from more autonomous forms of extrinsic motivation (i.e., identified regulation, interjected regulation) to less autonomous forms of motivation (i.e., external regulation). External regulation corresponds to the most extrinsic form of motivation, in which the athlete's behavior is controlled by external factors (e.g., obtaining a reward or avoiding punishment). The last form of regulation on the continuum is amotivation. When amotivated, athletes participate in an activity without valuing it, typically feeling that the action is beyond their control.

Four studies by Lemyre and colleagues (Lemyre 2005; Lemyre, Treasure, and Roberts 2006; Lemyre, Roberts, and Stray-Gundersen 2007; Lemyre, Hall, and Roberts 2008) illustrate the importance of motivation and perfectionism relative to the training workload. The purpose of the first study (2006) was to assess whether shifts along the self-determined motivation continuum, as well as variation in negative and positive affect in elite athletes, would predict signs of athlete burnout across the course of a competitive season in a sample of elite collegiate swimmers. Results indicated that swimmers who experienced a shift in motivation from more to less self-determined were more at risk of burning out than athletes whose motivational profile was more self-determined. Findings also indicated that variations in negative affect throughout the competitive season were significantly linked to signs of burnout in elite swimmers at the season's end. Finally, negative affect added to the prediction of all burnout subdimensions beyond that provided by shifts in self-determined motivation, while variation in positive affect only helped to better predict emotional and physical exhaustion at the season's end. This association with negative affect was also reported in other studies (Kellmann 2002).

Mood swings are thought to be linked to recovery states: Strong and acute negative emotions indicate inadequate recovery. Subsequently, incomplete or inappropriate recovery seems to cause a reaction in the athlete. When athletes' basic psychological needs are not fulfilled (see Ryan and Deci 2001), their motivation is affected, since they will typically look for external reasons for staying involved in sports. This shift in motivation puts athletes at risk for burnout over time if the situation is prolonged. Without adequate recovery, fatigue affects personal regulations. Without adequate recovery and self-determined motivation, athletes run a greater risk of following a training schedule without owning it, questioning it, or adapting it to their level of available resources. This loss in autonomous motivation is believed to hamper attainment of high-performance goals. To summarize, reduced self-determination and

strong emotional responses are markers of physical and emotional fatigue that may actively be used to prevent burnout in athletes.

A second study by Lemyre and colleagues (2007) confirmed these findings and added to our understanding of the link between overtraining and burnout in athletes. One hundred and forty-one athletes participating in winter sports (of which 45 participated in the winter Olympics) were enrolled in the study. The aim of the study was to predict the likelihood of experiencing athlete burnout when monitoring levels of self-determined motivation at the season's start and symptoms of overtraining at the season's end. Motivation was measured using the sport motivation scale and symptoms of overtraining were measured with the short overtraining symptoms questionnaire. Although the results must be interpreted with regard to the athlete's performance level, they globally show that a high level of self-determined motivation is negatively linked to burnout and that symptoms of overtraining are positively linked to burnout. While overtrained athletes can maintain their motivation, athletes suffering from burnout show signs of demotivation.

The third study illustrates the importance of motivation for various aspects related to training, recovery, and burnout in elite athletes (Lemyre, Hall, and Roberts 2008). Psychosocial variables were measured at the start of the season and compared to burnout scores at the end of the season. The set of questionnaires used focused on motivation, including tools to evaluate achievement goals, the motivational climate, perceived ability, and perfectionistic tendencies. Based on previous findings (Hall, Cawthra, and Kerr 1997; Lemyre, Treasure, and Roberts 2006), Lemyre and colleagues (2008) suggested that during their career, athletes suffering from burnout show a variety of ill-adapted dispositional characteristics, but that these only become apparent when they experience difficulties in their sporting career. A social-cognitive approach (Roberts 2001) was adopted to identify motivational profile variables that may put elite athletes at risk of experiencing burnout. Results revealed that elite athletes who had a coach and parents who emphasized performance outcomes and who were focused on normative comparisons, preoccupied with achieving unrealistic goals, and in doubt of their actions and own ability and were clearly more at risk of developing signs of burnout than athletes who used self-referenced and learning-oriented criteria for goal achievement. These findings underline the importance of emphasizing appropriate motivational patterns for athletes with a strong desire to improve their self-referenced ability and skill levels. This type of profile increases athletes' likelihood to seek out challenges or to persevere when facing achievement-related obstacles.

Finally, in a fourth study, Lemyre (2005) looked at motivational disposition, perception of the motivational climate, and the quality of motivation in elite athletes as well as variation in basal cortisol levels and how these variables predicted levels of burnout in elite swimmers at season's end. Lemyre hypothesized that maladaptive motivation shifts, together with variation in basal cortisol in elite athletes, could better predict overtraining and burnout at season's end in elite swimmers. Athletes were tested with a two-bout exercise test at three points during the year, corresponding to the easy, very hard, and peaking times of the season. A hard testing protocol was chosen, and venous blood drawn five times over a 24-hour schedule was assayed for cortisol by radio immune assays. Results indicated that basal cortisol variation as well as a perception of a high performance climate on the team predicted burnout at season's end. When looking at the quality of motivation in swimmers at season's end, intrinsic motivation was negatively correlated to burnout, while external regulation and amotivation were positively correlated to burnout levels. Lemyre suggested that enduring high basal cortisol variation was most likely detrimental to recovery and could lead to athletes overtraining and ultimately burning out.

These findings clearly demonstrate the advantage of assessing motivation and other dispositional sociocognitive variables when describing the phenomena linked to recovery and the risks for burnout in athletes. The role of motivational profiles is key, but other variables, such as mood swings or perfectionistic tendencies, are also important. While several physiological variables are generally measured to assess recovery, several psychological variables clearly must also be considered. Motivation is one of great importance.

Engagement and Motivation Pattern

From a psychosocial viewpoint, the work of Raedeke and Smith (2004) is significant. Through the extensive study of burnout and engagement in sport, they suggested that strategies preventing burnout may be more valuable over time than those aiming to treat burnout. A positive, rather than simply corrective, approach can be offered to attempt to reinforce engagement, which is the conceptual opposite of burnout. A definition of burnout has recently been adopted in sport psychology, based on an adaptation from general psychology. Raedeke considers that burnout is characterized by the persistent presence of three criteria: (1) physical and emotional exhaustion, (2) a reduced feeling of accomplishment, and (3) disinterest in sport. Physical and emotional exhaustion is linked to intense training and numerous competitions. A reduced sense of accomplishment relates to perceived ability and skills. Disinterest results in a lackadaisical attitude and ill-feeling toward the sport (Raedeke and Smith 2001). Thus, athletes who are under pressure and are restricted by the practice of their sport may notice the negative effects (both physiological and psychological) of overtraining on their performance.

In contrast, engagement is defined as a "cognitive-affective experience related to sport, characterised by confidence, involvement, and vigour" (Lonsdale, Hodge, and Jackson 2007; Lonsdale, Hodge, and Raedeke 2007). Confidence is a belief in one's own personal capacities to achieve a high level of performance and to attain goals. Confidence is the opposite of the reduced feeling of accomplishment experienced in burnout. Involvement is a desire to invest effort and time to meet the objectives considered important; it is the opposite of disinterest. Finally, vigor is an energetic, physical, mental, and emotional component that can also be termed *vitality*. It appears to be the opposite of physical and mental exhaustion.

Lonsdale and colleagues (Lonsdale, Hodge, and Raedeke 2007) showed that their model describing engagement in sport differs from the definition of engagement for professionals. They propose a four-factor model based on confidence, involvement, vigor, and enthusiasm, a factor characterized by feelings of excitement and joy. Like vigor, in sport, enthusiasm is the opposite of physical and mental exhaustion. The athlete engagement questionnaire was therefore developed to assess these four aspects of engagement. Participants are asked to evaluate how frequently they experienced feelings over the last 4 months.

Based on a 5-point Likert scale (ranging from "almost never" to "almost always"), they respond to affirmations such as the following: "I am capable of attaining my personal goals in my sport" (for confidence), "I want to train to attain my personal goals" (for engagement), "I feel full of energy when I practice my sport" (for vigor), and "I derive pleasure from practicing my sport" (for enthusiasm). However, this tool, which assesses a positive aspect of the athlete's environment, relates to a stable state, since it refers to feelings encountered over the last 4 months. In addition, as for most of these tools, translation and transcultural validation are necessary before it can be used in another country.

Finally, burnout may be the conceptual opposite of engagement, but this does not mean that the psychosocial processes leading to engagement are either opposed to or parallel to those leading to burnout. Further study is necessary to determine the relationships involved. The interactions between engagement and recovery must be examined. However, studies of engagement show that it is possible to study positive aspects of training, and not just the negative effects. It is probably wise to continue exploring these themes (engagement and burnout) simultaneously. Indeed, like perfectionism, which can lead to burnout, it would be interesting to assess the links between the two themes. Could excessive engagement also lead to burnout? This phenomenon exists for some employees, but there is a lack of data in the field of athletics.

Evaluation Methods for Psychological Recovery

Various tools can be used to assess the effects of different types of recovery. We have chosen to present the questionnaires that appraise the main psychological variables affected by recovery and excessive training.

Different scales are used in the medical community, such as the PRS (perceived recovery scale). The PRS is used in the medical department at the French Institute of Sports in a paper format. It consists of five subscales that allow for a graphical assessment regarding perception of (1) pain (delayed-onset muscle soreness, or DOMS), (2) well-being, (3) stiffness, (4) fatigue, and (5) heat. Answers are given by drawing a cross bar on a nongraduated 10 cm continuum between two extremes for the measured concept (e.g., extremely hot to extremely cold). In its paper form, results are measured with a ruler to the millimeter.

Assessment of Psychological Variables

In sport psychology, the following variables are assessed with self-administered questionnaires regarding (1) vitality, (2) motivational states, and (3) emotions. A single indicator of these variables may not be sufficient to detect a drop in vitality or motivation or a switch from positive to negative emotions. Hence, we suggest monitoring these variables to detect poor recovery as early as possible.

- *Vitality.* Vitality is related to well-being, a feeling of alertness and dynamism (Ryan and Deci 2001), and an impression that energy is available (Ryan and Frederick 1997). Two versions of the subjective vitality questionnaire exist. The trait version relates to (stable) personality traits, that is, a characteristic of the subject that is positively linked to self-esteem and negatively linked to depression and anxiety. The state version measures fluctuations in subjective vitality rather than its stable aspect. Vitality is negatively linked to physical pain, and it positively contributes to maintaining autonomy—an important factor in sport-related motivation (Nix et al. 1999). Recently, studies have shown that participants with low levels of autonomy were more likely to feel emotionally and physically drained by their investment in their sport. For example, the vitality (state) scale has been used as an indicator of well-being in sports (Adie, Duda, and Ntoumanis 2008). The seven-item version developed by Ryan and Frederick (1997) was improved by Bostic and associates (2000).

- *Motivation state.* Lemyre's studies show that a drop in motivation in elite athletes increases the risk of burnout at the end of the season (Lemyre, Treasure, and Roberts 2006). For this reason, motivation states should be measured regularly, for example, using a situational motivation scale (SIMS) (Standage et al. 2003). This questionnaire consists of 14 questions relating to the self-determination continuum (intrinsic and extrinsic motivation). It is recommended for the study of the motivational state associated with recovery. A shorter version, more targeted to a specific aspect of motivation, has been developed as an aide to training.

• *Emotions.* High rates of burnout are observed in athletes whose emotions become negative during a competitive season. Hence, as for vitality and motivation, it is relevant to monitor emotions, especially if they change from positive to negative. We recommend using the positive and negative affect schedule (PANAS) because of its solid reputation based on numerous publications. The original version in 20 items was created and validated by Watson and colleagues (1988).

Recently, a short version in 10 items was published by Thompson (2007). This version is suited to athletes because it is a fast way of monitoring emotions and their variation throughout a season.

Assessment Systems

Several computer assessment systems for psychological aspects have been commercialized. Of course, paper and pencil can still be used, but newer systems are more efficient. The Mind-eval system (www.mindeval.com) ensures that all data remain confidential. A key, or personal code, allows computerized files to respect anonymity. The system records all data, thus avoiding problems with missing information that can be encountered when using paper questionnaires. Once the personal code has been entered, athletes perform their self-evaluation on the computer (or on a compatible mobile telephone). This

---

PRACTICAL APPLICATION

Recommendation for the psychological component of recovery:

1. Motivation: Take care of the training climate and the personal goals of the athletes.

First, make sure athletes have planned their season with respect to the training and competitive workload. Here, the performance goals (or the outcome, such as medals or ranking) should be distinguished from practice goals (technical or tactical issues) that will target areas of improvement. Too much emphasis on competitive goals may contribute to burnout. Attention should be paid to excessive peer pressure, as well as pressure from sponsors, institutions, and parents.

As a coach, look for signs of motivation (engagement, determination) and demotivation.

2. Plan and respect recovery periods over the season.

Find appropriate times for recovery over the course of the season (during vacation, on weekends without tournaments). Indeed, these times must be anticipated and respected. They should not be considered time that could potentially be devoted to extra hours of training. Burnout may be a product of too much training or too little

recovery. Planning of the recovery component must also be respected.

As a coach, avoid asking for more training when the recovery period comes.

3. Appropriate recovery: Resting and sleeping is not enough.

Simply stopping training might be enough to compensate for the training workload, but the psychological aspect should also be considered. If goals help direct attention and sustain motivation, social activities must also be considered. Athletes should take time each week to plan or attend social events that are different from athletic events with family or friends from outside the sporting world. Extra hours of training can be compensated for by engaging in social or cultural activities, such as partying with friends, dancing, or going to the movies or the theater. If the practice of another sport as a social event is chosen, the coach in charge of the physical training program must be informed in order to adapt the quantity of training accordingly.

Coaches should watch for inappropriate behavior during parties, but must be tolerant of activities outside of the sport, since they balance the athlete's life.

system prevents errors that can be made during retranscription of paper questionnaires for entry into a computerized data-collection system. The paper version of the PRS has been transformed to present visual analogue scales (VAS): Athletes can use a computer mouse to move a slider to indicate their perceptions between the two extremes. Results are produced automatically and immediately, not just for the VAS (sliders), but also for calculations related to the vitality indicator, the scores of each subscale for motivation, and the global motivation index (self-determination index). This online system allows simultaneous data collection at different locations and times. The administrator, a researcher, or a coach can consult the results (graphs and raw data) at any time.

Summary

In sport psychology, most of the literature describing the effects of training workload relates to burnout, and not directly to recovery. To avoid fatigue and to perform consistently over time, it is believed that athletes must optimize their recovery. However, the main characteristics of burnout (and engagement, if we take a positive view) have been studied under a psychosocial paradigm. Interventions favoring recovery are centered on personal variables that are mainly linked to motivation (setting goals, annual schedule), but also to variables such as vitality, emotions, and perfectionism. Contextual variables, such as the motivational climate, are taken into account, but peer pressure, as well as pressure from sponsors, institutions (national governing bodies, clubs), and parents have yet to be taken into consideration.

First, a well-designed and validated questionnaire can help prevent overtraining. Indicators that are part of a training or research program can be measured.

Secondly, psychological aspects are often mentioned to explain what physiology cannot ("It must be a mental blockage, because from the physical point of view, everything was fine. There were no problems during training."). If the psychological variables of recovery are relevant, it is essential to appraise them to supplement the physiological data. Finally, we expect that significant advances will be made once the physiological and psychosociological processes have been studied together, rather than separately.

PART
III

Strategies for Optimizing Recovery

CHAPTER 6

Stretching

Giuseppe Rabita, PhD, and Anne Delextrat, PhD

With: Cédric Lucas, Arnaud Daufrène, Frank Métais, and Christophe Cozzolino

Various types of muscle stretching are usually performed before and/or after physical activity. Stretching is indeed a common practice among all sport participants, particularly for elite athletes.

Traditionally, muscle stretching is thought to be beneficial both for athletic performance and preventing injuries. However, over the last decade, many scientific studies have minimized these beneficial effects. Some have even shown stretching to have a negative effect in some circumstances. On the other hand, the specific effect of stretching on recovery is rarely described, although it has been extensively studied in recent years.

This chapter summarizes theoretical knowledge in order to give practical advice about stretching. To reach this goal, it analyzes scientific findings with regard to athletes' daily routines. Finally, it suggests practical recommendations.

Types of Muscle Stretches

To better understand the effects of stretching, we must know which physiological structures are involved in the different types of stretches. Thus, before describing the different methods for stretching the muscle–tendon system, we will briefly discuss the physiological structures involved. We will also describe the basic theories linking the effects of stretching to various performance indicators to explore the specific relationship between stretching and recovery.

Passive Stretching

Passive stretching of a muscle group refers to the use of external force, in the absence of voluntary contraction, to induce its lengthening. This external force may be applied by a third party or by the athletes themselves. In this case, the subject can use weights or changes in body position to stretch the muscle group in question. This type of stretching is the most commonly used in general sporting practice.

Passive stretching can be further subdivided into static and cyclic stretching. Static passive mobilization techniques involve first bringing the joint to an angle at which the muscle group is close to maximal lengthening, and then maintaining it at this angle (figure 6.1). During the first, dynamic phase, the passive torque (reflecting the resistive force of the muscle group being stretched) increases in a curvilinear way: Passive stiffness increases in proportion to the degree of stretching. In contrast, during the static phase, passive torque decreases. This is representative of the system's viscosity, which is easily characterized by the reduction in peak resistance between the start and end of the static phase.

Cyclic stretches, on the other hand, involve repeated stretching of a musculoarticular system, immediately followed by a return to the starting position. The two types of passive stretches have been compared on several occasions (Taylor et al. 1990; Magnusson et al. 1998; McNair et al. 2001; Nordez et al. 2010; Nordez, Casari, et al. 2009).

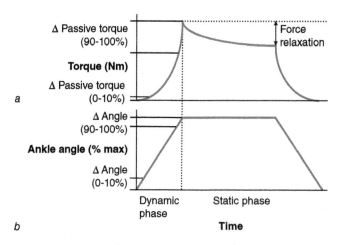

Figure 6.1 Diagram of *(a)* torque and *(b)* ankle angle versus time during passive dorsiflexion. Musculoarticular stiffness (calculated by the ratio of passive torque to variations in ankle angle) increases regularly between the start (0–10% of the imposed maximal angle) and end (90–100%) of the movement. The viscosity index is obtained from the force-relaxation parameter.

Reprinted, by permission, from P.J. McNair et al., 2001, "Stretching at the ankle joint: viscoelastic responses to holds and continuous passive motion," *Medicine and Science in Sports and Exercise* 33: 354-358.

In general, passive resistance of the musculoarticular system is greater during stretching (loading) than during return to the initial position (unloading). Thus, as shown in figure 6.2, the relationship between the passive torque and the joint angle is slightly modified. The hysteresis (i.e., the area between the two curves) results from dissipative properties, which are largely due to the viscosity of the musculotendinous structures involved. The hysteresis (or the energy dissipated) can be easily quantified by calculating the difference between the surface under the loading (stored energy) and unloading (released energy) curves.

The energy dissipated increases with stretching velocity (figure 6.3), and is an indicator of the viscoelastic characteristics of the structures involved. One of the practical interests of this figure is that it shows the effect of the number of stretch cycles: The dissipation coefficient (i.e., the ratio of energy dissipated to energy stored) drops with the number of repetitions. In other words, the more the muscle groups have been previously stretched, the lower the viscoelastic resistance (Magnusson et al. 1996; Nordez, McNair, et al. 2009). It is therefore important to avoid stretching muscles at high velocities during the first repetitions. Indeed, high-speed stretching requires the muscle to have a capacity to absorb energy.

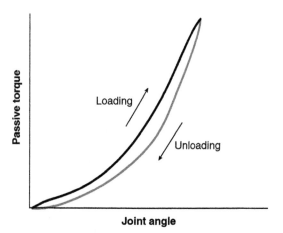

Figure 6.2 Viscoelastic hysteresis. The area of the hysteresis loop corresponds to the dissipated energy.

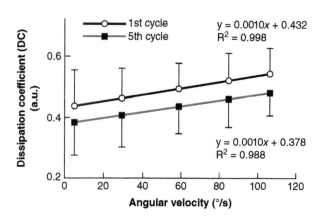

Figure 6.3 Dissipation coefficient (DC) as a function of the angular velocity for the first and fifth cycles. y: dissipation coefficient; R^2: determination coefficient.

Based on A. Nordez, 2010, "Static and cyclic stretching: Their different effects on the passive torque-angle curve," *Clinical Biomechanics* 13:156-160.

Figure 6.3 shows that in practice, cyclic stretching must start at slow velocities to allow the muscle to increase its capacity for energy absorption. Only a small number of repetitions (3 or 4) have been shown to be enough to substantially reduce this energy, regardless of the speed of passive mobilization. This partially protects the muscle from injury.

Active, Dynamic, and Ballistic Stretching

Various types of stretching involve muscle contraction. Stretches are considered *active* when the muscle (or muscle group) and its tendon are stretched prior to an isometric contraction of the same muscle group. They are *dynamic* when this isometric contraction in a stretched position is followed by a dynamic contraction of the same muscle group.

Ballistic stretching involves bouncing movements. In this case, a brief contraction of the agonist muscle groups is combined with the weight of the limb to stretch antagonistic groups. These latter groups must be fully relaxed at the time of stretching. This type of movement is usually repeated several times with no rest between repetitions, while regularly increasing joint amplitude, in order to gradually raise the level of stretching of the tissues.

Passive–Active Mode

Hold–release stretching involves placing the muscle group in a position for an almost maximal stretch, and then contracting it. The muscle is then stretched to increase articular amplitude. The action is then repeated in the new position. *Contracted–released* stretching is similar, except that rotation is left free. Finally, *contracted-released-contracted* stretching is based on the same principle, but the stretch is assisted by a voluntary contraction of the antagonistic muscle group.

Physiology of Muscle Stretching

During passive stretching (i.e., without any muscle contraction), the resistive force results from the mechanical properties of various anatomic structures as well as from mechanisms related to the physiology of the nervous system through reflex activities. Structures that affect the quality of stretching include connective tissue and tendons, elements of the cytoskeleton, actomyosin crossbridges, and pathways of the central nervous system.

Connective Tissue and Tendons

The endomysium, the perimysium, and the epimysium, connective matrices made up of elastic collagen fibers, surround and protect the muscle fibers, its fascicles, and the muscle as a whole, respectively (figure 6.4). The viscoelastic properties of these structures are greatly involved in passive stretching. Compared to the external connective tissue (aponeurosis), these structures are more compliant (i.e., less stiff).

Tendons are also composed of collagen fibers. They are mechanically arranged in series, interlaid with connective tissue, to transmit force to the skeletal system. Since tendons are stiffer than muscular conjunctive tissues, they were previously considered to play a minor role in stretching in passive conditions. However, recent results from ultrasound observations have questioned this assumption. It appears that the stiffness of both muscle and tendon tissues contributes to resistance during passive stretching.

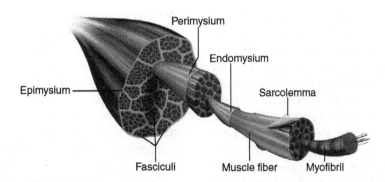

Figure 6.4   Cross-sectional representation of muscle.

Elastic Elements of the Cytoskeleton

Within the muscle fiber, several structures also contribute to muscle resistance during passive stretching. Titin (or connectin) is the main cytoskeleton protein mainly involved in the different types of stretching (figure 6.5). Its essential functions are to maintain myosin in the center of the sarcomere and, due to its elastic properties, to return the sarcomere to its resting length after stretching.

The desmin protein connects adjacent myofibrils at the level of the sarcomeres' Z-lines. It probably also contributes to passive resistance during large amplitude stretching, leading to transversal misalignments of the sarcomeres (Campbell 2009).

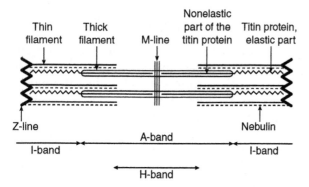

Figure 6.5 Diagram of interactions between the sarcomere and titin protein, consisting of an elastic isotropic (I-band) and a nonelastic part (A-band). The M-line in the middle of the sarcomere is formed of cross-connecting elements of the cytoskeleton. The H-band is the zone where the thick filaments are not superimposed. The Z-line represents sarcomere extremities.

Residual Actomyosin Crossbridges

Even in the absence of muscle contraction, a certain number of crossbridges between actin and myosin molecules are maintained. During stretching, these crossbridges provide a certain level of resistance, depending on the muscle's immediate history. Since stretching breaks these crossbridges, they are therefore less involved during subsequent stretches, provided that the stretch remains passive (Proske and Morgan 1999; Whitehead et al. 2001). However, a recent study showed that the role of this factor in humans might be negligible (Morse et al. 2008).

Nervous System Pathways

In addition to the mechanical properties of the structures mentioned previously, neuromuscular activity plays a decisive role during muscle stretching. By means of afferent pathways (i.e., returning information to the central nervous system), inhibitory and excitatory reflexes contribute to stretching through various effects. We recommend Guissard and Duchateau's review (2006) for comprehensive coverage of this topic. The authors describe how stretching the muscle–tendon unit can reduce spinal-reflex excitability. In acute conditions, this reduced excitability lowers passive tension and increases joint amplitude, thus allowing greater movement.

In the long term, stretching programs reduce tonic reflex activity, which significantly contributes to increased flexibility. Sensory theories have emerged, suggesting that the increased stretching capacity is mainly due to altered sensations (particularly reduced pain). These theories, which therefore highlight the major role played by the neurological system in stretching and its effects, are currently being debated (Weppler and Magnusson 2010).

Effects of Stretching on Performance

Before discussing the relationship between stretching and recovery, it is important to address the question of the effect of stretching on physical performance before fatigue has taken place. A large number of studies have quantitatively assessed this relationship. The acute effects, observed immediately after exercise, can be distinguished from the chronic effects, which result from stretching sessions scheduled over several weeks or months.

Acute Effects

Stretching is often recommended to improve physical performance. However, recent scientific reviews have shown that these recommendations are based on scientific evidence from specific cases. Although many studies have shown stretching to have a negative effect on performance, research aiming to explain the underlying physiological mechanisms often applies to stretching exercises that are, both in amplitude and duration, in excess of what is used during daily sporting activity.

The following paragraphs cover the most recent updates on the conclusions from scientific studies, specifically focusing on the results obtained with stretching similar to that used during normal practice.

Maximal Isometric, Isokinetic, and Isotonic Strength

Shrier (2004) and more recently, Rubini and associates (2007) systematically reviewed the literature on these performance indicators. The conclusions are obvious at first glance: For all performance indicators, stretching before exercise has a negative effect on performance. The results of these two reviews show that this effect is observed for all types of stretching: static, ballistic, or PNF (proprioceptive neuromuscular facilitation). However, stretching used during these studies varied in duration between 2 min and 1 h, which is significantly different from what is observed during daily

sporting activity. Therefore, table 6.1 is modified from the results of Rubini and colleagues (2007), only showing the studies for which stretches of one or two muscle groups were performed before exercise, for a total duration, including rest time, of less than 6 min.

From table 6.1, it can clearly be seen that stretching has an acute negative effect on strength, even when its duration is limited. Of the nine studies with a total stretching duration between 2 and 4 min, only one does not report reduced strength for a group that has stretched compared to a control group. In addition, of the 31 conditions tested in these studies, only 4 do not have a significant effect. Among these, 3 involve a total duration of stretching (including rest phases) that is less than or equal to 1 min, or is not indicated.

Table 6.1 Studies Investigating the Acute Effect of Stretching on Strength Performance

Reference (sample size)	Type of stretching	Duration of stretching	Muscles stretched	Muscles tested	Total duration (s)	Type of action	Results
Kokkonen et al. (1998) M (n = 15) F (n = 15)	Static (passive) [assisted/not assisted]	5 exercises 3 × 15 s 15 s rest	Hamstrings, hip adductors, plantar flexors	Hamstrings, quadriceps	450	Isotonic 1RM	↓ 7.3% flexion ↓ 8.1% extension
Nelson et al. (2001) M (n = 10) F (n = 5)	Static (active and passive)	3 exercises 4 × 30 s 20 s rest	Quadriceps	Quadriceps	360	Isokinetic	↓ 7.2% 60°/s PT ↓ 4.5% 90°/s PT
Nelson et al. (2001) M (n = 25) F (n = 30)	Static (passive)	2 exercises 4 × 30 s 20 s rest	Quadriceps	Quadriceps	240	Isometric	↓ 7% PT, at angle of 162°
Nelson and Kokkonen (2001) M (n = 11) F (n = 11)	Ballistic	5 exercises	Hamstrings, thigh adductors, plantar flexors, quadriceps	Hamstrings, quadriceps	450	Isotonic 1RM	↓ 7.5% flexion ↓ 5.6% extension
Tricoli and Paulo (2002) M (n = 11)	Static (active)	6 exercises 3 × 30 s 30 s rest	Quadriceps, hamstrings	Quadriceps, hamstrings	540	Isotonic 1RM	↓ 13.8% maximum strength
Garrison et al. (2002) (n = 29)	Static	NA	Quadriceps	Quadriceps	480	Isokinetic	No change in performance
Mello and Gomes (2002) M (n = 5) F (n = 8)	Static (passive)	2 exercises 2 × 15, 30 and 60 s 10 s rest	Hamstrings, quadriceps	Hamstrings, quadriceps	30 60 120	Isokinetic	No change in performance

(continued)

Table 6.1 *(continued)*

Reference (sample size)	Type of stretching	Duration of stretching	Muscles stretched	Muscles tested	Total duration (s)	Type of action	Results
Evetovich et al. (2003) M (n = 10) F (n = 8)	Static (2 active; 1 passive)	3 exercises 4 × 30 s 15 s rest	Biceps brachii	Biceps brachii	360	Isokinetic	↓ 30°/s PT ↓ 270°/s PT
Bandeira et al. (2003) F (n = 10)	Static (active)	6 exercises 1 × 15 s and 60 s	Hip flexors, hip extensors	Hip flexors, hip extensors	90 or 360	Isokinetic	↓ Flexors 60°/s
Avela et al. (2004) M (n = 8)	Static (passive)	2 exercises 60 min 2 weeks between exercises	Plantar flexors	Plantar flexors	360	Isometric	↓ 13.8% MCV (1st measure) ↓ 13.2% MVC (2nd measure)
Cramer et al. (2005) F (n = 14)	Static (1 active; 3 passive)	4 exercises 4 × 30 s 20 s rest	Quadriceps	Quadriceps	480	Isokinetic	↓ 3.3% 60°/s PT ↓ 2.6% 240°/s PT
Rubini et al. (2007) M (n = 18)	Static PNF (passive)	1 exercise 4 × 30 s or 4 × (3 × 10 s)	Hip adductors	Hip adductors	120	Isometric	45°: ↓ 8.9% and ↓ 12.3% 30°: ↓ 10.4% and ↓ 10.9% (static and PNF, respectively)
Cramer et al. (2004) M (n = 15)	Static (1 active)	4 exercises	Quadriceps	Quadriceps	NA	Isokinetic	No change in performance
Marek et al. (2005) M (n = 9) F (n = 10)	Static PNF (passive)	4 exercises 5 × 30 s 30 s rest	Quadriceps	Quadriceps	120 (static) 120 (PNF)	Isokinetic	↓ 2.8% (static and PNF)
Behm et al. (2006) (pre) M (n = 9) F (n = 9)	Static (passive)	3 exercises 3 × 30 s 30 s rest	Quadriceps, hamstrings, plantar flexors	Quadriceps	270	Isometric	↓ 6.5% MVC
Behm et al. (2006) (Post) M (n = 12)	Static (passive)	3 exercises 3 × 30 s 30 s rest	Quadriceps, hamstrings, plantar flexors	Quadriceps	270	Isometric	↓ 8.2% MVC
Brandenburg (2006) M (n = 10) F (n = 6)	Static (assisted/not assisted)	2 exercises 3 × 15 s or 30 s	Hamstrings	Hamstrings	90 or 180	Isometric concentric exercise	↓ 15s ↓ 30s NS difference between type of actions
Egan et al. (2006) F (n = 11)	Static	4 exercises 4 × 30 s 20 s rest	Quadriceps	Quadriceps	480	Isokinetic	NS 60°/s NS 300°/s 5 min after stretching

Only studies with a total duration of stretches lower than 6 min have been reported.

MVC: maximal voluntary contraction; NS: nonsignificant; PNF: proprioceptive neuromuscular facilitation; PT: peak torque.

For full reference information of studies cited in this table, please see Rubini, Costa, and Gomes (2007).

Springer and *Sports Medicine* 37, 2007, pgs. 213-224, "The effects of stretching on strength performance," E.C. Rubini, A.L. Costa, and P.S. Gomes, table II, with kind permission from Springer Science and Business Media B.V.

Jumping Performance

Table 6.2 presents the acute effects of stretching on performance for vertical jumps (VJ) with countermovement.

The effect of stretching on jumping performance is reduced compared to its effect on strength. However, of the 14 conditions described in the 9 studies analyzed by Rubini and colleagues (2007), 6 (or 43%) show a significant reduction in performance, 7 (50%) show no effects, and only 1 (7%) shows an increase in performance after stretching.

Chronic Effects

Few studies have focused on the chronic effects of muscle stretching on performance. However, their results are quite homogeneous.

In contrast with its acute effects, these results show that when muscle stretching is repeated over a period of a few weeks, it tends to increase strength, measured in both isometric and dynamic conditions. Currently, relatively few indicators allow a complete understanding of the mechanisms underlying muscle adaptation. A hypothesis of muscle hypertrophy has been suggested

Table 6.2 Studies Investigating the Acute Effect of Stretching on Jumping Performance

Reference (sample size)	Method	Sets and exercises	Muscles stretched	Stimuli duration (s)	Test	Results
Church et al. (2001) F (n = 40)	Static PNF	3 sets	Quadriceps, hamstrings	NA	VJ	↓ PNF NS static
Cornwell et al. (2001) M (n = 10)	Static (passive)	1 set 3 exercises	Hip extensors, knee extensors	90	VJ	↓ 4.4% (VJ) ↓ 4.3% (VJCM)
Knudson et al. (2001) M (n = 10) F (n = 10)	Static	3 sets 3 exercises	Quadriceps, hamstrings, plantar flexors	45	VJ	NS (VJ)
Cornwell et al. (2002) M (n = 10)	Static (passive)	3 sets 2 exercises	Triceps surae	180	VJ	↓ 7.3% (VJCM) NS (VJ)
Serzedêlo and Corrêa et al. (2003) F (n = 10)	PNF (CR)	3 exercises	Quadriceps, hamstrings, calf, gluteus	240	VJ LJ	NS (VJ) ↑ 10.7% (LJ)
Young and Behm (2003) M (n = 13) F (n = 3)	Static	4 exercises	Quadriceps, plantar flexors	120	VJ	↓ 3.2% (VJ)
Power et al. (2004) M (n = 12)	Static	2 sets 3 exercises	Quadriceps, hamstrings, plantar flexors	270	VJ	NS (VJ)
Unick et al. (2005) F (n = 16)	Static Ballistic	3 sets 4 exercises	Quadriceps, hamstrings, plantar flexors	180	VJ	NS (VJ) Static/ballistic
Wallmann et al. (2005) M (n = 8) F (n = 6)	Static (passive)	3 sets	Gastrocnemius	90	VJ	↓ 5.6% (VJ)

CR = contract/relax; F = females; LJ = long jump; M = males; NA = not available; NS = statistically not significant; PNF = proprioceptive neuromuscular facilitation; VJ = vertical jump; VJCM = vertical jump with counter movement; ↑ indicates significant increase; ↓ indicates significant decrease.

For full reference information of studies cited in this table, please see Rubini, Costa, and Gomes (2007).

Springer and *Sports Medicine* 37, 2007, pgs. 213-224, "The effects of stretching on strength performance," E.C. Rubini, A.L. Costa, and P.S. Gomes, table II, with kind permission from Springer Science and Business Media B.V.

based on animal studies, but the protocols used in these studies are so far from general practices found in sport activities that it does not seem reasonable to extrapolate. Whatever the case, and before making any generalizations with regard to recovery, we can say that although stretching is not recommended immediately before exercises requiring muscle strength, the regular practice of stretching is beneficial, and does not hinder strength performance.

Stretching and Recovery Research

Few studies have specifically investigated the effect of stretching on recovery from an athletic effort. However, the topic is now being explored more extensively, although no review articles have been published as yet. Therefore, the studies analyzed in the following paragraphs are recent.

Maximal Strength

Robey and associates (2009) compared the effect of different recovery methods in subelite and elite rowers. After a maximal effort over a 3.6 km running race, including a series of stairs (242 steps in total), 20 athletes recovered for 15 min using one of the following three methods, which were also repeated after 24 and 48 h: (1) stretching, (2) alternating immersion in hot and cold water, or (3) sitting (control group). This protocol was repeated three times over several weeks so that subjects were randomly assigned to the different recovery methods. Recovery was assessed using several tests before, immediately following, 24, 48, and 72 h after maximal effort.

The main results of the study show no difference between the two recovery methods and the control group in terms of perceived pain and maxi-

mal strength of the knee extensors. Table 6.3 gives the results for a 2 km test on a rowing ergometer.

The recovery method has no effect on performance in the 2 km ergometer test. For the elite group, a nonsignificant trend was observed for higher values after 72 h, including for the control condition. The results of this study, performed in almost normal training conditions show that, in elite athletes, the recovery time is a very important factor. In contrast, the different recovery methods, including stretching, do not enhance recovery.

Rate of Force Development

The rate of force development (RFD) is one of the parameters that most precisely describes the explosive nature of muscle activity. It can be calculated as the slope of the force/time relationship at the start of muscle contraction. It depends on nervous, contractile, and mechanical factors. Contradictory results have been reported by studies analyzing the acute effect of a stretching session on RFD (Costa et al. 2010; Gurjão et al. 2009; Bazett-Jones, Winchester, and McBride 2005; Maïsetti et al. 2007). However, among these studies, the protocols and objectives differed. In terms of population studied and duration of stretching, only Maïsetti and colleagues (2007) describe a situation similar to that encountered in normal sporting practice (five cycles of dorsiflexion maintained for 15 s at an angle corresponding to 80% of maximal dorsiflexion). These authors tested the rate of force development during maximal voluntary contractions (MVC) of the plantar flexors (isometrically) just before, immediately after, and 30 min after the series of stretches (figure 6.6).

Based on the results of this study, although the maximal plantar flexion strength is significantly reduced after stretching—not just immediately

Table 6.3 Mean (±SD) Differences Between Pre- and Posttreatment (72 h)

Group	Treatment	Mean pretime (s)	Difference (s)*
Club	Control	445.82 (45.71)	−0.21 (4.72)
	Stretching	453.03 (47.62)	0.34 (5.70)
	Hot/cold	451.65 (45.94)	−2.89 (4.22)
Elite	Control	415.97 (32.10)	2.55 (6.16)
	Stretching	415.75 (33.91)	6.92 (7.62)
	Hot/cold	415.20 (29.44)	3.00 (4.12)

Maximal 2 km rowing ergometer times.

*NB : Negative numbers reflect a slower post score and positive numbers reflect a faster post score. No significant differences were recorded either within or between recovery treatments.

Reprinted, by permission, from E. Robey et al., 2009, "Effect of postexercise recovery procedures following strenuous stair-climb running," *Research in Sports Medicine* 17: 245-259.

after (−9 [± 6]%), but also 30 min later (−10 [± 7]%)—the rate of force development does not vary significantly between tests. By definition, the maximal rate of force development during explosive contractions is a crucial factor for the level of force that can be generated at the start of the contraction (from 0 to 200 ms) (Wilson, Murphy, and Pryor 1994). It also determines the capacity to generate high speed if the joint is allowed free movement. These factors do not appear to be affected by stretching. However, this conclusion must be confirmed by further studies.

Time to Exhaustion at Submaximal Forces

Mika and associates (2007) studied the effect of various recovery techniques consisting of (1) *hold–release* stretching (see previous section, Passive–Active Mode) with the help of a therapist, (2) active recovery (on an ergocycle at moderate speed), and (3) passive recovery. The effects were assessed based on performance during a specific dynamic exercise. After a session of familiarization and determination of baseline values, 24 subjects took part in three additional testing sessions, composed of series of knee flexions–extensions at 50% of maximal strength, immediately followed by one of three randomly allocated recovery methods. To test the effects of the recovery method, participants performed a maximal isometric knee exten-

sion, as well as a contraction at 50% of maximal isometric knee extension until exhaustion.

Maximal strength measured after the three series was greater for the active recovery method than for the other two methods. Given the results presented previously, this was to be expected. The times to exhaustion during the fatigue test were significantly reduced compared to baseline values, but they were not affected by the recovery method.

Based on these data, the authors suggested that recovery is best when it involves light active exercise. Once again, stretching does not appear to affect recovery, since this method does not lead to a difference, compared to the control condition (passive recovery), for any of the parameters measured. Thus, stretching does not appear to have an effect on the capacity to maintain a low-intensity effort over a long period of time.

Cumulated Fatigue During Competitions

Recent studies by Montgomery and colleagues (Montgomery, Pyne, Hopkins, et al. 2008; Montgomery, Pyne, Cox, et al. 2008) are particularly interesting because the effect of different recovery strategies were tested in a competitive context (national basketball tournaments), where the different successive events led to cumulated fatigue. In basketball, each game involves many accelerations and decelerations of various intensities, from moderate to very high (Janeira and Maia 1998; McInnes et al. 1995), explosive jumps, and damage due to eccentric movements (Lakomy and Haydon 2004). The fatigue induced is accentuated by the rule adaptations which, to accelerate the game, lead to increased biomechanical and physiological constraints (Cormery, Marcil, and Bouvard 2008; Delextrat and Cohen 2008). These observations can be extended to other team sports, such as handball (Ronglan, Raastad, and Børgesen 2006). In this sort of competitive context, the choice of a particular recovery procedure used will determine performance and, eventually, the final ranking. The studies compare various parameters in 29 players split among three groups, depending on the recovery method (stretching, wearing full-leg compression garments, or cold-water immersion at 11.5 °C). The pre- and posttournament effects were normalized for the cumulated playing time over the 3 days. Table 6.4 summarizes some results of the first study (Montgomery, Pyne, Hopkins, et al. 2008). Several parameters selected based on

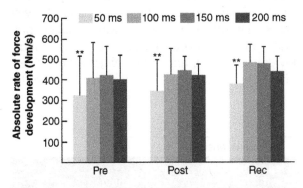

Figure 6.6   Absolute rate of force development (RFD) during maximal isometric plantar flexion before (pre), immediately after (post), and 30 min (rec) after repeated stretches. RFD was estimated as the slope of the torque-time curve for 50, 100, 150, and 200 ms.

**P < 0.01: RFD for 50 ms was lower than RFD calculated over longer durations for each condition (pre, post, rec).

**Statistically significant (P < 0.01).

Reprinted from *Isokinetics and Exercise Science*, Vol. 15, O. Maïsetti, "Differential effects of an acute bout of passive stretching on maximal voluntary torque and the rate of torque development of the calf muscle-tendon unit," pgs. 11-17, copyright 2007, with permission from IOS Press.

their relevance to this chapter are also presented in the table.

Based on these results, stretching is the least effective of the various types of recovery methods used between games. The negligible or low efficacy of stretching on parameters requiring speed and explosiveness (20 m sprint, basketball-related agility, vertical jumps) was expected (see table 6.2 for more on the acute effects of stretching on performance). Its effect on perceived pain is moderate. However, for this marker, wearing leg-compression garments has a strong effect, and cold-water immersion has a very strong effect. Surprisingly, after normalization for playing time, the effect of stretching on decreased flexibility is less significant for the compression group than for the stretching group. In addition, stretching has only a low incidence on perceived fatigue. This study suggests that recovery using passive stretching in a competition context, such as during a tournament, should be complemented by other recovery methods if the conditions allow.

Kinugasa and Kilding (2009) also determined the effects of different recovery methods on soccer players in a competition context. Matches were organized to test the effects of these various modalities on physical (vertical jumps) and physiological (heart rate) performance indicators, and on perceived recovery (ordinal scale). Twenty-eight soccer players played three 90 min games followed, in a random order, by (1) alternated immersion in cold (12 °C) and hot (38 °C) water, (2) immersion in cold water with active recovery (ergocycle at moderate speed), and (3) passive stretching. Tests were carried out before the games, immediately after the recovery session, and the next day. Vertical-jump performance did not vary significantly after a soccer game, although a trend for a reduction was observed for all recovery methods. Perceived recovery was substantially higher in the immersion with active recovery condition than for the other methods tested. However, this effect lasted only 24 h. Once again, stretching immediately after an event does not appear to affect recovery, either for physical and physiological indicators or in terms of subjective perceptions.

After Repeated Sprints

Favero and colleagues (2009) did not observe any significant effect of stretching on performance when used by trained athletes between three series of 40 m sprints. Remarkably, in the most flexible athletes in this study, this type of programmed stretching has a negative effect on mean running speed. This effect was confirmed by Beckett and colleagues (2009), who showed that a 4 min bout of static stretching of the lower limbs during recovery periods may hamper performance in repeated sprints. The negative effect on performance is lowered when the sprints are combined with changes of direction.

After Eccentric Exercises

Delayed-onset muscle soreness (DOMS) can appear within 12 and 48 h of intense or unusual muscular exercise involving eccentric contractions. Muscle microlesions resulting from this type of effort initiate an inflammatory process that leads to pain (Cheung, Hume, and Maxwell 2003). DOMS is associated with reduced proprioception, reduced articular range of motion, lowered strength, and maximal activation. These effects generally persist for 8 days after the dis-

Table 6.4   Effects of Different Recovery Interventions

Parameter	RECOVERY MODEL		
	Immersion	Full-leg compression garment	Stretching
Line-drill test (s)	Trivial	Moderate	Small
20 m sprint (s)	Small	Moderate	Trivial
Agility (s)	Small	Large	Trivial
Vertical jump (cm)	Moderate	Moderate	Small
Flexibility (cm)	Small	Moderate	Small
Fatigue (u.a.)	Small	Large	Small
Muscle soreness (u.a.)	Very large	Large	Moderate

Reprinted, by permission, from P.G. Pyne et al., 2008, "The effect of recovery strategies on physical performance and cumulative fatigue in competitive basketball," *Journal of Sports Sciences* 26: 1135-1145.

PRACTICAL APPLICATION

The effectiveness of stretching to improve recovery is still debated in the literature. The practice of passive stretching is mostly beneficial for well-being, and it contributes to athletes' cool-down process. Therefore, stretching after specific training sessions should be planned with moderation.

When and how should stretching be used in a context of recovery?

- It should not be done directly after training sessions or competitions.
- Duration of stretching: 15–30 s for each muscle group.
- Number of repetitions: 1–3 repetitions. However, subsequent repetitions should not focus on the same muscle group.

Contraindications for an optimal use of stretching in a recovery setting include the following:

- Avoid stretching following high-intensity sessions (specifically in the preparatory phase of the season) or after strengthening sessions. Stretching should not be performed on muscles groups experiencing DOMS.
- Avoid stretching before sessions that will involve high levels of muscle strength. For performances where joint amplitude is essential (martial arts, gymnastic, diving), alternate 15 to 20 s of stretching and contraction of the muscles previously stretched.
- Do not perform stretching with high amplitude when the muscle is in a resting state.
- Avoid body positions that involve muscle contractions of high intensity or induce elevated constraints on joints that are not involved in the stretch.

---

appearance of the pain itself. Therefore, this is a period when injuries are likely to occur. In this context, what effect does muscle stretching have on DOMS? This subject has been largely described. The scientific reviews compiling the main studies led to fairly homogeneous conclusions. Since the study by McGlynn and associates (1979) up to the review by Herbert and de Noronha (2007), the results are quite conclusive: Stretching performed before or after intense muscular activity does not reduce DOMS (Buroker and Schwane 1989; Dawson et al. 2005; Gulick et al. 1996; High, Howley, and Franks 1989; Johansson et al. 1999; Maxwell et al. 1988; McGlynn, Laughlin, and Rowe 1979; Terry 1985, 1987; Wessel and Wan 1994) or its indirect effects (Johansson et al. 1999; Lund et al. 1998).

Stretching and Strength

It is well known that muscle strength drops following eccentric exercise compared to initial levels. The timeline for recovery of this strength depends on various parameters, including exercise intensity. In general, it takes several days for subjects to fully recover their initial strength. How does stretching the muscle group involved during exercise affect the recovery of strength?

The following two figures provide a straightforward response to this question. Lund and colleagues (1998) performed the same eccentric exercises with the quadriceps muscle groups of both lower limbs. However, one of the limbs was also stretched (three 30 s repetitions) on the day of eccentric exercise and on each day of recovery. Strength recovery in the stretched quadriceps was lower compared to the limb that performed only the eccentric exercise. This result is observed with concentric exercise (figure 6.7), but it is particularly notable in tests involving the same exercise modality (i.e., eccentric) (figure 6.8).

DOMS is triggered by mechanical perturbations of the muscle fiber, particularly by damage to elements of the cytoskeleton (desmin, titin, and nebulin) (Yu, Fürst, and Thornell 2003; Fridén and Lieber 1998; Lieber, Thornell, and Fridén 1996), which are also involved in passive stretching. Some of these changes are observed within 5 min of the onset of the eccentric exercise (Lieber, Thornell, and Fridén 1996). This explains why stretching is particularly contraindicated when performed after exercise. In contrast, it seems that stretches performed before eccentric exercise have no effect on the drop in strength. Johansson and associates (1999) performed a study similar

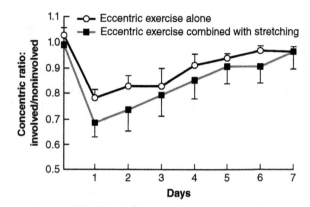

Figure 6.7 Changes in concentric maximum muscle strength following eccentric exercise alone or eccentric exercise combined with stretching.

Reprinted, by permission, from H. Lund, 1998, "The effect of passive stretching on delayed onset muscle soreness, and other detrimental effects following eccentric exercise," *Scandinavian Journal of Medicine & Science in Sports* 8: 216-221.

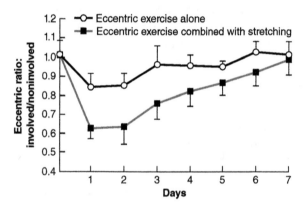

Figure 6.8 Changes in eccentric maximum muscle strength following eccentric exercise alone or eccentric exercise combined with stretching.

Reprinted, by permission, from H. Lund, 1998, "The effect of passive stretching on delayed onset muscle soreness, and other detrimental effects following eccentric exercise," *Scandinavian Journal of Medicine & Science in Sports* 8: 216-221.

to that of Lund and colleagues (1998), but they used only stretching before the eccentric exercise. They showed that, in their conditions, the drop in strength was not amplified by stretching. Thus, according to these authors, the increased drop in strength observed by Lund and associates (1998) would mainly be due to the stretching after the eccentric exercise.

Stretching and Pain

The effect of stretching on pain felt with or without palpation can be summarized by the results presented in the following tables (Herbert and de Noronha 2007). For each of the studies cited, the pain reported by the experimental group (which performed stretching exercises) is compared to that reported by the control group (which performed the same tests without stretching) 1 day (figure 6.9), 2 days (figure 6.10), or 3 days (figure 6.11) after exercise.

From these figures, it is obvious that pain is not affected by stretching, whether performed before or after eccentric muscular exercise. Stretching creates significant muscle tension, leading to microlesions within the muscle. Although some short-term analgesic effects cause the athlete to feel an immediate, subjective sensation of reduced soreness, pain is not reduced in the longer term (between a few hours and several days). Some studies have even shown that a passive stretching protocol used after strength training induced significantly greater pain 2 days later than the level measured when strength training was not followed by stretching (Wiemann and Kamphövner 1995). The inflammatory process, which leads to the pain, is induced by microlesions to elements of the sarcomere that are also involved in stretching. Passive stretching carried out after a strenu-

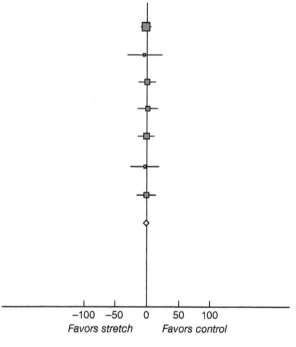

Figure 6.9 Perceived pain 1 day after eccentric exercise. The number of subjects (N) and the results (mean ± SD) are presented for each study analyzed. The differences between the experimental group, which performed stretches before or after exercise (stretch), and the control group (control) are expressed in %.

Reprinted, by permission, from R.D. Herbert and M. de Noronha, 2007, "Stretching to prevent or reduce muscle soreness after exercise," *Cochrane Database of Systematic Reviews* 17: 1-24.

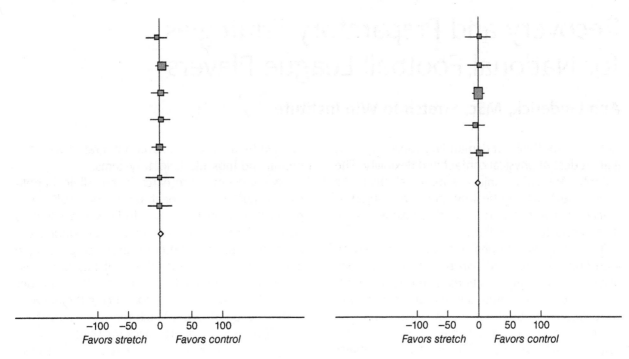

Figure 6.10 Perceived pain 2 days after eccentric exercise. The number of subjects (N) and the results (mean ± SD) are presented for each study analyzed. The differences between the experimental group, which performed stretches before or after exercise (stretch), and the control group (control) are expressed in %.

Reprinted, by permission, from R.D. Herbert and M. de Noronha, 2007, "Stretching to prevent or reduce muscle soreness after exercise," *Cochrane Database of Systematic Reviews* 17: 1-24.

Figure 6.11 Perceived pain 3 days after eccentric exercise. The number of subjects (N) and the results (mean ± SD) are presented for each study analyzed. The differences between the experimental group, which performed stretches before or after exercise (stretch), and the control group (control) are expressed in %.

Reprinted, by permission, from R.D. Herbert and M. de Noronha, 2007, "Stretching to prevent or reduce muscle soreness after exercise," *Cochrane Database of Systematic Reviews* 17: 1-24.

ous exercise session generates microlesions that amplify the muscle trauma and prolong the time necessary for full recovery.

Stretching and Edema

Muscle damage as a result of eccentric exercise leads to edema. It has been suggested that stretching can help reduce the extent of the edema (Bobbert, Hollander, and Huijing 1986). However, the swelling induced by inflammation is also part of the regeneration process that follows muscle damage. Therefore, according to Barnett (2006), dispersion of the edema by stretching should not be used as an aid to recovery.

Summary

The different aspects of stretching covered in this chapter converge to the following conclusions: Stretching alone is not generally effective as a recovery method, and it is even contraindicated in some conditions. It should be avoided immediately after muscle strengthening sessions or specific athletic activities that induce muscle soreness. When used as a recovery intervention, stretching does not improve performance indicators, including maximal strength, rate of force development, vertical jumps (indicators of explosiveness), or time to exhaustion. In the context of competitions, however, stretching can have a general effect on perceived pain and fatigue. Nevertheless, the influence of these factors on recovery after an event (which have unintended effects on the athlete's well-being and performance) is more successful when other recovery methods are also used. Consequently, while stretches are commonly used in recovery because of their ease of implementation, it is advisable to associate them with other recovery methods for optimal results.

Recovery and Preparatory Strategies for National Football League Players

Ann Frederick, MSc, Stretch to Win Institute

American football is a demanding sport requiring a great deal of physical contact and flexibility. The potential for serious injury is extremely high. As in any sport, staying healthy and free of injury is crucial for success, and increasing flexibility may be an important contribution.

This study was carried out in a clinical setting with one therapist and two professional football players in opposing positions in the sport. Both athletes had long and successful careers and were treated throughout the in-season and off-season. The two athletes have been receiving fascial stretch therapy (FST) for 14 years (subject A), and 11 years (subject B). Even retired athletes continue receiving FST to counteract the years of wear and tear from playing football.

Fascial stretch therapy is based on 10 principles that do not necessarily need to be executed in the order in which they are presented here; however, all the principles must be applied for optimal success:

1. Synchronize breathing with movement.
2. Tune the nervous system to current conditions.
3. Follow a logical anatomical order.
4. Increase range of motion without pain.
5. Stretch the fascia, not just the muscle.
6. Use multiple planes of movement.
7. Target the entire joint.
8. Use traction for maximal lengthening.
9. Facilitate body reflexes for optimal results.
10. Adjust stretching according to current goals.

FST differs from traditional stretching in many ways. It is performed on a specialized table using customized straps that keep one leg stabilized while the other leg is free to be moved through space. This allows greater leverage and a biomechanical advantage for the therapist and allows the athlete to relax. FST uses a global approach in assessment and treatment, emphasizing the connective tissue system (fascia) integrated with the neural and musculoskeletal systems.

As important as research is, the clinical application must prove repeatable in its results. The method used in this case study was applied to both subjects. There is a team effort during the entire treatment, and there is an ongoing dialogue between therapist and athlete. It has been discovered over the past two decades that in order to get the best outcome, "romancing the nervous system" is a better approach than attacking it.

Subject A was a quarterback for 13 years, which is a position on offense. He is 6 feet 2 inches (188 cm) tall and weighs 240 pounds (109 kg). He was required to throw the football, run, and avoid sacks. A quarterback's agility is critical, as is flexibility in recovering from sacks.

Subject B was a lineman who played defensive tackle professionally for 9 years. He is 6 feet 4 inches (193 cm) and weighs 315 pounds (143 kg). Linemen are responsible for tackling offensive players, so there was an ongoing strategy to manually decompress the spine and extremities and reestablish mobility and strength quickly during the in-season. He also was kiddingly termed "flexibility challenged."

FST was performed throughout the season as well as in postseason and the off-season. Because of the extremely high demands of the sport, the treatments needed to be adjusted for each session. If there was a specific problem from the most recent game played, the treatment plan was adjusted accordingly.

Goals of treatment are to address current challenges resulting from training or playing and to treat the entire body for balance and restoration. The most common requests from the athletes were to have the hips opened up and to remove the feeling of heaviness in the legs. Work was done in a preparatory manner if it was game day, and focus was on the sympathetic nervous system to ready the athletes for game time. On game day, the movement was at a much quicker pace than what was used on other days; little, if any, proprioceptive neuromuscular facilitation (PNF) was used. Parameters of FST were adjusted as follows for pregame: higher reps of low-force PNF, and decreased intensity and duration of stretches as indicated.

If it was postgame, then work was done on the parasympathetic nervous system to facilitate recovery in the restoration phase. The tempo of these sessions was much slower and corrective in nature. Attention was given to the areas of concern from the game. The goal was to get the athletes ready to go for the next game. Parameters of FST were adjusted as follows for postgame: fewer reps of low-force PNF, and increased intensity and duration of stretches as indicated.

Although much of the research generally shows negative effects of static stretching in sports, very few studies have been done on dynamic stretching methods. Research is needed on dynamic stretching methods (such as FST and others) to get a better indication of what is helpful. In our case study, common comments from the subjects were that the therapy made them feel faster, lighter, stronger, and significantly reduced their soreness. They also felt less restricted overall, in that their hips, back and shoulder "opened up" and the stretching greatly increased their freedom of movement. The athletes' flexibility increased through the off-season and was maintained during the competition seasons. Additional gains were made during each off-season. The longer they played the more important it was to maintain their flexibility, in order to stay healthy and at the top of their game. Compared to their teammates, they were healthier and remained uninjured for the majority of their careers (both subjects were completely injury free for 7 years).

Because of the partnership nature and customization to each athlete's needs, fascial stretch therapy has proved successful with hundreds of professional athletes over the past two decades. Improved performance and reduction in injuries from this type of stretching are evident from the long-term relationships between athletes and therapists. Those who are in the field, locker room, or treatment room and work with athletes of any caliber know that it must always be about the athletes.

Key Points

- Always follow the 10 principles.
- Set specific goals of what to achieve in terms of flexibility needs.
- Listen to and respect what the athlete is feeling, and treat accordingly.
- Give the type of treatment needed at the time.
- Restore balance to the body, either to pregame state or training state.
- Increase flexibility in the off-season.
- Restore current level of flexibility during the season.

CHAPTER 7

Hydration

Christophe Hausswirth, PhD

With: Véronique Rousseau

The detrimental effects of dehydration during exercise are well recognized. Performance will suffer if a fluid deficit is allowed to develop. Moreover, practicing sport at a high level results in a significant increase in energy needs, and maintaining the energy balance requires increased food intake. Providing adequate rehydration is one of the key factors for an athlete's fluid recovery. Different metabolic modifications have been noted during exercise: loss of electrolytes and proteins (possibly due to muscle-cell degradation), combined with depletion of glycogen reserves and mobilization of liquid reserves. The recovery phase therefore needs to compensate for the fluid losses resulting from the physical effort provided during training or competition by adapting hydration to pre-, per-, and posteffort periods.

The quality of the performance and the safety of the athlete depend largely on an adequate hydroelectrolytic balance. This balance is dangerously perturbed during exercise due to redistribution of water and electrolytes between various compartments, but also to losses through the skin through perspiration.

Water is a remarkable and omnipresent nutrient. In the body, water serves as a means of transport and as a reaction medium. Gas diffusion is often possible only through surfaces impregnated with water. Due to its incompressibility, water gives structure and shape to the body by increasing tissue volumes. In addition, water (because of its thermostatic qualities) is capable of absorbing a considerable amount of heat without significantly increasing its temperature. This quality, in combination with a high vaporization temperature, contributes to the maintenance of a relatively constant body temperature, despite external thermal constraints and significant increases in internal temperature during exercise.

The quantity of water in the body remains relatively stable over time. Although considerable quantities of water are lost during physical activity, appropriate absorption of fluids rapidly restores imbalances of the liquid levels in the body. An averagely active adult in a neutral environment requires approximately 2.5 L of water per day (table 7.1). In contrast, the water needs of an active person placed in a warm atmosphere can increase between 5 and 10 L per day. This water has three different origins: drinks, food, and metabolic processes.

Table 7.1   Water Balance in the Body

NORMAL ENVIRONMENT, SEDENTARY				HOT ENVIRONMENT, ACTIVE			
Daily water consumption		**Daily water usage**		**Daily water consumption**		**Daily water usage**	
Water in food	750	Skin	800	Water in food	750	Skin	8000
Water from beverages	1500	Lungs	350	Water from beverages	1500	Lungs	1900
Water from oxidation	250	Feces	100	Water from oxidation	250	Feces	100
		Urine	1250			Urine	500
Total input	**2500**	Total output	**2500**	Total input	**2500**	Total output	**10500**

Distribution of Water Loss During Exercise

Athletes are required to optimize their water needs, since water is the main constituent of the body. It represents between 40% and 70% of body mass (depending on age, sex, and body shape). It constitutes 65% to 75% of muscular mass and approximately 50% of fatty mass (Greenleaf 1992). Water is, indeed, indispensable for several physiological functions in the body: cellular activity, the cardiovascular system, regulation of body temperature, elimination through the kidneys, and so on.

Plasma is at the center of exchanges between all the organs of the body. Because of its easy accessibility, it represents an ideal source for the study of metabolism. However, study of a single plasma sample constitutes only a still image of the metabolic state, which may be subject to rapid variations. The volume of plasma reflects the physiological state of its source, the body. When not compensated for by adequate rehydration, sweating is responsible for a reduction in the total volume of body fluids, possibly leading to reductions in plasma volume. This can have an unintended effect on all the fluid reserves within the body. From a certain level of deficit, hypovolemia is responsible for a reduction in the systolic ejection volume that cannot be compensated for by increasing cardiac frequency. This leads to a reduction in the maximal cardiac output during exercise. Whatever the initial body temperature and speed of thermal storage, an internal temperature threshold determines the appearance of fatigue. This threshold is lowered in case of dehydration (González-Alonso et al. 1999). Hemo-concentration then occurs, often encountered during long-term exercise or exercise carried out in a warm atmosphere. This can lead to a certain number of difficulties that are incompatible with a high-quality performance:

- Increased viscosity
- Reduced elimination capacity
- Disimproved thermal exchange with the skin
- Difficulty irrigating the active muscles
- Poor function of cardiac pump
- Increase of mesenteric ischemia, which can be associated with colon motility problems

With regard to the latter, dehydration, and the resultant plasma hypovolemia, is responsible for an additional reduction of splanchnic blood flow. It also increases the risk of appearance of digestive lesions (Beckers et al. 1992). Prevention relies on regular, carefully designed physical training, on limiting sympathetic hypertonia, and on fighting hypovolemia, which can be done only through adequate rehydration.

Dehydration Processes

Dehydration is a process that originates during the practice of sport from sweating. If the loss of fluid remains modest (i.e., 0.5% of body weight), no side effects are observed. However, above this value, which many athletes exceed rapidly, numerous complications hinder correct exercise. This dehydration represents an imbalance between fluid movements, indicating that water loss is not compensated for by water intake (if the athlete is, to start with, normally hydrated or hyperhydrated). The simplest means of evaluating the percent dehydration is to use a double weigh-in. This limits the poor reliability of personal weighing scales (variations include ± 0.5 kg). The measurements must be carried out on an athlete who is naked and dry, so as not to count the water contained in clothes or remaining at the skin's surface. Athletes should always have an empty bladder for weigh-ins. The following formula will be applied to calculate dehydration:

$$\text{weight before (wB)} - \text{weight after (wA)} = \text{mass of water lost (ml)}$$

$$\text{ml/wB} \times 100 = \text{percent dehydration}$$

The sweating rate corresponds to the difference between the two weights, minus the amount of urine collected after exercise:

$$\text{ml} - \text{urine (A)} = \text{loss through sweating}$$

Measuring body weight before and after exercise allows the extent of fluid loss to be predicted, since the reduction in weight is due mainly to fluid loss through perspiration and evaporation as part of breathing. Despite additional losses of substrates and metabolic water loss, it is not, in fact, necessary to maintain exactly the same weight to preserve water balance. In addition, the extent of weight reduction can vary from one person to the next. It depends on losses through sweating and, thus, on the increase in core body temperature. During cricket matches conducted in various exterior temperature conditions, Gore and colleagues (1993) were able to observe weight losses varying up to 0.5 kg/h for outside temperatures of 22.1 °C, and up to 1.6 kg/h at temperatures of 27.1 °C.

Exposing oneself to heat or performing exercise in a heated atmosphere above 25 °C results in increased sweating, which depletes the body's fluid reserves, thus creating a relative state of dehydration. Because of this, the body pays a high price for the efficacy of its thermal regulation through the loss of intravascular and intracellular liquids. Indeed, when sweat is produced in significant quantities, plasma volumes are observably reduced (Claremont et al. 1976). This affects the body's capacity to cool itself. Loss of water by the body therefore affects both of its water compartments, with a variable distribution between extracellular fluid and tissue water.

The effect of this deficit on athletic performance differs, however, depending on the origin of this dehydration. During very sustained exercise, or exercise carried out in unfavorable thermal conditions, huge volumes of sweat may be eliminated over a very short time. In this case, the water comes mainly from plasma. Physical aptitudes are more affected here, at least until the athlete acclimatizes to the heat. After acclimatization, tolerance to this loss has been shown to improve (Pandolf 1998). Simultaneously, the increase in metabolic reactions linked to the energy demand results in a more significant reduction of water within cells: The successive reactions of glycogen and, to a lesser degree, lipid degradation lead to resynthesis of adenosine triphosphate (ATP) on the one hand and production of water on the other (Sawka, Wenger, and Pandolf 1996). However, this metabolic water production is insufficient to prevent dehydration. Any water deficit, or any dehydration, can negatively influence physiological processes and, more specifically, athletic performance.

Dehydration and Physical Performance

Small reductions in the body's fluid reserves significantly affect performance. The longer exercise continues, the worse it is tolerated if losses due to sweat evaporation are not compensated for. Thus, given the importance of water and the consequences of any deficit, athletes must compensate both for exercise-related losses and for those linked to general daily activities as rapidly and completely as possible. Dehydration is known to lower athletes' physical capacities by increasing their cardiac rhythm and core body temperature (Nielsen 1984). It has been frequently shown that a water deficit corresponding to just 2% of body weight reduces aerobic aptitudes by almost 20% (Armstrong et al. 1985). Numerous studies have shown that dehydrated subjects are intolerant of prolonged exercise (Claremont et al. 1976). Dehydration also affects how the thermoregulatory and cardiovascular systems work. Loss of liquid reduces plasma volume, which leads to a reduction in arterial pressure, in turn reducing blood flow through the muscles and skin, thus increasing cardiac frequency. In this context,

Sawka, Francesconi, and colleagues (1984) indicate that the alteration of cardiac rhythm depends on the drop in blood volume, and thus, on the level of dehydration. A study by Walsh and associates (1994) showed that 1.8% dehydration in cyclists riding at 90% of their maximum oxygen consumption led to a 32.8% reduction in the time to exhaustion when compared with cyclists who were rehydrated during the test. It is likely that the effect of dehydration on performance differs from one physical activity to another. How dehydration sets in, the climatic conditions during exercise, and the type of exercise are all key factors.

With regard to exercises of relatively short duration (requiring anaerobic metabolism) the effects of dehydration seem considerably less significant. During efforts lasting a few seconds, during which ATP comes exclusively from anaerobic metabolic systems (ATP-phosphocreatine and glycolytic), anaerobic performance does not appear particularly affected, to the extent that subjects can rehydrate adequately, in proportion with the observed fluid loss (Backx et al. 2000). However, and as we have previously shown, significant heat makes recovery from these short, intense efforts more difficult, particularly because of reduced liver catabolism of lactate and reduced muscle blood flow. It can therefore affect these short-duration activities (Sawka and Pandolf 1990). In addition, the effect of dehydration on anaerobic performances is not constant. The primary physiological mechanisms involved in reducing anaerobic performance relate to electrolyte imbalance, mainly due to increased potassium, and increased core body temperature (Sawka and Pandolf 1990). Regarding effort production, dehydration does not appear to affect maximum effort values (Mountain, Smith, and Mattot 1998; Greiwe et al. 1998), but it does reduce how long the effort can be maintained (Bigard, Sanchez, and Claveyrolas 2001). In a population whose body mass was reduced by 2.95%, these authors were able to show a reduction of 23% (P < 0.01) of the endurance time to exhaustion for isometric knee extension at 25% of maximal voluntary contraction (MVC) compared with a euhydrated situation (table 7.2). These authors also recorded a 13% reduction in dehydration when the intensity of the endurance time was fixed at 75% of MVC. However, this percentage is at the limit of the significance threshold (P = 0.06). Therefore, it should be taken with care. No other significant difference was recorded among the MVC of the various treatments (table 7.2).

Dehydration and Mental Performance

Objective data from cortical function tests confirmed the hypothesis that even moderate dehydration reduces mental performance in young adults (Maughan, Leiper, and Shirreffs 1997). Short-term and working memory seemed altered, although no effect was observed on long-term memory. According to Grandjean and Grandjean (2007), reduced performance is observed in tests involving psychomotor ability. These authors note an increase of 10.7% (P < 0.05) in the time to reaction for 2% dehydration and 21.4% (P < 0.01) for 4% dehydration (figure 7.1). In addition, a reduced percentage of correct answers, in the range of 10.6% for 2% dehydration and 22.4% for 4% dehydration, was recorded by the same authors using the same tests (Grandjean and Grandjean 2007).

Table 7.2 Variation of Maximal Voluntary Contraction (MVC) Values and Endurance Times to Exhaustion (ET)

Hydration status	MVC (N.M)			
	Right KEM	Left KEM	ET70 (s)	ET25 (s)[a]
Eu	416 ± 26	420 ± 20	46 ± 3	195 ± 19
Dhy	422 ± 20	404 ± 20	40 ± 4	150 ± 18‡
Rhy	428 ± 22	418 ± 16	47 ± 4	163 ± 17†,*

N.M: Newton meter. KEM: knee extension movement. Global significant effect of the treatment. Three conditions: Eu = normal hydration, Dhy = dehydration, and Rhy = rehydration after a period of dehydration. Values are expressed as mean ± standard deviation. [a] P < 0.01: significant difference with global effect of treatment. † P < 0.05, ‡ P < 0.01: significant difference with the Eu condition. * P < 0.05: significant difference with the Dhy condition.

Adapted, by permission, from A.X. Bigard, H. Sanchez, G. Claveyrolas, 2001, "Effects of dehydration and rehydration on EMG changes during fatiguing contractions," *Medicine and Science in Sports and Exercise* 33: 1694-1700.

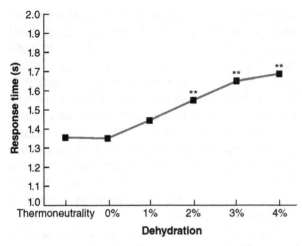

Figure 7.1 Variation in response times as a function of the percent dehydration induced. Values are expressed as means.

** P < 0.001: significant difference with the neutral condition at 0%.

Adapted, by permission, from H.R. Lieberman, 2007, "Hydration and cognition: A critical review and recommendations for future research," *Journal of American College of Nutrition* 26(5): 5495-5545.

Ongoing animal studies show that dehydration inhibits production of nitric oxide synthase—nitric oxide is considered a true neurotransmitter between nerve cells—and thus acts directly on the process of memorization in the medium to long term. Grandjean and Grandjean (2007) reviewed all the deleterious effects of dehydration above 1% of body weight on mental faculties (table 7.3).

From 2% of lost body weight, notable reductions in cognitive performance appear that are independent of how dehydration was induced (e.g., warm atmosphere, prolonged exercise) (Cian,

Koulmann, and Barraud 2000). This is very significant in sports where physical activity requires short reaction times and very precise movements. In addition, this appears even more problematic in exercise involving combined events, such as the Nordic biathlon (rifle shooting and cross-country skiing) or the modern pentathlon (pistol shooting, fencing, swimming, show jumping, and running).

Hydration Recommendations

The recommendations of the French Agency for Food Safety (AFSSA) for the general French population (Martin 2001) state that the hydration needs of an average 60 kg adult are between 2.1 and 3 L/day (35 to 50 ml · kg^{-1} · day^{-1}). In 2008, the European Food Safety Authority (EFSA) made these recommendations even more precise. Among adults (anyone over 14 years of age), a woman should consume 2 L/day and a man 2.5 L/day, or 1.4 to 2 L/day from drinks and 0.4 to 0.75 L/day from food. Since the work of Koulmann and colleagues (2000), the daily liquid needs required to avoid acute dehydration have been known to be between 1.5 L and 1.8 L. It must be noted that 87% of 12- to 19-year-olds and 70% of 20- to 55-year-olds drink less than 1.5 L of water per day.

In addition, it appears that quantity is not the only factor influencing hydration. According to the Research Center for the Study and Observation of Living Conditions (CREDOC) in France, distribution (figure 7.2) is equally important (Hoibian

Table 7.3 Variation of Cognitive Parameters as a Function of States of Hyperhydration and Dehydration due to High External Temperature or Exercise-Related Dehydration

Cognitive elements	Hyperhydration	Dehydration due to external temperature	Exercise-related dehydration
Perceived fatigue	↔	↑	↑
Mood	↔	↔	↔
Reaction time and accuracy	↔	↔	↔
Number of errors	↔	↑	↑
Comparison of the average reaction time	↔	↔	↔
Short-term memory	↑	↓	↓
Long-term memory	↔	↔	↔

Adapted, by permission, from A.C. Grandjean and N.R. Grandjean, 2007, "Dehydration and cognitive performance," *Journal of American College Nutrition* 26(5): 549S-554S.

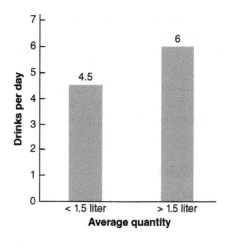

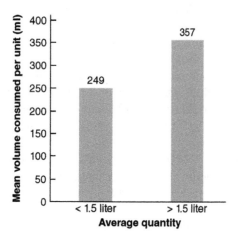

Figure 7.2 Recapitulative histograms of the number of drinks taken per day and the average quantities consumed at each intake.

Based on CREDOC data 2007.

2007). Their survey indicates that the adults who drink most also drink larger volumes more frequently throughout the day.

Preventing Dehydration for Improved Recovery

Preventing dehydration is a primary objective for athletes, but structured preventive measures are not always easy to implement. It is important to know that, once it has set in, the dehydrated state may not be compatible with athletic performance. Its prevention is difficult because the body is incapable of storing adequate water. Any water absorption before the competition systematically leads to increased urination, which can be inconvenient in a competitive situation. Euhydration, or normal hydration, of the athlete is generally

advised. Drinking 150 to 300 ml before a competition allows hydration to be maintained, without causing urination in the following hours. In addition, the passage through plasma is very rapid, since the water absorbed appears in sweat 9 to 18 min later (Armstrong et al. 1987). However, no real standard to evaluate water imbalances exist. A study by Brandenberger and colleagues (1989) showed that, despite other factors, active prehydration with 500 ml of water 40 min before exercise tended to increase the sweating rate, limiting the perturbations induced by the absence of hydration during exercise.

Thirst is not a good indicator of the need to rehydrate during exercise: It is therefore necessary to drink before it is felt, in greater volumes than the feeling would appear to indicate. The volume to consume should be determined individually and adjusted according to individual tolerance, while trying to compensate for losses. To do this, the latter should be estimated from the characteristics of the exercise performed (intensity, duration), the environmental conditions (ambient temperature, relative humidity, reflected heat), and specific individual factors (level and state of training, acclimation to heat or not). To favor stomach emptying, it seems preferable to ingest large volumes of fluid in line with individual tolerance levels (Rehrer 2001).

Electrolyte Replacement During Recovery

Variations in body weight indicate the quantity of water lost during exercise and the adequacy of rehydration during and after exercise. Significant levels of sweating result in loss of hydroelectrolytes. However, the rate of sweating and the amount of electrolytes in sweat (Na^+, K^+, Cl^-, Ca^{2+}, Mg^{2+}), which can vary greatly among different people, influence its composition. When perspiration increases, the level of sodium (Na^+) and chloride (Cl^-) ions increases, while that of potassium (K^+) and magnesium (Mg^{2+}) ions remains unchanged and that of calcium (Ca^{2+}) ions drops. As a result, the loss of extracellular compounds, like sodium and chloride, is most likely to be the source of the deficit induced by sweating. Supplementation can contribute to adequate daily levels in some circumstances, particularly during or after an extended period of exercise, or exercise conducted in high external temperatures. Sweat is hypotonic compared to other body fluids, and it becomes even more so through training and acclimatization to heat. This is why plasma-electrolyte concentrations (and osmolarity) can increase during exercise if the athlete does not absorb liquid to replace lost fluids. It is, however, very difficult to establish a precise evaluation of some electrolytes based on their plasma concentrations. This is because of exercise-induced movements of both fluids and electrolytes.

Temporary mineral imbalances can be encountered during prolonged exercise. Noakes and associates (1985) were able to show that compensating for substantial fluid loss by drinking flat water led to hyponatremia (up to 25%) in some athletes. As a result, athletes should compensate for soluble sodium losses when participating in sports over extended periods of time. Sodium tablets are not advisable, whatever the level of deficit, because they can secrete water into the intestinal lumen, leading to discomfort and other deleterious effects on the gastro-intestinal tract. Because it stimulates absorption of water in the intestine, the addition of sodium to a sugary drink is generally justified (Noakes, Goodwin, and Rayner 1985). Rehydration drinks, which have added sodium chloride, have been shown to be better at correcting water imbalance and plasma volume (through water retention) than plain water. Among other things, they allow the concentrations of hormones involved in fluid regulation to normalize much faster (Nose et al. 1988; Brandenberger et al. 1989). If the rehydration solution does not contain enough sodium, the excess liquid absorbed will simply increase the volume of urine excreted; it will not benefit rehydration at all (Shirreffs et al. 1996). These authors were able to compare several rehydration schemas on subjects who lost 3% of body mass. In group L (addition of 23 mM sodium, or 1.2 g/L), only rehydration at 150% and 200% led to a significant increase in urine volumes. Increasing the sodium concentration (61 mM, or 3.3 g/L) in the rehydration solution contributes to restored fluid levels, but rehydration with this solution must be controlled through precise ingestion volumes. Indeed, the three conditions (B, C, and D) for group H all differ significantly from each other. The authors concluded that drinking a large volume of water after exercise does not compensate for dehydration if the sodium concentration is not high enough, and vice versa.

In addition, moderate exercise results in only moderate loss of potassium through sweating (Cunningham 1997). Whereas, during more intense exercise (for example in competition), potassium lost through sweat is at levels reaching 5 to 18 mEq. More significant losses of potassium can be compensated for by consuming foods rich in potassium (lemons, bananas). A glass of orange or tomato juice replaces practically all the potassium, calcium, and magnesium lost in approximately 3 L of sweat (Cunningham 1997).

Alkaline Drinks During the Early Recovery Phase

It is well documented that during short but intense exercise, anaerobic production of lactic acid in the active muscles and its rapid diffusion into extracellular fluids leads to increased proton concentrations (Sahlin et al. 1978). From a physiological point of view, we have observed that intramuscular proton accumulation reduces the contractile capacity of muscles (Hausswirth et al. 1995) by inhibiting enzymes such as phosphofructokinase (PFK), by perturbing calcium fluxes, and by reducing the number of actin–myosin bridges (Vollestad and Sejersted 1988), as well as by perturbing membrane excitability and the transmission of related incoming nerve signals. In the case of athletic activity, the acid–base imbalance in the blood perturbs the endocrine system, the availability of energy substrates for the muscles and blood (Sutton, Jones, and Toews 1981), and the cardiocirculatory and ventilatory systems (Jones et al. 1977), as well as subjective perception of fatigue (Swank and Robertson 1989). This is why, for about 20 years, athletes who participated in what can be termed *anaerobic* sports were advised to use alkaline drinks before and after intense effort. However, even if scientific interest has often been linked to improvements in endurance times between 120 and 240 s (McNaughton and Cedaro 1992), absorbing bicarbonates before an effort is often associated with gastro-intestinal perturbations (McNaughton et al. 1999), especially when the bicarbonate concentration exceeds 300 mg/kg of body weight. Because of this, we strongly advise the introduction of alkalinizing drinks during the early recovery phase. The acid–base homeostasis recovery time is best reduced consuming bicarbonate drinks postexercise (see table 7.4).

Table 7.4 Review of the Different Bicarbonate and Sodium Concentrations in Mineral Water Brands Commercially Available in France

Mineral water/electrolytes	Bicarbonates (mg/L)	Sodium (mg/L)
Donat Mg	7800	1600
Hydroxydase	6722	1945
Saint-Yorre	1368	1708
Vichy Célestin	1989	1172
Rozanna	1837	493
Quézac	1685	255

Glucose Drinks and Rapid Recovery

A certain level of dehydration is inevitable during physical activity, particularly when it is carried out in a warm atmosphere. Recovery must therefore optimize restoration of water and mineral balances. This is particularly problematic for athletes involved in events where repeated effort at short intervals is required (like for a triathlon, decathlon, or modern pentathlon, or when combats or assaults are repeated during the day). It is, however, well established that after a period of dehydration, a prolonged period exists during which it is not possible to restore fluid resources to 100%.

Drinking water causes rapid urination, even if the athlete remains dehydrated, hindering return to normal hydration levels. Drinking water alone is also responsible for slaking thirst, by acting on the dipsogenic osmotic stimuli, all the more so when the water is cold (Rehrer et al. 1992). It seems that consuming carbohydrate (CHO) solutions during the recovery period allows exercise capacity to be restored more efficiently than when water is consumed alone (water serves as a vector for the carbohydrates, which are indispensable for the resynthesis of hepatic and intramuscular glycogen stores). This becomes important when the recovery interval between two events is short or when exercises are repeated over several days (Maughan, Leiper, and Shirreffs 1997).

Concerning the type of sugar or carbohydrate, glucose is often favored during recovery. It has been shown to be more effective (Nadel, Mack, and Nose 1990) because its absorption involves an active process. In addition, this active transport is associated with sodium absorption (Crane 1962). Fructose, on the other hand, is not absorbed by active transport, but by facilitated diffusion. It is also independent of sodium. The active transport of glucose associated with sodium leads to increased water absorption following intake of a solution containing glucose (sugar, glucose polymers, or starch). Indeed, several authors have noted an increase in net water absorption (after subtraction of secretions) of solutions containing less than 7% sugars and sodium. In contrast, plain flat water induces a net secretion of electrolytes into the gastro-intestinal lumen (Leiper and Maughan 1986).

To complement this, clinical studies have shown that oral rehydration solutions (those containing 5% glucose and 30 to 60 mM sodium) improved the net absorption of soluble sodium and reduced stool frequency in athletes suffering from dehydrating diarrhea (Farthing 1988). In order to better understand the different rates of net water absorption, Beckers and associates (1992) proposed an experiment (its results are illustrated in figure 7.3). The glucose-containing postexercise rehydration solutions used in this study all contained 20 mEq/L Na$^+$, but had variable concentrations and osmolarities. As indicated in figure 7.3, perfusion of a glucose solution at 4.5% results in better net fluid absorption than plain water alone. A glucose and electrolyte solution at 7% (mainly sucrose) also led to significantly higher fluid absorption compared to plain water. A hypertonic glucose solution at 17% resulted in net fluid secretion only. On the other hand, a maltodextrin solution at 17% (17 Md), whose osmolarity was close to that of the 4.5% solution, led to net fluid absorption, but this absorption was revealed to be much slower than for the solution at 4.5%. These authors point out, however, that the best sugar absorption during recovery was obtained with more concentrated glucose solutions, with

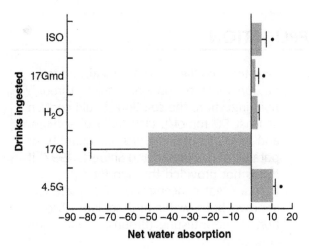

Figure 7.3 Net intestinal fluid movements (negative figures = net absorption; positive figures = net secretion) after absorption of water, of an isotonic glucose solution at 4.5% (4.5 G) [301 mOsm], of an isotonic glucose and electrolyte solution at 7% (ISO), of a hypertonic glucose solution at 17% (17G) [1,223 mOsm], or of an isotonic solution of glucose polymers at 17% (17 Gmd).

a corresponding reduction in fluid absorption. A compromise must therefore be found by the athlete for postexercise glucose-based rehydration strategies. The type of drink needed depends on the exercise performed, the rate of sweating, environmental constraints, and requirements for rapid resynthesis of glycogen used.

Hyperhydration

To avoid the adverse effects of hypohydration, some studies suggested that pre-exercise hyperhydration might delay or prevent hypohydration during exercise and, therefore, the potential negative consequences that are sometimes associated with this condition (Freund et al. 1995). Although many studies have examined the topic of hyperhydration, much controversy continues to surround this issue. It has been suggested that hyperhydration might improve the thermoregulatory function through expansion of blood volume, therefore enhancing performance (Lyons et al. 1990). Several studies have examined the effects of hyperhy-

dration using water, or water in combination with electrolyte solutions (Grucza, Szczypaczewska, and Kozlowski 1987). However, both methods led to only a transient expansion of total body water, since fluid overload is promptly excreted by the kidneys (Riedesel et al. 1987). Other studies have demonstrated that, compared to water ingestion alone, the ingestion of glycerol along with a large volume of fluid led to higher fluid retention (Lyons et al. 1990). Glycerol ingestion has been associated with hyperhydration for periods of up to 4 h (Riedesel et al. 1987). Peak hyperhydration is determined by the relationship between absorption of glycerol and fluid and clearance. It is difficult to determine when peak hyperhydration occurs because the timing of measurements of fluid retention varies greatly between studies. Because the absorption of a large volume of fluid occurs quickly when fluid intake is rapid, exercise should start 30 min after final fluid consumption. Waiting for an extended period results in unnecessary water loss and reduces overall hyperhydration. Waiting 30 min after fluid intake should allow sufficient time for the sensation of stomach fullness to subside, while ensuring that exercise begins when hyperhydration is close to maximal.

Pre-exercise glycerol hyperhydration will be most advantageous when sweat losses cannot be replaced during exercise. However, if euhydration can be maintained during exercise, then hyperhydration may not provide any additional advantage. Theoretically, water hyperhydration with or without the use of glycerol seems to possess great potential for improving performance and reducing the risk of experiencing problems associated with hypohydration (Convertino et al. 1996). Some studies reported that, in a subsequent exercise following a hyperhydration procedure, core temperature was lower when compared with a control trial (Grucza, Szczypaczewska, and Kozlowski 1987). In addition, some of these studies also reported higher sweat rates following the hyperhydration procedures (Lyons et al. 1990). On the other hand, other studies have demonstrated that hyperhydration does not provide any thermoregulatory-related advantage (Hitchins et al. 1999). In addition, rapid ingestion of large quantities of water (or other low-sodium fluids) can lead to hyponatremia, which has very serious clinical consequences (Von Duvillard et al. 2004).

PRACTICAL APPLICATION

- Replace body fluids lost during exercise as rapidly as possible. Weigh athletes before and after the event to determine their fluid needs. Full replacement of fluids and their retention can only be achieved by ingesting 150% to 200% of the measured fluid loss (Shirreffs et al. 1996).

- Using thirst to guide rehydration of athletes always results in insufficient compensation for fluids lost during exercise. In this case, slightly flavored, cool water at 12–15 °C can truly enhance the increase in fluid replacement as compared with other drinks (cold water, tepid water, mineral water) (Hubbard et al. 1984). Palatability of rehydration solutions is important to stimulate intake, but whenever possible, electrolytes should be added through solid nutrition in order to reduce the sodium concentration of drinks (Beckers et al. 1992).

- The water balance is more rapidly corrected by the ingestion of large (rather than small) volumes of liquid, over a longer period of time. After this ingestion, it is advisable to split intakes to about 200 ml every 15 min (Kovacs et al. 2002).

- If the athlete has at least 6 h before the next training session, fluid balance can be corrected by a combination of water and solid foods (Galloway 1999).

- If the time lapse between two training sessions or competitions is too short to allow ingestion and digestion of a meal, and rehydration must be obtained solely through fluid ingestion, the solution should contain at least 50 mmol/L (1.15 g/L) of sodium, and as far as possible, a small quantity of potassium (Maughan and Shirreffs 1997) if this is not provided through food. In addition, a sugar concentration about 2% can improve both water and sodium absorption in the intestine (Brouns, Kovacs, and Senden 1998).

- The ingestion of salt tablets is not advisable since they induce hypertonicity in the intestinal lumen, leading to increased digestive secretions and potentially causing digestive problems (Bigard and Guézennec 2003).

- The ingestion of caffeine during recovery is to be avoided, since it increases urinary excretion of electrolytes (Mg^{2+}, Na^+, and Ca^{2+}) (Brouns et al. 1998).

- Consuming drinks with more than 2.5 g/100 ml glucose slows the speed of stomach emptying. The addition of sodium chloride or potassium, often in excess, to exercise drinks has no effect on the speed of stomach emptying (Owen et al. 1986).

- In order for plasma volumes to be optimally restored with stable osmolarity, mineral concentrations should be between 0.5 and 0.6 g/L for Na^+, 0.7 and 0.8 g/L for Cl^-, and from 0.1 to 0.2 g/L for K^+ (Lamb and Brodowicz 1986).

Summary

Body weight variations indicate the quantity of water lost during exercise and the adequacy of rehydration during and after exercise. Significant levels of sweating result in loss of hydroelectrolytes. Supplementation can contribute to adequate daily levels in some circumstances, in particular during or after an extended period of exercise or exercise conducted in high external temperatures. Glucose is often favored during recovery and has shown its efficacy because its absorption involves an active process. It seems that consuming carbohydrate (CHO) solutions during the recovery period allows exercise capacity to be restored more efficiently than when water is consumed alone. This becomes important when the recovery interval between two events is short or when exercises are repeated over several days.

Recovery Between Ironman and XTERRA Triathlon World Championships

Iñigo Mujika, PhD, University of the Basque Country

Every year in the month of October, the sport of triathlon turns its attention to Hawaii for two of the sport's biggest non-Olympic events: the Ironman World Championship on the Big Island and the XTERRA World Championship on the island of Maui. The Ironman World Championship, also known as Ironman Hawaii, is a long-distance triathlon consisting of a 3.8 km ocean swim, 180 km of road cycling, and a 42.2 km marathon run. The XTERRA World Championship is a grueling off-road triathlon event that includes a 1.5 km ocean swim, followed by a 30 km mountain bike ride and an 11 km trail run. Both are mass-start events performed without a break and raced in that order, in hot and humid conditions.

A special prize, known as the Hawaiian Airlines Double, is awarded to the male and female triathletes with the fastest combined Ironman and XTERRA World Championships time. Before 2007, these two major events took place 1 week apart. Although they are now separated by 2 weeks, recovering from Ironman Hawaii to race again at the XTERRA Maui represents a major challenge for participating athletes and coaches alike.

When it comes to the Hawaiian Airlines Double, Basque triathlete Eneko Llanos is the most successful athlete of all time, since he has won this prime consecutively from 2006 to 2010 (table 1). To be successful, Eneko needs to find a reasonable balance between maximizing recovery from an exhausting event, such as the Ironman, and maintaining his training adaptations despite the necessary period of training cessation (Mujika and Padilla 2000). In 2006, when the two events were still separated by only 8 days, Eneko achieved this goal by resting completely for 4 days after the Ironman, with a focus on recovery from the muscle damage induced by the race (Suzuki et al. 2006), then getting some training stimuli in the 3 days prior to the XTERRA (table 2). He finished fifth in both events.

Eneko's typical recovery and training plan for the years 2007 to 2010, when the two events were separated by 15 days, included a period of complete inactivity, a period of active rest or cross-training (Loy, Hoffmann, and Holland 1995), and a period of training and race preparation (table 3). Note that Eneko's Ironman performance in 2009 was below average (14th place). However, that relative counterperformance was followed by his third XTERRA World Champion title, his first since taking part in the event after racing Ironman Hawaii. That particular year, Eneko's training program included an overreaching period in the lead-up to Ironman Hawaii (Coutts, Wallace, and Slattery 2007), but the subsequent 2-week taper probably provided insufficient recovery. Indeed, a longer taper seems to be necessary for optimal performance after an overload training period (Thomas, Mujika, and Busso 2008). After the Ironman race, a recovery phase followed by a moderate training phase probably induced supercompensation similar to that elicited by a two-phase

Table 1 Performance History of Triathlete Eneko Llanos Over a 5-Year Period at the Ironman and XTERRA World Championships

	IRONMAN WORLD CHAMPIONSHIP		XTERRA WORLD CHAMPIONSHIP	
Year	Time (hh:mm:ss)	Place	Time (hh:mm:ss)	Place
2006	08:22:28	5	02:46:49	5
2007	08:26:00	7	02:51:17	13
2008	08:20:50	2	02:42:01	6
2009	08:37:55	14	02:37:22	1
2010	08:22:02	7	02:40:44	6

taper, which is characterized by training reduction followed by a moderate increase in the training load (Thomas, Mujika, and Busso 2009).

Based on the experience accumulated by triathlete Eneko Llanos over the years, it seems that 2 to 4 days of complete rest, followed by 3 or 4 days of active rest or cross-training, and another 6 to 8 days of moderate training and race preparation is an effective recovery strategy to be competitive at Ironman and XTERRA World Championships.

Table 2 Recovery and Training Plan of Triathlete Eneko Llanos During the 8-Day Period Between Ironman and XTERRA World Championships in 2006

Days after Ironman	Swim (km)	Bike (min)	Run (min)
1-4	Rest	Rest	Rest
5	1.5	45	
6	1.2	60	
7		30	15

Table 3 Typical Recovery and Training Contents of Triathlete Eneko Llanos During the 15-Day Period Between Ironman and XTERRA World Championships From 2007 to 2010

Days after Ironman	Cross-training (min)	Swim (km)	Bike (min)	Run (min)
1-3	Rest	Rest	Rest	Rest
4	Walking/hiking 90–120			
5	Beach kayaking 60–90			
6	Snorkeling/surfing 60–90			
7		2.0		60
8			90	30
9		2.0	60	
10		2.2		50
11		2.7	45	20
12		2.0	60	40
13	Rest	Rest	Rest	Rest
14	0	1.0	40	15

CHAPTER 8

Nutrition

Christophe Hausswirth, PhD

With: Véronique Rousseau

Nutrition factors play an essential role in athletic success. One of the basic rules for health is that athletes must maintain a balance between their nutrition needs and diet in order to restore biological constants. This balance must be understood both in terms of calories (quantitative balance) and in terms of macro- and micronutrients (qualitative balance).

The aims of nutrition recovery are specific to each athlete and to each training period, and appear thus to be determined by a group of factors:

- Physiological and homeostatic modifications resulting from training:
  - Depletion of energy substrates (mainly glycogen)
  - Dehydration (see chapter 7)
  - Muscle injury or protein catabolism
- The goals in terms of performance improvement or adaptation of training sessions:
  - Increase in muscle size or strength
  - Reduction of percent body fat
  - Increase in enzymes, functional proteins, or synthesis of functional cells or tissues (e.g., red blood cells, capillaries)
  - Level of substrates ingested and the hydration status before the next exercise

- The duration of the period separating two exercises:
  - Total recovery time
  - Other obligations or needs during the recovery period (e.g., sleep, travel)
- The availability of food for ingestion during the recovery period:
  - Immediate availability of food after training
  - Athletes' appetite and the opportunity to consume food and beverages during the recovery period

Protein Metabolism and Recovery

Physical exercise has been shown to affect protein metabolism. Regular strength or aerobic training sessions influence protein metabolism by different processes depending on the type of activity. We know that strength training increases muscle mass, while endurance training increases oxidative enzymes. In both cases, protein synthesis is favored. This protein synthesis is essential for development and growth, but also for the maintenance of body mass. While sugars represent the main source of energy supplied during exercise, regular exercise significantly increases daily needs in nitrogen-containing compounds. Indeed, Rennie and colleagues (1981) showed, during experiments on human subjects, that reduced protein synthesis combined with a concomitant increase in proteolysis and amino acid oxidation inevitably led to a nitrogen debt within the muscle. As shown in figure 8.1, proteolysis during exercise is greater than protein synthesis. The difference between degradation and synthesis is about 5 g of protein per hour of activity. This value relates to that measured through urea production due to amino acid oxidation. The protein balance is therefore negative during exercise: Nitrogen expenditure is not covered by protein intake.

In specific conditions, some amino acids are susceptible to oxidation. Thus, they constitute energy substrates in themselves. However, all the proteins present in the body play a precise functional role. In contrast to carbohydrates and lipids, no reserve of amino acids is available. Where necessary, amino acids derived from structural or functional proteins are used. Among the amino acids that make up the proteins of the body, only substituted amino acids (leucine, isoleucine, valine), alanine, glutamate, and aspartate are oxidized in muscles (Goldberg and Odessey 1972).

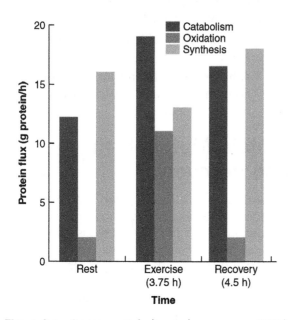

Figure 8.1   Protein metabolism in humans over 3.75 h of moderate exercise. Protein flux is expressed in g of protein and amino acids metabolized per hour, during resting or active periods.

Adapted with permission, from M.H. Rennie et al., 1981, *Clinical Science*, Vol. 61, pgs. 627-639, © The Biochemical Society and Medical Research Society.

Only these amino acids are capable of contributing to the energy needed for ATP resynthesis during exercise. In addition, some experimental evidence indicates that, among these amino acids, those harboring a substituted group play a bigger role in energy metabolism (Lemon 1991). It is clear, however, that the contribution of amino acid oxidation to energy provision depends strictly on the type of exercise, its relative intensity, its duration, and the athlete's fitness and nutrition status (Lemon 1991).

We know very little about the level of protein synthesis in athletes during recovery. It is, however, well established that an increase in amino-acid levels in the sarcoplasm is observed in the hours following exercise, of whatever type. Goldley and Goodman (1999) were able to show that this accumulation is observable starting the fourth hour after application of the workload. Intrasarcoplasmic accumulation of amino acids during recovery from physical exercise has been attributed to a specific effect of muscle use, resulting in increased transmembrane transport (Goldberg et al. 1975). The increased membrane transport of amino acids occurs at the same time as glucose transport. Despite the absence of direct experimental proof, postexercise amino-acid accumulation in muscle fibers creates favorable conditions for protein resynthesis. It is now well established that the sarcoplasmic accumulation of amino acids and protein synthesis are strongly correlated (Goldley and Goodman 1999).

Contractile Activity and Muscle Development

It is remarkable to note that, in regularly trained athletes, consuming excess dietary protein during recovery has a double result: The primary function is to ensure the repair of morphological lesions as a result of exercise and the second is to allow the synthesis of structural proteins. However, it is now well established that the anabolic effect of proteins depends strictly on muscle contraction. According to Décombaz (2004), three factors must coincide to activate net muscle protein synthesis: muscle contraction, amino acid availability, and insulin circulation.

The fact that the very nature of nutrients influences growth hormone release is essential in the relationship between recovery and endocrinology. In humans, it has been shown that the absorption of 500 calories (cal) as carbohydrate (maltodextrin) or 500 cal as protein (commercially available supplement made up of a complex of several amino acids) was followed by a reduction in the release of growth hormone (Matzen et al. 1990). After this initial drop, protein-based nutrition has the particularity of inducing a peak of growth-hormone secretion. This starts 90 min after absorption of the supplement and extends into the fourth hour (figure 8.2). These results show that the composition of the daily diet plays an important role in the control of growth-hormone release. We can easily hypothesize that this stimulation of growth hormone release, observed after the ingestion of dietary protein, is a factor favoring anabolism of contractile and structural proteins in skeletal muscles. However, there seems to be a plateau for protein synthesis. Above this threshold, amino acids from excess dietary protein are more likely to be oxidized than stored (above 1.5 g/kg each day) (Tarnopolsky, MacDougall, and Atkinson 1988). These results cast a significant shadow of doubt over the advantages of consuming very large quantities of protein (i.e., above the needs induced by exercise) for the development of muscle mass during recovery.

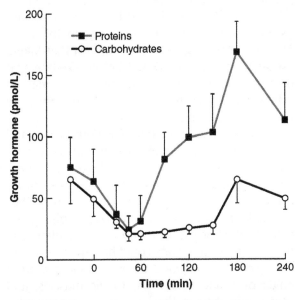

Figure 8.2 Growth hormone release over time after ingestion (at time 0) of a bolus of carbohydrate or protein.

Adapted, by permission, from L.E. Matzen et al., 1990, "Different short-term effect of protein and CH intake on TSH, growth hormone, insulin, C-peptide, and glucagon in humans," *Scandinavian Journal of Clinical Laboratory Investigation* 50: 801-805.

Temporal Recovery

In terms of protein metabolism, results are qualitatively similar for a very large range of exercise types, from brief, very intense exercise to long-term endurance work. Thus, all exercise links reduced protein synthesis and increased degradation, while the reaction is reversed during the recovery period (Poortmans 1993). Therefore, the notion of temporal recovery with regard to nutrition, most specifically for protein, is an essential factor in the rapid restoration of energy substrates.

In animal experiments as early as 1982, muscular protein synthesis was shown to increase over the 2 h following completion of high-intensity exercise (Booth, Nicholson, and Watson 1982). The increase in protein content in the muscle, both total and myofibrillar proteins (25% for the gastrocnemius muscle), depends on the intensity and duration of exercise (Wong and Booth 1990). Similar results were reported in humans after 4 h of running on a treadmill (Carraro et al. 1990) or after repeated strength training at 80% of maximal capacity (Chesley et al. 1992). In order to compensate for protein loss during exercise—60 min at 60% of maximal oxygen consumption—one recent study investigated the importance of the timing of postexercise protein supplementation. Levenhagen and colleagues (2001) were able to show that efficiency of protein synthesis in the whole body was improved by ingesting 10 g of protein rapidly after exercise, with significant effects observed up to t + 3 h (figure 8.3). When protein was ingested more than 3 h after exercise, recovery did not allow complete restoration of proteins to basal levels. The authors concluded that insulin may play a central role in the regulation of protein synthesis. Thus, it appears important to consume a source of protein immediately after exercise of long duration, or that of brief duration but very high intensity.

The athlete's carbohydrate (CHO) reserves at the end of a period of exercise must also be taken into account during recovery. Although they are difficult to measure in the muscle, it seems important to be able to estimate their level from the extent of energy expenditure. Indeed, the availability of energy substrates, and particularly CHO, is known to be an essential factor in the level of amino-acid oxidation. In this context, Lemon and Mullin (1980) measured urea excretion as

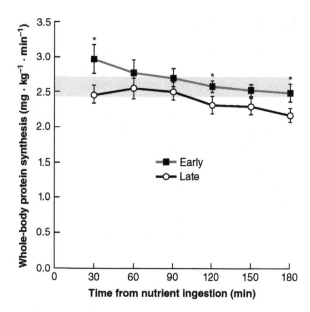

Figure 8.3 Protein synthesis over time for a group of 10 athletes who ingested protein either immediately after or 3 h after exercise.

* Significant difference between the early-intake and late-intake conditions (P < 0.05).

Adapted, by permission, from D.K. Levenhagen et al., 2001, "Post exercise nutrient intake timing is critical to recovery of leg glucose and protein homeostasis," *American Journal of Physiology* 280: E982-E993.

a marker of oxidation of nitrogen-containing compounds. They compared glycogen depletion during exercise and while resting, and found that during exercise, it induced a greater increase in urea excretion. The availability of glycogen therefore contributes to regulating amino-acid oxidation. These observations appear essential to the understanding of recovery: Athletes with low glycogen stores during exercise will see an increase in their nitrogen balance.

Tarnopolsky and associates (1988) estimated the minimal protein intake to avoid a negative nitrogen balance in endurance athletes to be 1.2 to 1.4 g/kg each day. Given the significant levels of interindividual variability in terms of protein digestibility and assimilation, a daily intake of 1.5 to 1.7 g/kg of body weight is recommended for endurance athletes (Martin 2001). This corresponds to between 12% and 16% of total daily energy intake. For athletes performing strength sports and for those who must maintain muscle mass, sufficient protein intake to equilibrate the nitrogen balance is estimated at between 1.4 and 1.6 g/kg each day. This so-called safety intake is indicated for protein with high nutritional value. For athletes wishing to increase their muscle mass,

increased dietary protein intake, varying between 2 and 2.5 g/kg each day, may be offered for limited periods.

However, high protein intake must not be prolonged. It cannot exceed 6 months per year (Martin 2001). Given our current knowledge, it seems difficult to justify intakes sometimes exceeding 3 g/kg each day. It is also logical to suppose that excess dietary protein may be bad for the athlete's health, particularly for renal function. It is especially important to note that urinary excretion of nitrogen induces increased fluid loss. This is why fluid intake must be closely monitored and adjusted in these populations. The absence of visible alarm signals should not be used to encourage consumption of abnormal protein quantities, particularly since we now know that there is no proven scientific justification for this practice.

The importance of an anabolic environment was indicated by a recent study in which immediate intake (5 min after muscle reinforcement training) of a dietary supplement including 19 g of milk protein (rich in essential amino acids) over 12 weeks produced muscular hypertrophy and strength gain at increased levels compared to when intake was delayed for 2 h. This was observed in a population of master athletes (Esmarck et al. 2001). In parallel to this study, the authors investigated the relevance of food intake just before exercise, and its incidence on athletes' recovery. In this case, Tipton and colleagues (2001) suggest that taking essential amino acids (EAA) before exercise that is mainly based on resistance has a more marked effect on later protein synthesis than if they are taken after exercise (figure 8.4).

These authors compared situations before and after exercise and showed higher protein synthesis when athletes were given a solution of 6 g EAA and 35 g glucose before performing resistance exercises. The indicator of muscle-protein synthesis used was based on the net rate of phenylalanine production from amino acids circulating in the blood. These results could be explained by an increased rate of blood flow, favoring a greater influx of amino acids (AA) to the muscles, and thus a reduction in the lag time for protein synthesis. The authors also showed that this protein synthesis persists for 1 h after exercise when EAA were taken 5 min before (in comparison with the situation in the post group).

In agreement with this result, it is of note that the effect of increased AAs on protein synthesis

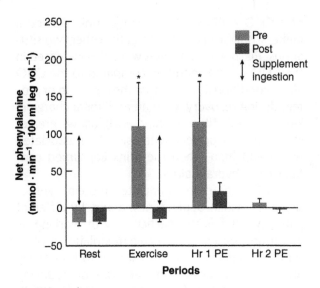

Figure 8.4 Net phenylalanine production in the blood over four time periods (rest, exercise, Hr 1 PE: 1 h postexercise, and Hr 2 PE: 2 h postexercise). *Pre* indicates intake of essential amino acids 5 min before exercise. *Post* indicates intake of essential amino acids 5 min after exercise.

* Significant difference with the post condition (P < 0.01).

Reprinted, by permission, from K.D. Tipton, 2001, "Timing of amino acid carbohydrate ingestion alters response of muscle to resistance exercise," *American Journal of Physiology: Endocrinology and Metabolism* 281:E197-206.

is of short duration, despite efforts to maintain consistently high blood concentrations (Bohe et al. 2001). Thus, it is more practical to have several small intakes of dietary protein at regular intervals (lunch, snack, dinner), which allows for several protein concentration peaks during the recovery period. The acute effect observed by Tipton and associates (2001) over the first 2 h of recovery seems to be prolonged (positive nitrogen balance) over 24 h (Tipton et al. 2003). This suggests that muscle mass could grow if the protein intake, preferably of dietary protein, was repeated over an extended period.

We lack convincing evidence that exercise combined with supplements induces a more positive protein balance in young athletic adults over the long term than exercise without supplements. We must also take into account the possible influences of physical exercise on protein metabolism over the remainder of the day (various meals, sleep, and so on). However, most of the results described here were obtained for a single variable (i.e., protein synthesis), which, although necessary, is not sufficient for muscular hypertrophy. Measurements in real life, possibly showing a benefit in terms of strength or muscle diameter after several weeks, are difficult to carry out.

Combining Protein, Carbohydrate, and Leucine for Muscle Recovery

The availability of energy substrates, particularly of carbohydrates, is an essential determinant in the level of amino-acid oxidation. Indeed, glycogen depletion during exercise induces a greater increase in urea excretion than during rest. As we have seen, urea excretion is a reflection of the use of nitrogen-based compounds (Lemon 1997). It seems clear that oxidation of amino acids during exercise is closely related to the availability of other energy substrates. The enzymatic complex, branched-chain alpha–keto acid dehydrogenase (BCKA-DH), is a major player in this process. It is the limiting enzyme in the leucine-catabolism pathway, and its activity is controlled by factors such as intensity and duration of exercise. Experiments based on animal models have shown that BCKA-DH activity in the muscle increases with running speed (Kasperek and Snider 1987). Similarly, endurance training induces an increase in leucine oxidation during exercise by increasing the activity of muscle BCKA-DH.

A first study by Anthony and colleagues (1999) showed the effects of leucine supplementation on recovery. In various rat populations, leucine was found to significantly stimulate protein synthesis following exercise on a treadmill. More recently, studies carried out with humans highlighted the fact that ingestion of carbohydrate (CHO) combined with protein (PRO) and amino acids often affected how plasma insulin levels were regulated (Pitkanen et al. 2003). It is assumed that these high insulin concentrations can stimulate uptake of selected amino acids along with the rate of protein synthesis (Gore et al. 2004). In addition, insulin is known to inhibit proteolysis (Biolo et al. 1997).

Howarth and associates (2009) got long-distance athletes to perform 2 h of cycling and then ingest various solutions over the following 3 h. The solutions contained a medium concentration of carbohydrates (1.2 g · kg^{-1} · h^{-1} L-CHO), a very high concentration of carbohydrates (1.6 g · kg^{-1} · h^{-1} H-CHO), or a medium concentration of carbohydrates in addition to proteins (1.2 g · kg^{-1} ·

h⁻¹ CHO and 0.4 g · kg⁻¹ · h⁻¹ PRO). Consuming PRO–CHO during recovery allowed a significant increase in the net protein balance 4 h after the end of aerobic exercise, as well as an increase in the rate of protein synthesis. The results from the group consuming PRO–CHO during recovery led the authors to conclude that there is a possible adaptation of the muscle, on the one hand, and muscular anabolism, on the other, to repair the damage induced by long duration exercise.

Another study was carried out on several high-level athletes running for 45 min on a treadmill (Koopman et al. 2005) (figure 8.5). Immediately after exercise, the athletes consumed energy drinks composed of carbohydrate (0.3 g · kg⁻¹ · h⁻¹ CHO); carbohydrate and protein (0.3 g · kg⁻¹ · h⁻¹ CHO and 0.2 g · kg⁻¹ · h⁻¹ PRO); or carbohydrate, protein, and free leucine (0.3 g · kg⁻¹ · h⁻¹ CHO, 0.2 g · kg⁻¹ · h⁻¹ PRO, and 0.1 g · kg⁻¹ · h⁻¹ Leu). The results show, primarily, that the net protein balance was significantly higher in the CHO + PRO + Leu condition compared to the other two. In addition, as indicated in figure 8.5, ingestion

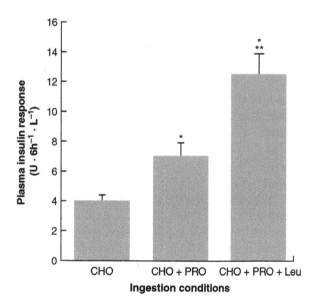

Figure 8.5 Plasma insulin concentration response to 2 h of recovery and three ingestion conditions (CHO: carbohydrates; CHO + PRO: carbohydrate + protein; CHO + PRO + Leu: carbohydrate + proteins + leucine).

* Values significantly different from the CHO condition (P < 0.01).

** Values significantly different from the CHO + PRO condition (P < 0.01).

Adapted, by permission, from R. Koopman et al., 2005, "Combined ingestion of protein and free leucine with carbohydrate increases postexercise muscle protein synthesis in vivo in male subjects," *American Journal of Physiology: Endocrinology and Metabolism* 288: E645-653.

of a CHO + PRO + Leu energy drink allowed a higher insulin response than in the other two situations. Plasma insulin levels were also increased in the CHO + PRO condition compared to the CHO alone condition. It seems that the ingestion of protein during recovery, with an additional charge in leucine (a substituted amino acid), allows greater stimulation of protein synthesis when associated with CHO than when the drink consumed contains carbohydrate alone.

Leucine therefore stimulates protein synthesis, in an insulin-dependent manner, by different pathways. It has the particularity of working as a nutrition-signaling molecule modulating protein synthesis. Leucine was also shown to potentially affect muscle protein metabolism by reducing degradation (Nair, Schwartz, and Welle 1992). This is most likely achieved through increasing circulating insulin and phosphorylating key proteins involved in regulating protein synthesis (Karlsson, Nilsson, and Nilsson 2004). However, while most in vivo or in vitro studies on animals report that administering leucine can inhibit protein lysis and stimulate its synthesis, in vivo studies in humans show that the ingestion of leucine or branched-chain amino acids (BCAAs) reduces proteolysis, but does not otherwise stimulate protein synthesis. The maximal rates of protein synthesis during postexercise recovery probably require signaling from these amino acids (i.e., substituted and branched), but also from the anabolic signal provided by exercise.

Finally, no studies of the benefits of other substituted amino acids (such as isoleucine and valine) on protein synthesis during recovery have been carried out. The complexity of, and metabolic interrelationships within, pathways involving amino acids do not promote analysis of their transformation.

Branched-Chain Amino Acids and Postexercise Mental Performance

The central fatigue hypothesis states that branched-chain amino acids (BCAAs) competitively inhibit tryptophan transfer across the blood-brain barrier. Oral supplementation aiming to regulate the BCAA concentration in the blood would reduce the conversion of tryptophan to serotonin in the brain. Serotonin is a neurotransmitter that plays a role in sensing fatigue. The serotonin response and evolving prolactin concentrations accompany modifications of substituted amino acid (SAA)—valine, isoleucine, leucine—concentrations in plasma. From this observation, Newsholme (1986) formulated the hypothesis that changes to the concentration of SAA can regulate the central mechanisms of fatigue by increasing the speed of serotonin synthesis. These arguments led researchers to propose SAA or tryptophan supplementation in humans, in order to test Newsholme's hypothesis. Different studies, not very numerous, showed an improvement in mental performances during a football match after ingesting a 6% CHO solution and BCAAs (Blomstrand, Hassmén, and Newsholme 1991). To our knowledge, a single recent study of top-level athletes investigated the influence of BCAAs and their effect on mental performance during the recovery phase. Portier and colleagues

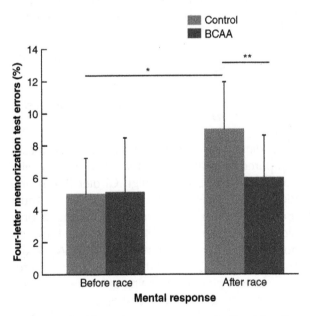

Figure 8.6 Mental responses symbolized by the percentage of errors made during the four-letter memorization test just after a sailing competition. During the competition, subjects ingested either a control CHO solution, or a CHO + branched-chain amino acid (BCAA) solution.

* Significant difference between the values obtained for the before-race and after-race conditions (P < 0.05).

** Significant difference between the values obtained for control and substituted amino acid conditions (P < 0.05).

Springer and *European Journal of Applied Physiology* Vol. 104, 2008, pgs. 787-794, "Effects of branched-chain amino acids supplementation on physiological and psychological performance during an offshore sailing race," H. Portier et al., figure 2. With kind permission from Springer Science and Business Media.

(2008) studied the effects of a diet enriched in SAA (50% valine, 35% leucine, and 15% isoleucine) on a short-term memory test taken just after a sailing competition (figure 8.6). The results show that a diet enriched in SAA during a competition allows better conservation of mental performances during the recovery phase than the carbohydrate-based diet generally observed in this sport. This could be an interesting application when, in cases like sailing, competitive legs follow each other throughout the day with very limited recovery periods.

Sugar and Recovery

The majority of energy needs are met by sugars (carbohydrates) and lipids. In the sugar category, glucose plays a predominant role, since it is immediately available. It is transported in the blood, and its catabolism supplies cells with energy. All cells therefore use blood glucose. For example, it covers half the energy needs of the central nervous system, the remainder coming from the degradation of ketone bodies. However, the body's glucose reserves are low (25 g), which requires the human body to permanently import new glucose molecules from food sources. When they are not used to renew the glycogen stores of various tissues, excess dietary sugars are converted into lipids in the liver and in adipocytes. Glucose alone represents 80% to 90% of the energy supplied by sugars. In aerobic conditions, complete oxidation of a mole of glucose leads to the formation of 38 moles of adenosine triphosphate (ATP). When resting, during postprandial periods, glucose absorption is discontinuous. For 100 g absorbed during a meal, it is estimated that 60 g are oxidized over the following 3 h. This use of glucose allows relative lipid savings. The cost of glucose storage (for later use) is 5%.

Glycogen Resynthesis

Reductions in muscle glycogen as a result of prolonged exercise stimulate the metabolic pathways leading to glycogen synthesis during recovery. Ingesting foods that contain sugar during this recovery phase leads to two phenomena: on the one hand, an increased rate of resynthesis, and on the other, an increased level of glycogen above those present prior to exercise. Glycogen resynthesis capacities differ depending on the nature of the sugars available. The speed of muscle glycogen resynthesis is identical during the recovery phase after ingestion of glucose or glucose polymers, but it is slower with fructose (Blom, Hostmark, and Vaage 1987). In contrast, fructose increases the resynthesis rate for hepatic glycogen, to the extent that glycogen synthesis is promoted by insulin activity. It is therefore more efficient, during recovery, to consume carbohydrate with a high glycemic index.

These data were confirmed a few years later, taking the notion of glycemic index (GI) into consideration. The GI allows the physical response to oral intake of carbohydrate (CHO) to be characterized. It is defined as the area under the curve for glycemic response after the ingestion of a food containing sugar. This curve reflects, on the one hand, the speed of appearance of sugar in the blood, and on the other, its speed of capture by the tissues using it. The GI thus allows foods to be compared. In this way, for a population of highly trained cyclists, Burke and associates (1993) were able to show that the maximum muscle glycogen resynthesis (vastus lateralis muscle) observed 24 h after exercise (2 h at 75% of the $\dot{V}O_2$max) was obtained when postexercise alimentation was based on high GI carbohydrates (figure 8.7). This result cannot be totally explained by variations in insulin and glucose concentrations. Indeed, the extent of glycogen resynthesis postexercise is about 30% of the pre-exercise level. This cannot be explained by smaller variations in insulin and glucose levels over 24 h. These data are essential when the athlete must train repeatedly during the day or when competitions are repeated.

In order to better explain the mechanisms relating to the lower efficacy of low-GI CHOs, some authors have pointed to poor intestinal absorption (Wolever et al. 1986; Jenkins et al. 1987). Indeed, Joszi and colleagues (1996) showed that the low digestibility of a mixture with a high concentration of starch amylose (low GI) was responsible for a lower restoration of glycogen observed during 13 h of postexercise recovery compared with the ingestion of glucose or maltodextrins (high GI). These authors therefore show that the poor digestibility of some CHOs leads to overestimation of their availability in the intestine (Joszi et al. 1996). We think, in addition, that these studies should be reinforced by others, during which real food should be ingested. Nevertheless, a longitudinal study, carried out over 30 days, showed that an active population exposed to a low GI

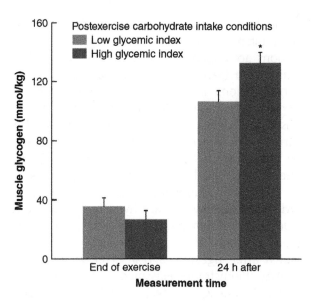

Figure 8.7 Muscle glycogen concentrations, immediately and 24 h after performing prolonged exercise in subjects having consumed 10 g of carbohydrate (CHO) per kg body weight over the 24 h following exercise. These CHO are sugars with a low or high glycemic index.

* Significantly different from the use of sugars with a low glycemic index (P < 0.05).

Adapted, by permission, from L.M. Burke, G.R. Collier, and M. Hargreaves, 1993, "Muscle glycogen storage after prolonged exercise: Effect of the glycemic index of carbohydrate feeding," *Journal of Applied Physiology* 75: 1019-1023.

daily diet showed reduced glycogen synthesis, compared to initial values and values for a similar population consuming a high GI diet (Kiens and Richter 1996). From this observation, we must therefore be careful when we recommend only diets with a low GI, to the extent that these do not always favor glycogen resynthesis. While the composition of CHO solutions appears crucial for the recovery phase, the timing of intake also influences muscle glycogen resynthesis.

Timing of Postexercise CHO Intake

Intake of CHO and its precise timing during the recovery phase largely influence the quality of glycogen resynthesis. These strategies are very important during unique restrictive situations (such as triathlon or marathon), but also during events where competitive legs are repeated throughout the day (such as swimming, middle-distance racing, or repeated judo combats). The sooner carbohydrate is consumed after completing exercise, the higher the amount of muscle glycogen resynthesized. Thus, when some CHO is ingested immediately after exercise, the quantity of muscle glycogen measured 6 h later is higher than when the intake of CHO is delayed for 2 h after the end of exercise (Ivy, Lee, et al. 1988). It is now accepted that exercise increases both sensitivity to insulin (Richter et al. 1989) and permeability of the muscle's cell membrane to glucose, and that the highest rates of muscle glycogen resynthesis are recorded during the first hour (Ivy, Lee, et al. 1988). This is mainly due to the fact that the enzyme glycogen synthase is activated by glycogen depletion (Wojtaszewski et al. 2001). Sugar-based nutrition immediately after exercise takes advantage of these effects, as reflected by the higher rates of glycogen storage (7.7 mmol · kg^{-1} · h^{-1}) over the first 2 h of recovery. The usual rates of glycogen storage (4.3 mmol · kg^{-1} · h^{-1}) are judged insufficient in this context (Ivy, Lee, et al. 1988).

This study showed the basis for recovery with regards to glycogen: Ingestion of too little CHO immediately after exercise induces very low rates of glycogen resynthesis, rates that are not inclined to promote repeated performances (training or competition). In addition, delaying food-based CHO intake for 4 h after the end of exercise does not allow high rates of glycogen resynthesis, in contrast with immediate postexercise intake. These results are particularly relevant for relatively short recovery periods between exercises (between 6 and 8 h). When recovery is longer (between 8 and 24 h), dietary intake of CHO immediately after exercise does not result in accelerated glycogen resynthesis (Parkin et al. 1997).

As part of twice-daily training sessions for high-performance athletes, it is preferable to eat soon after exercise, with a view to promote replenishment of glycogen stores and thus avoid penalizing the second training session. In the case of athletes who do not train more than once per day, it is not so much a question of rushing to consume CHO just after exercise as of favoring consumption of a meal or snack with adequate CHO before the next training session. However, in the case of exercises of long duration and those that require a high rate of energy expenditure, sugar sources should be provided during training.

Several CHO ingestion strategies are often noted: Either the athletes prefer to eat solid, sugar-rich foods as part of their main meals, or several snacks are offered during the different days of training. Studies interested in the 24 h recovery period have shown that large meals based on complex carbohydrate twice a day or carbohydrate-based snacks repeated seven times per day, have equivalent power to reconstitute muscle glycogen stores (Costill et al. 1981). More recently, similar results were found for high-performance athletes ingesting 4 complex carbohydrate-based meals per day or 16 snacks, 1 per hour (Burke et al. 1996). In this last study, although the glycogen resynthesis rates were similar in both conditions, the blood glucose and insulin concentrations were different over the course of the 24 h (Burke et al. 1996).

In addition, very high rates of glycogen synthesis have been reported over the first 4 to 6 h of recovery when high quantities of CHO were ingested at 15- to 30 min intervals (Doyle, Sherman, and Strauss 1993; van Hall, Shirreffs, and Calbert 2000; van Loon et al. 2000; Jentjens et al. 2001). These high rates were attributed to the maintenance of insulin and blood glucose levels, as a result of this dietary protocol. The apparent conflict between these last results seems to reside in the fact that the concentrations are not compared to those obtained in protocols where several CHO-based snacks are offered to athletes. It seems, however, that the maximal rate of glycogen resynthesis measured during recovery is obtained for athletes consuming 0.4 g/kg of body weight every 15 min (i.e., 120 g CHO per h for a 75 kg subject) over the 4 h immediately following completion of exercise (Doyle, Sherman, and Strauss 1993) (figure 8.8). In addition, the authors indicate in this study that glycogen resynthesis is not comparable depending on the type of exhausting exercise performed (concentric or eccentric contraction) over the previous 48 h. Indeed, glycogen restoration is 25% lower when exercise involved eccentric contractions compared to concentric contractions.

In line with the importance of the timing of ingestion of CHO, it is of note that glucose penetration into cells is insulin dependent, requiring specific transporters, such as GLUTs (Williams 2004) (figure 8.9). During exercise, insulin and muscle contraction stimulate glucose capture in the muscles by means of GLUT-4 transporters (Holloszy and Hansen 1996). Even if a dissociated and cumulative effect of insulin and muscle contraction exists, the mechanisms leading to translocation of GLUT-4 transporters seem to be distinct (Nesher, Karl, and Kipnis 1985). Muscle

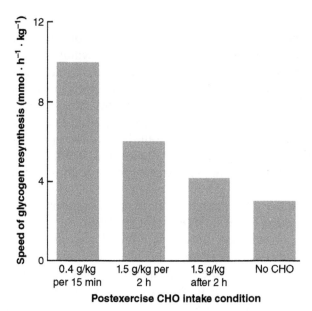

Figure 8.8 Rate of muscle glycogen resynthesis over the 4 to 10 h following the end of exercise. Low-osmolarity maltodextrin (approximately 1.6 g · kg^{-1} · h^{-1}) was used.

Adapted, by permission, from J.A. Doyle, W.M. Sherman, and R.L. Strauss, 1993, "Effects of eccentric and concentric exercise on muscle glycogen replenishment," *Journal of Applied Physiology* 74: 848-1855.

contraction and insulin promote the recruitment of GLUT-4 from different intracellular pools (Thorell et al. 1999). In rats, it has been suggested that the increase in muscle GLUT-4 protein levels after a strenuous day's exercise is responsible for the improved glycogen synthesis capacity of the muscle, as compared with sedentary rats (Richter et al. 1989). More recently, in a population of 11 cyclists McCoy and colleagues (1996) investigated the influence of GLUT-4 transporters on restoring glycogen stores after 2 h of pedaling at 70% of $\dot{V}O_2$max, followed by 4 × 1 min bursts at 100% $\dot{V}O_2$max, with 2 min recovery between bouts. These authors showed that high concentrations of GLUT-4 correlated significantly with the highest levels of glycogen resynthesis during the 6 h postexercise recovery period (r = 0.63, P < 0.05).

In addition, the increase in permeability of the muscle membrane to glucose in postexercise conditions is due to the number of glucose transporters integrated in the plasma membrane and, probably, to the increase in intrinsic transporter activity (Ivy and Kuo 1998). In this context, Goodyear and associates (1990) found that, on an isolated skeletal muscle plasma membrane, the rate of glucose transport was quadrupled immediately after exercise, while the number of glucose transporters was only doubled. Thirty minutes after exercise, glucose transport was approximately halved, while the number of glucose transporters associated with the plasma membrane decreased by just 20%. After 2 h of recovery, the transport and number of transporters at the level of the plasma membrane had returned to basal pre-exercise levels. Very interestingly, the decreased time for glucose transport observed by Goodyear and colleagues (1990) is identical to the initial triggering time for insulin in the glycogen synthesis phase described by Price and colleagues (1996) (see figure 8.9).

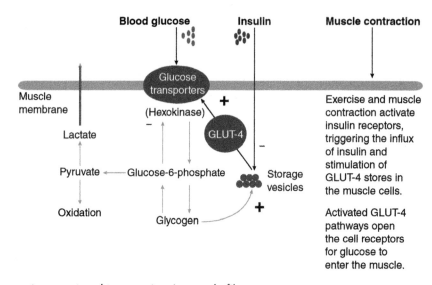

Figure 8.9 Glucose transport and transporters in muscle fibers.

Quantity of Carbohydrate Consumed

The first studies investigating quantities of carbohydrate consumed during postexercise recovery date from 1981 (Costill et al. 1981). These authors report that consuming 150 to 600 g of CHO per day induced greater replenishment of glycogen stores over a 24 h period than lower CHO quantities. A few years later, it was shown that the intake of 1.5 g CHO per kg of body weight over the 2 h following an exhausting exercise induced an appropriate rate of glycogen resynthesis. This rate was not improved when the CHO quantity was doubled (i.e., 110 g CHO per hour for a 75 kg subject) (Ivy, Katz, et al. 1988). In addition, Sherman and Lamb (1988) were similarly able to show that no difference was recorded with postexercise CHO quantities of 460 g or 620 g per day. In contrast, doses of 160 g and 360 g per day induced significantly lower resynthesis when comparing muscle glycogen levels before and after exercise (figure 8.10).

The quantity of carbohydrate to be consumed during recovery after exercise is often questioned. Thus, different energy drinks have been marketed

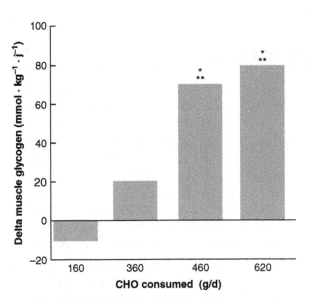

Figure 8.10 Glycogen resynthesis after exhausting exercise. Using four concentrations, low level of CHO (160 and 360 g/day) or very rich in CHO (460 and 620 g/day).

* Significantly different from the 160 condition (P < 0.05).

** Significantly different from the 360 condition (P < 0.05).

Adapted, by permission, from W.M. Sherman and D.R. Lamb, 1988, Nutrition for prolonged exercise. In *Perspectives in exercise science and sports medicine, Vol. 1: Prolonged exercise*, edited by D.R. Lamb and R. Murray (Traverse, MI: Benchmark Press), 213-280.

to maintain athletes' plasma and blood volumes. As part of this, Criswell and colleagues (1991) tested the influence of a drink at 7% glucose (and containing electrolytes) on the levels of hematocrit and hemoglobin after a football match in comparison with a drink containing no glucose. The data obtained from 44 football players showed that the energy drink allowed plasma volumes to be stabilized during recovery, while the nonglucose drink with electrolytes did not allow maintenance of blood volume (i.e., a 5% reduction in plasma volume). However, the energy drink did not influence the drop in anaerobic performance.

Coingestion of Carbohydrate and Protein

For a few years, the association of carbohydrates (CHO) with proteins (PRO) has been suggested to improve the speed of glycogen resynthesis. This seems to be linked to induction of higher levels of insulin secretion by the combination than those provoked by CHO alone (Pallotta and Kennedy 1968). Currently, somewhat contradictory results have been obtained regarding the potential benefit of associating CHO and PRO, the differences obtained might be attributable to the experimental protocol itself, to the frequency of supplementation, or to the quantities of CHO and PRO offered to athletes. For example, in studies where benefits with regard to glycogen resynthesis were found for the combination of CHO with PRO, subjects were fed every 2 h (Zawadzki, Yaspelkis, and Ivy 1992; Ivy et al. 2002). Studies that did not show any effect of the addition of proteins on glycogen resynthesis often used renutrition intervals of between 15 and 30 min (Tarnopolsky et al. 1997; Carrithers et al. 2000; van Hall, Shirreffs, and Calbert 2000; Jentjens et al. 2001). In addition, in some of these studies, very high quantities of CHO were administered (van Hall, Shirreffs, and Calbert 2000; Jentjens et al. 2001), while in others, a low proportion of protein was added (Tarnopolsky et al. 1997; Carrithers et al. 2000). Independent of the experimental procedures used, it seems obvious that CHO-rich postexercise nutrition at very frequent intervals reduces the benefit of protein supplementation during recovery (see summary in table 8.1).

Adding protein to carbohydrate during recovery aims primarily to increase insulin production. This hormone is a determinant for glycogen synthesis, acting both on glucose penetration into muscle

Table 8.1 Comparison of Studies Where Protein (PRO) Was Added to Carbohydrate (CHO) in the Diet Consumed During Recovery

Scientific study	Postexercise supplementation schedule	Calorie distribution between sources (%)	Quantities offered during recovery (g · kg^{-1} · h^{-1})	Recovery time (h)	Glycogen resynthesized (µmol · g^{-1} · h^{-1})
Zawadzki, Yaspelkis, and Ivy (1992)	Immediately and 2 h after	73% CHO 27% PRO	0.8 CHO 0.3 PRO	4	35.5 *
Tarnopolsky et al. (1997)	Immediately and 2 h after	66% CHO 23% PRO 11% LIPIDS	0.37 CHO 0.05 PRO 0.01 LIPIDS	4	24.6
Carrithers et al. (2000)	Immediately after and every 30 min	70% CHO 20% PRO 10% LIPIDS	0.71 CHO 0.20 PRO 0.04 LIPIDS	4	28.0
van Hall, Shirreffs, and Calbert (2000)	Immediately after and every 15 min	77% CHO 23% PRO	1.0 CHO 0.3 PRO	4	39.0
van Loon et al. (2000)	Immediately after and every 30 min	67% CHO 33% PRO	0.8 CHO 0.4 PRO	5	34.4 *
Jentjens et al. (2001)	Immediately after and every 30 min	75% CHO 25% PRO	1.2 CHO 0.4 PRO	3	25.0
Williams et al. (2003)	Immediately and 2 h after	74% CHO 26% PRO	0.8 CHO 0.4 PRO	4	159 *
Berardi et al. (2006)	10 min, 1 h, and 2 h after	67% CHO 33% PRO	0.8 CHO 0.2 PRO	6	28.6 *

* Indicates that the addition of protein to carbohydrate allowed significant muscle glycogen resynthesis during postexercise recovery.

PRACTICAL APPLICATION

There is some evidence to note that in trained athletes, consuming adequate protein during recovery has two effects: first, to ensure the repair of morphological lesions as a result of exercise and, second, to allow the synthesis of structural proteins and then increase muscle recovery. However, some other results need to be highlighted:

- For the stimulation of glycogen storage in muscles after exertion, all the data underline the importance of early food intake (from the end of exercise) and somewhat reduce the role of increased insulin, as well as the relevance of ingesting insulinogenic protein and amino acids (Ivy et al. 2002).

- Intake of substituted amino acids—mainly leucine (0.1 g · kg^{-1} · h^{-1})—associated with carbohydrate (0.3 g · kg^{-1} · h^{-1}) and protein (0.2 g · kg^{-1} · h^{-1}) is recommended in order to stimulate postexercise protein synthesis and, thus, recovery (Koopman et al. 2005).

- The importance of an anabolic environment was shown by an immediate intake after muscle reinforcement exercise of 19 g of milk protein (rich in essential amino acids) over 12 weeks: The strength gain was increased compared to when intake was delayed for 2 h (Esmarck et al. 2001).

- The availability of glycogen regulates amino-acid oxidation. Thus, athletes whose glycogen stores are depleted during exercise will see an increase in their nitrogen balance. In order to avoid a negative nitrogen balance in endurance athletes, the minimal protein intake seems to be between 1.2 and 1.4 g · kg^{-1} · d^{-1} (Tarnopolsky, MacDougall, and Atkinson 1988).

- Even if an SAA appears to help maintain postexercise mental performance, to date no quantitative evidence affirms that supplements targeting essential amino acids would be useful, on a continuous basis, to athletes in good health (Bigard and Guézennec 2003).
- Dietary composition plays an important role in the control of growth hormone release. Stimulation of this, observed 1 h after the ingestion of dietary protein, promotes anabolism of contractile and structural proteins in skeletal muscle (Matzen et al. 1990).
- There is a threshold for protein synthesis: Above it, amino acids from excess dietary protein are oxidized rather than stored (above $1.5 \, g \cdot kg^{-1} \cdot d^{-1}$) (Tarnopolsky, MacDougall, and Atkinson 1988).

It is often written that there is a metabolic window for refueling. However, the most important key factor in opening a window for refueling is a carbohydrate supply. We have to remember the following points:

- Immediate recovery after exercise (0–4 h): 1.0 to $1.2 \, g \cdot kg^{-1} \cdot h^{-1}$ should be consumed at frequent intervals (Jentjens and Jeukendrup 2003).
- Recovery over a day with moderate to strenuous endurance training: 7 to $12 \, g \cdot kg^{-1} \cdot day^{-1}$ should be consumed (Tarnopolsky et al. 2005).
- Recovery over a day with severe exercise program (>4 to 6 h/day): 10 to $12 \, g \cdot kg^{-1} \cdot day^{-1}$ should be consumed (Ivy, Katz, et al. 1988).
- Carbohydrate-rich meals are often recommended during recovery. The addition of protein (meat, fish, eggs, and so on) to the usual diet inevitably allows faster muscle glycogen resynthesis. Add 0.2 to 0.5 g of protein per day and per kilo to carbohydrate, in a 3:1 ratio (CHO:PRO). This is critical when an athlete trains twice a day or over a very prolonged amount of time (Berardi et al. 2006; Ivy et al. 2002).
- When the period between training sessions is less than 8 h, the athlete must consume carbohydrate as soon as possible after exercise in order to maximize glycogen reconstitution. There is an observable advantage to splitting CHO intakes over the early recovery phase, particularly when the mealtime is not soon (Ivy, Lee, et al. 1988).
- For longer recovery periods (24 h), athletes must organize the contents and timing of meals, as well as the CHO content of snacks, in line with what is practical and comfortable relative to their usual habits and schedule. It must be remembered that no difference was observed, with regard to glycogen replenishment, when carbohydrate was taken either in solid or liquid form (Burke, Kiens, and Ivy 2004).
- Carbohydrates with moderate or high glycemic indices supply energy rapidly for glycogen resynthesis during recovery. Therefore, they should constitute the main choice in postexercise energy-restoring menus (Burke, Collier, and Hargreaves 1993). The quantities ingested are crucial for muscle glycogen resynthesis. Athletes who do not consume enough calories often have difficulties reconstituting their glycogen stores: These populations must be carefully observed so as not to induce too high a deficit (Kerksick et al. 2008).

fibers and on the activity of glycogen synthase, the limiting enzyme in glycogen synthesis. In most cases, consuming a mixture of CHO and PRO should allow a more marked insulin response. In this context, the first study showing the efficacy of the combination was carried out in athletes after 2 h of ergocycle training following 12 h of fasting (Zawadzki, Yaspelkis, and Ivy 1992). Glycogen resynthesis was evaluated 4 h after the consumption of solutions containing 112 g CHO, 40.7 g PRO, or a mixture of CHO and PRO. The results of this study showed that the amount of glycogen formed per hour is significantly greater in athletes consuming the mixture of CHO and PRO. These results have, however, been questioned, since the quantity of CHO absorbed during recovery was not optimal. Thus, when using optimal CHO concentrations, van Hall and colleagues (2000) were not able to show a beneficial effect of adding protein to CHO on glycogen resynthesis. In 2002, Ivy and colleagues showed that after exhausting ergocycle training, intakes (spaced every 2 h) of a mixture containing both CHO and proteins induced an increase in muscle glycogen resynthesis. In the same context, Williams and colleagues (2003) got eight endurance-trained cyclists to perform 2 h of ergocycle at 65% to 75% $\dot{V}O_2$max, followed, after 2 h recovery, by a limit time at an intensity of 85% of $\dot{V}O_2$max. During recovery from this, CHO alone or a mixture of CHO and PRO was administered immediately after exercise and 2 h later. Results show a significant improvement (128%) in muscle glycogen resynthesis for the CHO-PRO condition. The solution composed of CHO and PRO allowed improvement of the time to exhaustion. A single cyclist showed lower performance in the CHO-PRO condition. All the others significantly improved their performances: The time to

exhaustion was 31 min in the CHO-PRO condition and only 20 min with CHO alone.

A final recent study (Berardi et al. 2006) studied the effect of 1 h of against-the-clock cycling in six experienced cyclists. The study subjects then ingested different meals and solutions, immediately, 1 h, or 2 h after exercise. The nutrition composition was as follows: (C + P: carbohydrate + proteins; CHO: carbohydrate; placebo: solid-food placebo). After 6 h of recovery, a second 1 h against-the-clock trial was performed. Various muscle biopsies were taken pre- and postexercise to quantify glycogen resynthesis. Although cycling performances were similar over the two trials (P = 0.02), the rate of resynthesis of muscle glycogen was greater (+ 23%) in the C + P condition than in the CHO alone condition. This final result has an undeniable influence: When postexercise nutrition must be spaced out and when two exercise periods are to be performed with only a small recovery period, the combination of carbohydrate and protein probably has great advantages with regard to increasing the speed of glycogen resynthesis.

Summary

In order to compensate for protein loss during exercise, the timing of postexercise protein supplementation is important. The efficiency of protein synthesis in the whole body is improved by ingesting 10 g of protein rapidly after exercise, with significant effects observed up to t + 3 hours. Another challenge is the refueling with carbohydrate supply. When time is short between fuel-demanding events, it makes sense to start refueling as soon as possible. The immediate target should be around 1 g of carbohydrate per kg of body mass, repeated every hour until meal patterns take the achievement of daily fuel needs.

Recovery Strategies for International Field Hockey Event

Ian McKeown, PhD,
Australian Institute of Sport

Major international field hockey events (e.g., Olympic Games, Champions Trophy, European Championships) lasting 7 or 8 days can typically involve at least 5 games. Normally, a few rest days are spaced near the end of the tournament. Recovery between games can range from less than 24 h to 48 h. Field hockey involves intense bouts of sprinting and maneuvering in often difficult and demanding body positions. Sprinting or high-intensity running can take place in any direction, with movements similar to any other ball sport. The following is an insight into the typical planning and delivery required to gain optimal recovery and regeneration between games during the European Championships, A division, 2009.

It is vital to plan well in advance of tournaments. Using the team hotel as the team base means that the facilities in the hotel and surroundings must be explored prior. Depending on variables such as cost, it would be optimal to have a swimming pool, ideally with a plunge pool, a gym with cardiovascular training equipment, and a team room.

As part of the planning, suggested menu options should be communicated to the management of the hotel, ideally the head chef, so that adequate meal options, size, and timing can be arranged. Timing for meals should be planned in the lead up to and throughout the tournament, covering every eventuality, depending on the degree of success during the competition.

When planning the timetable for each day, knowing when each of the group games will take place is imperative. Training prior to tournament should be used to practice all elements of the preparation, including the recovery. Issues such spacing between meals, exclusive use of swimming pool times, and any other possibility should be ironed out.

Another element to consider when planning for a tournament is the timing for rest and regeneration. Depending on the squad, this time should be compulsory for athletes to put their feet up and rest. With younger squads, this can be par-ticularly important, since some athletes may not realize the importance of rest and recovery. Our squad sometimes used this time to review individual player clips on laptops. However, this option was largely dictated by previous performance, and it was only open to some. This review process could cause higher levels of stress, depending on the players' mindset.

Scheduling systems to monitor the response to training and competition, and accumulation of fatigue should be put in place before major tournaments. As a quality assurance method or best practice, elements of monitoring each player should be implemented year-round. The format of data input and reporting to coaches must be developed to give worthwhile information to the coaches, with practical suggestions on how to best handle the various levels of fatigue. Fatigue during the tournament is inevitable, but it must be minimized, giving the players the best opportunity to perform in the next game.

Immediately postmatch and after a short debrief by the coaching staff, light active recovery involving low-intensity running and other dynamic movements (side stepping, skipping, backward running, and so on) lasting about 10 min is performed. During this time, each player has water and an electrolyte replacement drink. Small snacks and sweets are also available. Although this process is usually monitored for excessive intake, at this time, there was no need to ration.

Depending on time constraints and after ensuring immediate intake of food and fluid, the squad splits into groups to perform low-intensity pool recovery. This could involve very light swimming. In general, simple movements such as walking, skipping, carioca, and sidestepping are typically used. A circuit of movement, light dynamic stretching, swimming, and light active stretching selected from a bank of exercises for around 15 min worked well, but some activities were individually prescribed. Varying levels of water confidence can have a large influence on the intensity and stress of a seemingly light session. Nonswimmers must be identified, and the session altered appropriately.

Although recovery is very well planned, it must be noted that there has to be a large element of flexibility here. Depending on the performance and the result of the match, the attitude and atmosphere around the squad can be significantly different. It is the coaches' imperative to adjust the delivery to ensure that each player takes a role in recovery. As suggested previously, the use of schedules can provide some valuable feedback. Players' schedules should be monitored daily, and players should provide input before breakfast. Feedback on subjective feelings of stress, lethargy, and muscle soreness tend to give the most applicable information to influence coaching decisions and aid in prescribing appropriate interventions. By completing questionnaires first thing in the morning, sleep patterns can be assessed. This also allows enough time for the coaches to react. Extra investigation could be sought if scores were significantly deviated from normality.

Group self-massage and flexibility were extremely popular. After some player education, they could largely be left to administer massage themselves. Therefore, the only form of prescription needed here was time, specific areas of attention, and level of intensity. Typically, intensity was low to moderate due to the muscle damage from games and the short recovery time period available. Foam rolls and other implements were available for light trigger-point and self-massage work. Any specific therapy was to be prescribed and performed by the team physiotherapist. Other specific active recovery options, such as light stationary cycling, were available to cater for injuries or chronic issues.

As a side note, there is always the desire to watch other matches in the tournament, typically against sides that the team may have to play at later stages in the tournament. This was permissible as long as everything else was taken care of and recovery was not compromised. Other considerations, such as heat and sun exposure, standing time, uncomfortable sitting positions, availability of adequate food and fluid, and other aspects of the players' performance must be considered here. The rule applied was that any compromise to the players' well-being would not be tolerated.

To ensure that the support team members were proactive with all aspects of the squad's performance, daily staff meetings took place. Discussion around individual player feedback, scheduling, timetables, any alternatives to the plan, and any player requests were addressed. This proved valuable to ensure that every eventuality was covered and that the whole coaching and support staff had a specific and clear understanding of the day's plans.

Key Points

- Plan well: Investigate facilities available, menu suggestions, access to training facilities, and timetabling.
- Practice all strategies to be implemented before major events.
- Use simple diaries for player monitoring.
- Active recovery immediately postmatch has to be followed with additional work later.
- Be flexible and sensitive to the group dynamics.
- Use group sessions for self-massage and flexibility.

CHAPTER 9

Sleep

Yann Le Meur, PhD, Rob Duffield, PhD, and Melissa Skein, PhD

Sleep is a naturally recurring state that involves the controlled and regulated loss of consciousness, without concurrent loss of sensory perception (Beersma 1998). The process of sleep occurs at regular intervals throughout a 24 h cycle, with the development of sleep accompanied by a gradual reduction in muscle tone and explicit loss of conscious control of physiological regulation (Frank 2006). Sleep can be distinguished from unconsciousness (i.e., a coma) due to the continued control over reflexes, the ability to awake from the loss of consciousness (i.e., sleepers can open their eyes), and the ability to react to speech, touch, or any other external stimuli. Despite the regularity and presumed importance of the sleep process, to date, the explicit role of sleep remains equivocal (Beersma 1998). Accordingly, the question of why we sleep remains poorly understood, since scientists to date have described *how*, rather than *why*, we sleep. Nevertheless, given that sleep duration is tightly regulated, it is commonly accepted that engaging in sleep ensures multiple psychological and physiological functions (Beersma 1998).

In a recent review of the literature, Frank (2006) identified several theories as to the function of sleep. The first theory relates to somatic function, emphasizing sleep's restorative effect on the immune and endocrine systems, respectively. The second, neurometabolic theory suggests that sleep assists in the recovery of the nervous and metabolic cost imposed by the waking state. The final theory is based on cognitive development, supposing that sleep has a vital role in learning, memory, and synaptic plasticity. Recent scientific studies show support for the latter theory, sug-

gesting there are greater benefits for the brain than for the body (Frank 2006). Regardless, it seems probable that sleep ensures multiple physiological and psychological functions across a wide array of cognitive and physical aspects for either regenerative or preparatory purposes (Frank 2006).

Physical exercise is a source of significant physiological and psychological stress for the body, be it muscular, energetic, cognitive, or other. Almost all athletes attempt to apply regular training and competitive stimuli to overload their physiological and cognitive systems in order to promote adaptation and improve performance (Issurin 2010). The demands of such training and competitive stimuli are normally large, and the importance of recovery to ensure adequate adaptation is deemed critical to both short-term (Barnett 2006; Mujika and Padilla 2003) and long-term performance outcomes (Halson and Jeukendrup 2004; Issurin 2010; Roose et al. 2009). As highlighted previously, engagement in sleep is a regular and important aspect of human physiological functioning. It is viewed as important to both recovery and athletic success (Halson 2008). Accordingly, it is recommended that the postexercise recovery phase provide conditions conducive for high-quality sleep to ensure sufficient and adequate recovery and maintenance of exercise performance. The present chapter will accordingly discuss the process of the sleep cycle; the effects of sleep or lack of sleep on physiological, perceptual, and performance responses; and the interaction of sleep, hydration, and nutrition for postexercise athletic recovery.

Stages of the Sleep Cycle

Humans spend almost one-third of their lives asleep, equating to approximately 6 to 10 h per daily cycle of 24 h. Intense physical activity leads to significant increases in physiological and cognitive demands, which may then require increased recuperative time for physical restoration, part of which relates to increased sleep requirements (Walters 2002). Such increased sleep demands may ensure the continuation of genetic processes, allowing adaptation to training (Kraemer, Fleck, and Evans 1996) while facilitating the return to resting values of exercise-induced metabolic (Skein et al. 2011; Trenell, Marshall, and Rogers 2007), nervous (Takase et al. 2004; Zhong et al. 2005), and immune (Lange, Dimitrov, and Born 2010; Walsh, Gleeson, Pyne, et al. 2011) responses. Sleep is a regulated part of a series of cycles. Schematically, sleep cycles represent the respective stages of light sleep, deep sleep, and paradoxical sleep (during which dreams may be experienced). These stages are classified according to parameters such as electrical brain activity, blood pressure, heart and breathing rates, muscle activity, and eye movement (Lashley 2004).

It is believed that effective sleep requires the presence of all the aforementioned stages, which may then promote good physical and mental conditions on waking (Lee-Chiong 2006). Each cycle lasts approximately 90 min. Depending on individual needs, between four and six cycles are necessary for sleep to be deemed adequate (Lee-Chiong 2006). Accordingly, the total duration of sleep can vary between 6 and 10 h, depending on the person, location, and physical demands. These interindividual differences are considered normal. It is important to determine sleep periods and tailor to individual needs; the early cycles must be recognized to allow an athlete to retire when the first signs of sleepiness are felt. If the first indicators (yawning, heavy eyelids) are missed, it may be necessary to wait another 90 min before sleepiness is felt once again (Walters 2002). Consequently, athletes and sport scientists should be familiar with the signs of individualized sleep cycles to assist the process of ensuring sufficient sleep and regularity of sleeping patterns.

• *Stages 1 and 2 (light sleep).* These stages normally exist once a person has entered the bed and turned out the light. During this time, the athlete may start to drowse. The level of consciousness may proceed from active to passive standby, as thoughts and images of the day are reviewed. Conscious awareness of the surrounding environment is slowly reduced, but a return to wakefulness can occur rapidly and without difficulty. When the sleeper reaches this stage of drowsiness, the muscles start to relax, while blood pressure and heart rate concurrently lower (Lee-Chiong 2006).

• *Stages 3 and 4 (deep sleep).* Following the movement into light sleep at the start of the night, the athlete may rapidly transit into stages 3 and 4, considered the deep sleep phases. Physiological responses include slowed respiration and reductions in blood pressure and heart rate (Lee-Chiong 2006). During this stage, sleepers may start to snore due to muscle relaxation: The supple tissue of the palate vibrates from the movement of air, leading to the characteristic sound of snoring. While the cardiovascular system operates under a reduced load during deep sleep, in contrast, the hormonal system increases its activity. In particular, during the early phases of the sleep cycles, large quantities of growth hormone are secreted, which is thought to contribute to tissue regeneration (VanHelder and Radomski 1989). This activation of hormonal responses during deep sleep often commences simply by lying down and starting to drowse. However, if sleep is delayed or interrupted, athletes should avoid becoming irritated due to further delays in the development of deep sleep.

• *Paradoxical sleep phases.* Following progression into deep sleep phases, which may occur after approximately 80 to 100 min, the first phase of paradoxical sleep commences, often noted by the experience of dreams (figure 9.1). During this phase, brain activity is increased, showing some similarity to activity levels noted during wakefulness (Lee-Chiong 2006). Further, the sleeper's circulation is increased due to increased vagal tone, resulting in increased heart rate and respiration. The sleeper's body is immobilized during the paradoxical sleep phase, thus preventing the sleeper from acting out the actions in the dreams. The first phase of paradoxical sleep at night lasts only a few minutes. Following this, a new 90 min cycle takes place, with a phase of light sleep followed by deep sleep. During the second cycle, the paradoxical sleep phase is generally longer than the first (Lee-Chiong 2006). These 90 min cycles follow each other throughout the night. As the cycles continue, the deep sleep phase shortens after the second

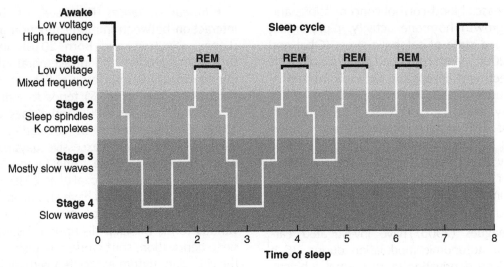

Figure 9.1 Sleep cycles during a normal 8 h night (hypnogram). The REM stage is mixed voltage and low frequency, with rapid eye movement and muscle atonia.

cycle, and it eventually disappears completely in the final sleep cycle occurring in the morning. In contrast, the light and paradoxical sleep phases lengthen as the night progresses, and the probability of the sleeper awaking gradually increases (Lee-Chiong 2006). Despite the widely held view, a healthy sleeper does not spend the whole night in deep sleep; even for a healthy young sleeper, deep sleep accounts for barely more than 20%, while a further 20% is spent in paradoxical sleep. The remaining volume of nightly sleep is split between light sleep, drowsiness, and wakefulness (figure 9.1). Accordingly, it is suggested a healthier sleeper awakes spontaneously after a long phase of very active paradoxical sleep, and may have less difficulty in starting the day.

Physiology of Sleep Deprivation

The findings discussed as follows may explain the reason some studies have shown that periods of intensive training, combined with overreaching or even overtraining, might alter the quality of sleep (Halson et al. 2006; Jurimae et al. 2004; Taylor, Rogers, and Driver 1997). Whether sleep disturbances lead to a state of overtraining, or are simply a symptom of it, still remains to be determined. Regardless, it appears that disrupted sleeping patterns may lead to altered physiological states, which may further affect training quality and adaptation (Halson 2008). Consequently, evalu-

ating sleep quality may be an important means to detect the potential for transition to an over-reached, or even an overtrained, state (Urhausen and Kindermann 2002).

• *Storage of energy substrates.* Skein and colleagues (2011) showed recently that 30 h (overnight) sleep deprivation with standardized food intake results in reduced muscle-glycogen concentrations compared to a normal night sleep in male team-sport athletes. It was hypothesized that the additional energy expenditure from being awake retarded full muscle-glycogen synthesis (Skein et al. 2011). Accordingly, the increased energy expended during prolonged hours awake may affect athletes' capacity to resynthesize the depleted muscle glycogen (Skein et al. 2011; VanHelder and Radomski 1989). While the volume of sleep deprivation used in these studies may seem extreme for most athletes, such findings suggest disrupted sleep to have some physiological consequences. Given that muscle glycogen content is important for any prolonged intermittent exercise (Krustrup et al. 2006), reduced pre-exercise glycogen content may adversely affect endurance capacity during intensive training over extended periods. Moreover, muscle glycogen deficits can affect both muscle and brain function, ultimately leading to reduction in work capacity (Costill et al. 1988).

• *Endocrine response and muscular recovery.* Similarly, sleep deprivation may slow the recovery process due to alterations to endocrine responses (VanHelder and Radomski 1989). A reduction in sleep increases blood-cortisol concentrations and reduces growth hormone activity, representing an increased state of catabolic stress (Obal and Krueger 2004). With continued presence of insufficient sleep and an ensuing catabolic hormonal state, muscle catabolism may increase, while protein synthesis can decrease. The long-term effects of such a state remain unknown; however, it is likely that continued lack of sufficient sleep may affect muscle function and the ability to recover from acute exercise bouts (VanHelder and Radomski 1989).

• *Autonomic nervous system balance.* Zhong and colleagues (2005) have investigated cardiovascular autonomic modulation during 36 h of total sleep deprivation in 18 normal subjects. Their results highlighted that sleep deprivation was associated with increased sympathetic and decreased parasympathetic cardiovascular modulation. To our view, this finding is interesting for athletes and coaches because autonomic nervous system (ANS) dysfunction or imbalance has been presented as one explanation for the signs and symptoms of the overtrained state (Hedelin et al. 2000; Hynynen et al. 2006; Uusitalo, Uusitalo, and Rusko 2000). Numerous studies have shown that intense physical activity is likely to disrupt this system by affecting the sympathetic/parasympathetic balance, and it may ultimately result in an overreached or overtrained state (Achten and Jeukendrup 2003). The ANS plays an essential role in maintaining body function; hence, a successful training period with appropriate recovery leads to increased resources of the ANS (i.e., increased parasympathetic and decreased sympathetic modulation) (De Meersman 1993; Garet et al. 2004; Iellamo et al. 2002; Pichot et al. 2000). Garet and colleagues (2004) reported that the reduction of the parasympathetic activity during intensive training was correlated with the loss in performance of well-trained swimmers, and the rebound in parasympathetic modulation during tapering paralleled the gain in performance. In this perspective, one of the main goals of recovery is to increase the magnitude of parasympathetic reactivation after training (Aubert, Seps, and Beckers 2003). Considering that sleep helps to regain the autonomic balance (Takase et al. 2004), thus allowing a return to intensive training, getting sufficient sleep is potentially important to ensure the maintenance of the autonomic balance during intense training periods.

• *Immune defenses.* There seems to be a close interaction between immune function and sleep (Lange, Dimitrov, and Born 2010). In laboratory animal studies, the intracerebral infusion of interleukin (IL)-1, interferon-γ (IFN-γ), or tumor necrosis factor-α (TNF-α) tends to induce sleep (Krueger and Madje 1990; Opp, Kapas, and Toth 1992). Further, studies of circulating cytokine levels in patients with excessive daytime sleepiness suggest that these same factors influence human sleep patterns (Gozal et al. 2010; Vgontzas et al. 1997). Associations have also been observed between abnormalities of immune function and various forms of sleep disruption. Issues include sleep deprivation, shift work, and disturbances of the circadian rhythm associated with global travel (Bishop 2004; Richmond et al. 2004). However, it has been difficult to determine whether the observed changes in immune responses reflect a disturbance of sleep per se, disturbances of the circadian periodicity of hormone secretions (Kusaka, Kondou, and Morimoto 1992; Lucey, Clerici, and Shearer 1996; Nakachi and Imai 1992), a general stress response, or a cognitive reaction to loss of sleep. Collectively, these results suggest that regular sleep may be important for optimal recovery from regular training and competition due to the highlighted relationship between insufficient sleep and immune function (Halson 2008).

• *Cardiovascular and thermoregulatory systems.* The examination of sleep loss on cardiorespiratory and thermoregulatory responses remains equivocal, although hemodynamic responses to steady-state exercise are unchanged following substantial (30–72 h) sleep deprivation. Horne and Pettitt (1984) examined the effect of 72 h sleep loss on steady-state cycling, reporting minimal differences in maximal and submaximal oxygen consumption, gross mechanical efficiency, and respiratory quotient. Additionally, Martin and Gaddis (1981) reported $\dot{V}O_2$, $\dot{V}CO_2$, $\dot{V}_E$, HR, and arterial blood pressure were not different between 72 h sleep deprivation or baseline over 3 days of exercise, including cycling protocol for 8 min at 25%, 50%, and 75% $\dot{V}O_2$max, respectively. Minimal differences in cardiorespiratory (HR, $\dot{V}O_2$, $\dot{V}_E$) and core-temperature responses were evident during lower intensity treadmill walking for 3 h following 36 h of sleep deprivation. Moreover, Mougin and associates (2001) reported an increase in HR and $\dot{V}_E$ during submaximal exercise (75% $\dot{V}O_2$max) following a night of partial sleep disruption. Finally, thermoregulatory responses to exercise following sleep deprivation have demonstrated equivocal results. Sawka, Gonzalez, and Pandolf (1984) reported during 40 min of cycling in warm conditions, sleep deprivation reduces evaporative and dry heat loss, while Martin and Chen (1984) and Kolka and Stephenson (1988), respectively, reported that during steady-state exercise, core and skin temperatures were not affected by sleep loss. Alterations in core temperature have been related to the flux of circadian rhythms, with a reduced ability to dissipate heat in the morning compared to the evening resulting in heat gain (Aldemir et al. 2000). Therefore, the disruption to circadian rhythm due to sleep loss may affect thermoregulatory responses to an exercise bout for a given time of day and, thus, potentially affect exercise performance if thermal load is unduly exacerbated (Waterhouse et al. 2005).

Sleep Deprivation and Exercise Performance

Due to competition and training schedules, travel commitments, and psychological stress and anxiety, athletes often experience disruption of sleep patterns, partial sleep loss, or complete overnight sleep deprivation (Bishop 2004; Richmond et al. 2004). The extent of this sleep disruption or loss can range from minor (2–4 h) to quite extensive (overnight), depending on the circumstances. Since athletes are often required to compete and train at high intensities for prolonged periods of time, they may be subjected to increased physical, physiological, and metabolic stress during exercise, exhibiting consequent delays in recovery (Skein et al. 2011). Furthermore, the reality of many competition and training schedules results in athletes performing exercise bouts on consecutive days. If adequate sleep is not achieved following a bout of exercise, subsequent recovery may be further dampened, since sleep deprivation or disruption increases energy expenditure and metabolic demands (Berger and Phillips 1995). Sleep deprivation or disruption, which is often experienced by athletes, may have negative implications on subsequent exercise performance or capacity. Although the effects of sleep deprivation on various physiological, thermoregulatory, metabolic, and psychological parameters at rest and during exercise remain relatively equivocal, some studies have highlighted the effects of a lack of sleep on exercise performance (table 9.1).

The comparison of previous studies examining the effects of sleep deprivation on subsequent exercise performance is difficult due to the considerable variation in study design, including exercise type and duration, duration of sleep deprivation, alterations in diets, and the use of sleep medications (Martin and Haney 1982; Oliver et al. 2009; Skein et al. 2011). However, the general overview of the literature suggests sleep disruption or deprivation can negatively affect exercise performance. Differences in findings relating to exercise performance have been associated with various physiological and metabolic perturbations and negative effects on perceived exertion and mood states. Incremental exercise protocols assessing exercise capacity have highlighted that 36 h sleep deprivation decreases time to exhaustion by 11% during heavy treadmill walking (Martin and Gaddis 1981). Further, partial sleep loss due to late bedtimes and early rising, which are common practices among athletes (Sargent, Halson, and Roach 2013), reduce maximal work rate compared to baseline values during an incremental test to exhaustion (Mougin et al. 2001). In contrast, Mougin and colleagues (1991) have reported no effect on time to exhaustion following a night of disrupted sleep, highlighting the mixed research findings on the effects of sleep deprivation on exercise performance.

Self-paced exercise protocols during which athletes can manipulate workload throughout a given exercise bout have shown sleep deprivation to reduce exercise performance. Oliver and associates (2009) reported reductions in distance covered during a 30 min self-paced treadmill run preceded by one night of sleep loss. Conversely, Martin and Haney (1982) have previously reported 30 h sleep deprivation to have minimal effect on the manipulation of workload at a set rating of perceived exertion (RPE) during treadmill exercise. These findings suggest that during self-paced exercise, the conscious regulation of exercise intensities due to the knowledge of sleep disruption may contribute to the performance alterations, rather than physiological or metabolic variables alone. While some investigations have focused on self-paced endurance exercise, few have examined sleep deprivation and intermittent-sprint performance.

Recently, Skein and colleagues (2011) examined the effect of 30 h sleep deprivation on free-paced, intermittent-sprint performance and reported a

Table 9.1 Effects of Sleep Deprivation on Performance in Various Athletic Activities

References	Sleep deprivation duration	Type of exercise	Effects of sleep deprivation
Bulbulian et al. (1996)	30 h (plus computer tasks and walking)	45 consecutive contractions at 3.14 rad/s	Peak torque reduced and with added exercise No effect on fatigue index
Horne and Pettitt (1984)	72 h	Cycling at 40%, 60%, and 80% $\dot{V}O_2$max for total duration of 40 min	Greater variability of mechanical efficiency Minimal effect on $\dot{V}O_2$
Martin and Chen (1984)	50 h	Steady-state walking following by walking until exhaustion	Reduced time to exhaustion No effect on $\dot{V}O_2$, $\dot{V}CO_2$, $\dot{V}_E$, HR, blood lactate, epinephrine, norepinephrine, or dopamine
Martin and Gaddis (1981)	30 h	8 min each at 25%, 50%, and 75% $\dot{V}O_2$max on a cycle ergometer	No effect on $\dot{V}O_2$, $\dot{V}CO_2$, HR, $\dot{V}_E$, or BP Increased RPE during moderate to heavy exercise
Martin, Bender, and Chen (1986)	a) Two nights of partial sleep loss b) 36 h	a) 30 min heavy treadmill walking b) 3 h treadmill walking	No effect on HR, core temperature, $\dot{V}O_2$ or $\dot{V}_E$, cortisol, or β-endorphins Disturbed mood states
Martin and Haney (1982)	30 h	Constant treadmill speed, adjust grade to maintain RPE of very hard	No effect on treadmill grade (17.1% vs 17.5%)
Mougin et al. (1991)	Partial sleep loss (3 h during the night)	20 min at 75% $\dot{V}O_2$max, followed by incremental exercise to exhaustion	Increased HR, VE at submaximal and maximal exercise Increased blood lactate during maximal and submaximal exercise
Oliver et al. (2009)	30 h	30 min at 60% $\dot{V}O_2$max and 30 min self-paced treadmill running	Less distance covered No effect on RPE, core and skin temperatures, or HR Raised $\dot{V}O_2$ at end of fixed-paced exercise
Plyley et al. (1987)	64 h	Pre- and post-sleep deprivation $\dot{V}O_2$max test. Additional 1 h treadmill walking every 3 h during the 64 h SD.	Reduced $\dot{V}O_2$max and $\dot{V}_E$max No effect on HR, RER, and blood lactate Additional exercise had no effect
Reilly and Piercy (1994)	Partial sleep loss (3 h during the night)	Maximal and submaximal strength: bicep curl, bench press, leg press, dead lift	Greater strength declines during submaximal lifts
Skein et al. (2011)	30 h	50 min self-paced intermittent-sprint exercise	Slower sprint times and distance covered during first and final 10 min of exercise
Symons, VanHelder, and Myles (1988)	60 h	Maximal and submaximal isometric strength; Wingate anaerobic capacity test; reaction time; 20 min treadmill exercise at 70% $\dot{V}O_2$max	Increased HR and RPE during steady-state exercise No effect of muscular strength or power output during Wingate, reaction times

significant decline in distance covered during the first and final 10 min of the exercise protocol and slower mean sprint times following sleep deprivation. Furthermore, the authors reported declines in voluntary force and voluntary activation during 15 maximal isometric contractions of the knee extensors following sleep deprivation. These findings concur with previous findings that dis-

rupted sleep for three consecutive nights reduces maximal and submaximal strength during both upper and lower body exercises (Reilly and Piercy 1994). However, 60 h sleep deprivation has been reported to have minimal effect on force during 25 maximal isokinetic contractions of the upper and lower body, again highlighting the contradictory findings in the literature.

Sleep Deprivation and the Perceptual State

Sleep deprivation can be a source of irritability, mental fatigue, and loss of motivation (Meney et al. 1998). As expected, sleep disruption also results in increased feelings of sleepiness (Blagrove, Alexander, and Horne 1995; Waterhouse et al. 2007). Further, sleep loss has also been associated with negative mood states including fatigue, loss of vigor, and confusion (Bonnet 1980; Meney et al. 1998; Reilly and Piercy 1994), and it can reduce tolerance to pain (Onen et al. 2001). A lack of sleep is detrimental to a wide range of psychological functions, with the severity dependent on the complexity of the subsequent task (Ikegami et al. 2009). Furthermore, the negative effects of reduced sleep are exacerbated when the tasks to be carried out are relatively simple or monotonous, and involve minimal environmental stimulation (Folkard 1990; Rosekind 2008; Van Dongen and Dinges 2005; Waterhouse et al. 2001). These simple skills affected by sleep deprivation include vigilance, simple and choice reaction times (Scott, McNaughton, and Polman 2006; Waterhouse et al. 2007), and short-term memory (Kim et al. 2001; Waterhouse et al. 2007). Additionally, previous studies have highlighted sleep deprivation to negatively affect auditory vigilance, logical reasoning, mental addition, and visual search tasks (Angus, Heslegrave, and Myles 1985; Bonnet 1980).

Sleep disruptions seem to have lesser effects on tasks of greater complexity and those that are rule based, but people become more sensitive to sleep deprivation when a task increases in familiarity (Harrison and Horne 2000). However, there is evidence that one night of sleep deprivation can lead to deterioration of more complex skills. This may be due to the fact that these respective tasks predominately require the prefrontal region and cerebral cortex, which are both affected by sleep deprivation (Harrison and Horne 2000). During periods of intense training, athletes face particular psychological and cognitive challenges and based on the aforementioned findings should monitor their sleep quality carefully. Beyond the reduced performance level, these degradations reduce proprioceptive capacities, a common source of injury in athletes (leading to sprains, for example) (Ivins 2006). Finally, lack of sleep affects memory, which may be problematic for training sessions involving learning tactics or new motor skills (Brawn et al. 2008). Laureys and colleagues (2002) further highlight the importance of sufficient sleep on motor and cognitive tasks, explaining that perceptive and motor learning continues after the training session through mnesic processes occurring during sleep. Of note, athletes may be able to alleviate the negative effects of sleep deprivation on sprint times, level of alertness, and short-term memory by taking after-lunch naps (Waterhouse et al. 2007).

Tryptophan and Sleep Cycles

Given the reported benefits of sleep to the recovery process, in combination with replenishment of fuel stores, the interaction between both the timing (Dollander 2002; Roky et al. 2001) and macronutrient content (Phillips et al. 1975; Wells et al. 1997) of meals is known to influence sleep. A number of macronutrients may influence the quality of sleep, particularly tryptophan (Trp), which serves as a precursor for brain serotonin and is a reported sleep-inducing agent (Hartmann 1982; Hartmann and Spinweber 1979). Tryptophan is an essential amino acid that is mainly converted into serotonin (also known as 5-hydroxytryptophan, or 5-HT) (Jouvet 1999). Conversion occurs when the ratio between free tryptophan and branched-chain amino acids increases, leading to a rise in brain tryptophan levels. Through 5-HT, free tryptophan is converted into serotonin, which in turn is transformed into melatonin (Jouvet 1999). The relationship between Trp and 5-HT and ensuing sleep processes suggests that it may be possible to assist improved sleep cycles through dietary manipulation, specifically by altering the ratio of free tryptophan and branched-chain amino acids (Markus et al. 2005).

A factor that promotes the entry of Trp into the brain is its plasma concentration relative to that of the other large neutral amino acids (LNAAs: tyrosine, phenylalanine, leucine, isoleucine, valine, and methionine) (Wurtman et al. 2003). It is now known that high glycemic index carbohydrates have the ability to increase the ratio of circulating Trp to LNAAs (Trp:LNAA) with a direct action of insulin, which promotes a selective muscle uptake of LNAAs (Berry et al. 1991). Afaghi and colleagues (2007) reported that a meal with a high glycemic index promotes sleep by means of an increase in brain Trp and serotonin as the ratio of plasma Trp to LNAA increases. These authors reported that a high glycemic index meal given 4 h before bedtime significantly shortened the sleep-onset latency by 48.6% compared with a low-glycemic index meal given 4 h before bedtime, or by 38.3% compared with the same meal provided 1 h prior to bedtime. Moreover, Hartmann (1982) reported that consuming tryptophan is associated with a 45% reduction in sleep-onset latency. Further, Arnulf and associates (2002) also report that increases in fragmented sleep and rapid eye movements over several nights are associated with a state of tryptophan depletion. Accordingly, a relationship between the availability of Trp and 5-HT and quality of ensuing sleep suggests the sleep cycle may be affected by dietary intake.

However, as well as the quality of ensuing sleep, Markus and colleagues (2005) studied the effects of Trp intake at dinner on vigilance and attention the next morning. The Trp source supplied was lactalbumin (i.e., milk proteins), known to be the richest source of Trp of all dietary sources. Within 2 h of meal consumption, this diet increased the ratio of plasma Trp to branched-chain amino acids by 130%, compared to the placebo group. This resulted in reduced drowsiness and greater cognitive activity the following morning (Markus et al. 2005). Such findings suggest a positive effect of Trp intake in the evening (through foods such as milk, meat, fish, poultry, beans, peanuts, cheese, and leafy green vegetables) on the following night's sleep. However, further research is required to determine the specifics of how the timing and quantities of tryptophan ingested can be optimized to provide a positive effect on ensuing sleep. To date, the recommendation seems to be to provide foods containing tryptophan, rather than consuming tryptophan alone as a food supplement.

Hyperhydration and Sleep Disturbances

Halson (2008) reported that sleep quality for athletes may be reduced by hyperhydration. Given the nutrition focus on ensuring adequate consumption of fluid volume following training, athletes may attempt to ingest larger quantities in bolus amounts later in the evening. Such practice may be particularly problematic if the water deficit was compensated late, rather than early, in the day. Athletes may ingest large volumes of liquids low in sodium (e.g., water) after training up until the time of going to bed. This hydration strategy may affect sleep by causing athletes to wake several times during the night to urinate (Halson 2008). From this point of view, the balance of fluid consumption for hydration versus improved sleep quality for recovery needs to be considered. Consequently, it may be wiser to ensure intake of fluid throughout the day and consume water rich in bicarbonates rather than large volumes late in the evening.

Effects of Alcohol and Stimulants on Sleep Quality

The consumption of alcohol is a regular and popular practice among athletes of both recreational and professional nature. The use of alcohol continues to be the most commonly consumed drug among athletic populations (O'Brien and Lyons 2000). Nevertheless, moderate alcohol consumption 30 to 60 min before bedtime may result in sleep disturbances (Stein and Friedmann 2005). Specifically, it seems that the alterations to sleep quality following alcohol consumption are explicitly mediated by blood alcohol concentration

Specificities for Ramadan

Ramadan, observed by hundreds of millions of Muslims worldwide, is based on 1 month's abstinence from drinking, eating, and smoking between sunrise and sunset. Roky and associates (2001) reported a reduction in deep sleep phases in 8 Muslims practicing Ramadan. Such findings appear to correlate with increased plasma cortisol or body temperature. These results may add further understanding to the elevated state of fatigue reported by Muslim athletes observing Ramadan. More generally, this study also highlights the difficulties faced by athletes restricting food intake as part of weight-control diets. Good sleep habits appear to be particularly important for these athletes. As a result, care should be taken during any dietary restriction plan to ensure high-quality sleep, in either qualitative or quantitative terms.

(O'Brien and Lyons 2000). Moreover, disruptions to sleep maintenance are seemingly most marked once alcohol has been completely metabolized from the body (O'Brien and Lyons 2000). Under conditions of moderate alcohol consumption where blood alcohol levels average 0.06% to 0.08% and decreases 0.01% to 0.02% per hour thereafter, an alcohol clearance rate of 4 to 5 h would coincide with disruptions in sleep maintenance in the second half of an 8 h sleep episode (O'Brien and Lyons 2000). Accordingly, if such metabolism coincides with later sleep cycles, an athlete may sleep fitfully during the second half of sleep cycles, although following awakening, may find returning to sleep difficult (O'Brien and Lyons 2000). Such findings demonstrate that alcohol consumption may affect both the quality and quantity of sleep for athletes. In addition, other studies have reported that alcohol consumption leads to increased heart and breathing rates, intestinal disorders, and headaches, all of which can affect sleep (Roehrs, Yoon, and Roth 1991). Therefore, it is recommended that this population limit the volume of alcohol consumed, particularly during intensive training periods when adequate good quality sleep is essential for appropriate recovery.

Despite the importance of sleep for acute and chronic recovery from exercise, during a normal day of training or competition, many athletes may make use of external stimulants to improve performance (caffeine, taurine, and so on). Caffeine is an ergogenic aid that stimulates the central nervous system. It is found in numerous food sources, such as coffee and tea. Several studies have shown that caffeine consumption can affect sleep, although many people may tolerate use once they become habituated (Roehrs and Roth 2008). In a literature review on the effects of caffeine on sleep, Bonnet and Arand (1992) suggest that in the hours before going to bed, consuming doses greater than 100 mg (i.e., one or two cups of coffee) can increase sleep-onset latency and reduce deep sleep and total sleep duration. Karacan and associates (1976) observed a dose-response effect: The greater the quantity of caffeine ingested, the greater the sleep degradation. Further, Shilo and colleagues (2002) also reported that melatonin secretion, which is part of the sleep cycle regulation, is significantly decreased by caffeine intake. Accordingly, these findings highlight that while caffeine may be reported as

beneficial for exercise performance during the day (Graham 2001), consumption later in the evening may negatively affect the quality and quantity of ensuing sleep patterns. However, Hindmarch and colleagues (2000) demonstrated that these deleterious effects on sleeping patterns were mainly noted in people who were not used to consuming caffeine, highlighting the process of habituation of caffeine consumption. Collectively, these findings suggest that caffeine consumption should be limited later in an evening (i.e., after 6 p.m.) and that energy drinks containing high concentrations of caffeine or taurine should also be avoided to limit ensuing effects on the athlete's sleep quality.

Advantages of Napping

In many cultures, particularly in Europe, the early afternoon is used for short naps and sleeping or resting practices. It is well known that daytime sleepiness occurs and performance declines in the afternoon (Dinges 1989, 1992). Given that meal consumption may not affect afternoon sleepiness (Stahl, Orr, and Bollinger 1983), it is therefore considered to be a part of the biological rhythm (Broughton 1989). Napping is one countermeasure to this sleepiness (Broughton 1989; Dinges 1989, 1992). While the afternoon nap has positive effects, such as refreshing mood and improving performance levels, it also has negative effects, such as sleep inertia (i.e., impaired alertness experienced on awaking) (Dinges 1989, 1992). These effects of napping depend on nap duration, prior wakefulness, and the time of day when napping (Naitoh, Englund, and Ryman 1982). Several studies have shown that short daytime naps (30 min) were effective for maintaining daytime arousal level (Hayashi, Ito, and Hori 1999; Horne and Reyner 1996; Naitoh, Kelly, and Babkoff 1992; Stampi et al. 1990).

Waterhouse and colleagues (2007) investigated the effects of a post-lunch nap on subjective alertness and performance following partial sleep loss. Ten healthy male participants either napped or sat quietly from 1 to 1:30 p.m. after a night of shortened sleep (from 11 p.m. to 3 a.m.). Their results indicated that alertness, sleepiness, short-term memory, and accuracy at the eight-choice reaction-time test were improved by napping, but mean reaction times and grip strength were not affected. Further, mean 2 m sprint time was reduced from 1.060 s to 1.019 s, and mean

PRACTICAL APPLICATION

Despite the importance of sleep for recovery of physiological and cognitive function, sleep hours are often too easily sacrificed when time is short (Sargent, Halson, and Roach 2013). Such lack of sleep may result from both voluntary or involuntary demands and actions. For example, voluntary actions may reduce duration of sleep when working late into the night or spending part of the night in alternate activities. Alternatively, lack of sleep may also be involuntary, such as with insomnia when attempts to sleep are unsuccessful. Such uncontrolled responses can often appear surreptitiously, after a few bad nights caused by anxiety (Uhde, Cortese, and Vedeniapin 2009) or overreaching in training (Halson 2008). Further, going to bed fearing ensuing difficulties at getting to sleep may also be sufficient to keep sleep at bay and continue a vicious cycle of insomnia. Nevertheless, most aspects of insomnia can be reduced, if not cured, by alterations in lifestyle choices, attitudes, and habits. Accordingly, a solution is therefore to relearn the processes involved with sleep. The following list provides a general checklist of strategies for dealing with issues related to sleep. The website Helpguide.org (Smith et al. 2011) presents a thorough list of suggestions for better sleep, including advice on when symptoms of sleep problems should be evaluated by a physician.

The following lifestyle choices should assist athletes aiming to improve their sleep:

1. **Set intense training before 6 p.m.**

2. **Keep a regular sleep schedule.**
   - Set a regular time for going to bed and waking up.
   - Nap for 30 minutes in the early afternoon to make up for lost sleep.

3. **Eat and drink correctly.**
   - Stay away from big meals at night.
   - Avoid caffeine after lunch.
   - Avoid drinking alcohol.

4. **Create a relaxing bedtime routine.**
   - Turn off your television and screen devices.
   - Take a hot shower or leisurely warm bath before bedtime.
   - Reserve your bed for sleeping.

5. **Make your bedroom more sleep friendly.**
   - Keep noise down.
   - Keep your room dark and cool.
   - Make sure your bed is comfortable.

6. **Manage anxiety and stress.**

7. **Utilize basic techniques to get back to sleep.**
   - Remain in bed in a relaxed position.
   - Make relaxation your goal, not sleep.

time for 20 m sprints was reduced from 3.971 s to 3.878 s. These results indicated that a post-lunch nap improved alertness and aspects of mental and physical performance following partial sleep loss, which may have implications for athletes who have had restricted sleep during training or before competition. Additionally, these advantages to performance may be further assisted when combined with an improved previous night's sleep (Takahashi et al. 2004).

Summary

Ensuring that athletes achieve an appropriate quality or quantity of sleep may have significant implications for performance and recovery and may reduce the risk of overreaching or overtraining. Indeed, sleep is often anecdotally suggested to be the single best recovery strategy available to elite athletes.

Particular attention should be given to pertinent strategies for improving sleep quality in athletes in order to optimize physiological adaptations for improving performance and maintaining health. The strategies refer mainly to training schedule and stress management, nutrition, and hydration.

Implementation of Recovery Strategies for International Rugby Union Tournaments

Campbell Hanson, Physiotherapist to ARU Junior Wallabies Squad for 2011 Tour

Rugby union is a sport that involves intense physical contact and requires repeated submaximal efforts over an 80 min period. It is unique in that 15 players on the field all have different roles and individual skill sets to carry out. This is reflected in the wide array of physiques and physiological makeups of the players.

When attending the annual International Rugby Board Junior World Championships, the premier junior rugby tournament in the world for male athletes aged 17 to 20, teams are faced with the task of playing five competition games in the space of 17 days. This is a large fixture load for any team sport, even more so for one with a high degree of physical contact and the associated increase in soft-tissue trauma. At tournaments like these, the teams that recover the quickest or limit their accumulated fatigue the most have a better chance of being successful as the tournament progresses.

The preceding factors present some unique situations to the medical staff for optimizing recovery and performance. We use some key strategies to combat levels of fatigue and muscle soreness. We believe that implementing the most highly technical methods or having complicated recovery procedures is not as crucial as focusing on consistently getting the basics right and developing good professional habits in these young athletes. With only 3 full days between matches, our recovery program generally looks as follows:

- *Day 1.* Recovery protocols: pool session, stretching, ice baths, injury assessment and treatment, and time spent off feet. Training involves video analysis of off-field match, individual player reviews, and so on. Athletes who played fewer than 20 min do a gym session.

- *Day 2.* On-field team session is held in the morning, usually with no contact. Athletes may run through set pieces. Massage is administered in the afternoon.

- *Day 3.* Teams run in the morning, visit match venues in the afternoon, and spend more time off their feet.

From a recovery viewpoint, three factors are deemed most important: sleep, nutrition, and an effective athlete-monitoring program.

Sleep is the most important recovery factor, so planning ahead for optimal sleep during travel and on arrival is crucial, especially when flying and crossing multiple time zones (this occurs when flying long haul from Australia to the Northern hemisphere). A travel strategy that we have found effective includes the following:

- Issue all players with an eye mask to facilitate sleeping on the plane.

- Advise them when they should attempt to sleep while flying so that they become synchronized with the local time zone as quickly as possible on arrival.

- Avoid caffeinated or sugary drinks and stay hydrated while flying to assist hydration levels and prevent negative effects on sleeping patterns.

- Run a team flex-and-stretch session in a vacant space in an airport lounge during transit or stopovers.

- Have a team walk and an outdoor stretch-and-flex session after arriving at the hotel to help adjust the players' circadian rhythm through exposure to natural light.

- Discourage sleeping before the local bedtime when arriving at the destination unless some players are struggling to stay awake.

- Plan with hotels in advance to ensure that all rooms have reliable working air conditioning. This is essential to provide the best sleep environment for athletes.

- For effective recovery, ensure that younger athletes spend time off their feet between training sessions when traveling in a different environment.

Daily hydration (urine) testing is carried out before breakfast, and players who are identified as being at risk of dehydration are advised how much fluid they should consume. Snacks are available immediately posttraining. Player education is carried out prior to travel so athletes have a clear idea of how many grams of carbohydrate and protein they should be consuming. Scales are available so posttraining fluid consumption can be calculated. Establishing basic nutrition protocols is an important part of an education process for developing athletes.

We run a daily athlete wellness-monitoring program that records players' visual analogue scale (VAS) scores on factors such as muscle soreness, general fatigue, sleep, emotional stress, and energy levels. We have found this to be an effective way of gathering information that helps coaching staff make decisions on training load and identify athletes at risk of impaired performance or injury.

The significant amount of physical contact and positional differences in rugby union requires an effective monitoring program that allows staff to identify players who are fatigued or suffering from significant muscle soreness and to adjust their training load. It is not uncommon for some players in the latter stages of a tournament to have a significantly limited training load due to fatigue and soft-tissue trauma. Effective athlete monitoring also allows better implementation of active recovery strategies.

We must not forget the mental side of recovery. A small team outing that doesn't require players to be on their feet for too long can be a valuable addition. It gets athletes out of the hotel environment and takes their minds away from rugby for a few hours.

Pretravel planning to establish sleep and nutrition strategies, as well as an effective athlete-monitoring program within a structure that has the flexibility to change individual training loads is essential to allow rugby union players to perform at their optimum during a tournament with short turnaround periods. If these three factors are optimized, they will provide an environment for the full potential and maximum recovery in the time frame available.

Key Points

- Do pretravel planning.
- Ensure sleep and nutrition strategies are in place.
- Have an effective monitoring system.
- Allow flexibility within these structures to cope with change.

CHAPTER 10

Massage and Physiotherapy

Antoine Couturier, PhD

With: Yann Le Meur, Cécile Huiban, Marc Saunier, and François-Xavier Férey

The contribution of massage to personal well-being has been recognized since antiquity. The effects experienced through their action on the skin, muscles, tendons, and ligaments are thought to provide many physical and psychological benefits, including improvements to circulation and the musculoskeletal system, pain reduction, and relaxation. Although they are the most requested and used method in sport recovery, their real effects are often controversial, and many authors question the benefits of massage for athletes.

Following technical developments, numerous therapeutic procedures based on the use of physical agents and grouped together under the heading *physiotherapy* (physical therapy) have emerged and have been developed in the last century. While physiotherapy techniques such as pressotherapy, electrostimulation, luminotherapy, or aromatherapy were initially reserved for medical applications, these techniques have gradually been adopted by a wider public, particularly athletes, for whom it was hoped the benefits would be similar to those seen in patients.

In a bid to distinguish between beliefs and reality, this chapter looks at the current knowledge on the use of massage (either manual or mechanical) and physiotherapy used to improve recovery and performance. We will first explain the physiological rationale behind the use of those various techniques, and then we will analyze the data in the scientific literature. While there are many published articles relating to the treatment of diseases and disorders, we have deliberately focused only on studies involving healthy or athletic subjects, which are the most relevant in our chosen context.

Massage Techniques

Massage is the result of a natural and intuitive need to rub a painful area of the body. Descriptions of massages to improve performance or favor recovery after strenuous exercise, particularly to treat injuries due to sport or war, can be found in texts dating from antiquity (Goats 1994a).

It is widely accepted that massages provide numerous benefits, such as increased blood flow, reduced muscle tension and excitability, and an increased feeling of well-being. It is therefore assumed that they will have a positive effect on performance and the risk of injury. They are, indeed, widely used to improve recovery between training sessions, or as part of preparation for and following competitions (figures 10.1 and 10.2).

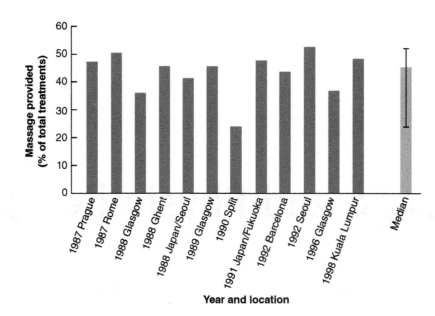

Figure 10.1 Massages provided, expressed as a percentage of total number of treatments, at each of the major athletic events between 1987 and 1998. Median and range values are given for the combined data over the period assessed.

Adapted, by permission, from S.D. Galloway and J.M. Watt, 2004, "Massage provision by physiotherapists at major athletics events between 1987 and 1998," *British Journal of Sports Medicine* 38: 235-236; discussion 237.

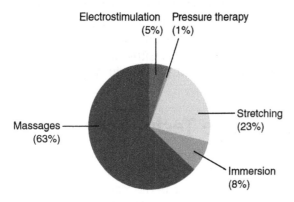

Figure 10.2 Recovery procedures used by athletes in the French delegation during the Olympic games in Beijing 2008 (CNOSF data).

Data courtesy of French Sport National Committee (CNOSF).

Many hypotheses have been formulated attempting to explain how massage acts on the circulatory, musculoskeletal, and nervous systems. However, the effects of massage on recovery (both physiological and psychological) and performance remain controversial (Weerapong, Hume, and Kolt 2005). After defining the classical massage techniques, this chapter will review their hypothetical mechanisms of action and their effects on performance and recovery.

Massage consists of manipulating tissues by pressure and kneading, with a view to improving health and well-being (Dufour et al. 1999). These manipulations can be performed manually or mechanically.

Manual Massage

Manual massages can be split into a succession of basic steps comprising the following:

• *Effleurage:* Light, continuous movement with the fingers and hands. Effleurage is often used at the start and end of a session. Graded pressures are generally thought to help reduce muscle tone and induce general relaxation, preparing the subject for more vigorous manipulations.

• *Pressure:* These can be light or strong compressions of the soft tissue against a more resistant underlying layer, such as bone. Pressure can be gliding or static. The expected effects include increased blood and lymph circulation, improved drainage, and reduced edema (Goats 1994b).

• *Kneading:* This technique involves isolating a tissue, skin, or muscle volume, and submitting it to repeated pressure and release, generally following a gradual progression (Dufour et al. 1999). Kneading may be good for stretching contracted or adherent tissues, and for relief of muscle cramps. It is thought to act more deeply than pressure to stimulate flow of blood and lymph, thus favoring drainage and reducing edema.

- *Friction:* Friction causes sliding of one anatomic plane over an underlying one, without rubbing. These movements are generally used to improve sliding of the different anatomical planes, to induce defibrosis, or to create a relaxing effect (Dufour et al. 1999).

- *Tapotement:* The aim of tapotement is to induce skin reflexes and local vasodilation. These are believed to increase muscle tone and dispersion of interstitial fluid, which collects as a result of muscle injury and inflammatory processes.

- *Vibrations:* These involve intermittent static pressure of variable intensity and frequency. This technique can be relaxing when the pressure is applied with low intensity and high frequency, and stimulating when using strong pressure and low frequency.

- *Rolling massage:* This technique mobilizes the skin with rolling and palpating movements. It may allow excess water and fat to be eliminated and may promote localized drainage.

Massages for drainage (venous and lymphatic return) and Swedish-type massages are the most used on athletes. This type of massage lasts approximately 30 min and includes a succession of effleurage, kneading, friction, and tapotement, and then ends with another series of effleurages. Swedish massage is also the most frequently cited technique in the scientific literature relating to recovery.

Mechanical Massage

Various devices imitate the action and reproduce the effects of manual massages. An example is the roller-type device, which mimics rolling massages. Through the application of independent motorized rollers mounted on a head and coupled with an air-depression system, these devices form a fold of tissue and act directly on the tissues (aponeuroses, muscle fascia, muscles, tendons, and so on). The roller is applied directly to the skin or through technical garments. Water beds are also used, where water projected toward the skin through a rubber sheet creates a sensation of massage. The area to be massaged, the pressure applied, and how the nozzles move can all be programmed in these systems.

Finally, intermittent pneumatic compression (IPC) devices improve venous return and lymphatic drainage. The limb to be treated is encased in an inflatable sleeve made up of several compartments that can be inflated and deflated sequentially, progressing from the extremity toward the torso.

Physiological Effects of Massage

Physiological responses to massage have been measured by range of motion, skin and intramuscular temperature, pulsed ultrasound Doppler, electromyography, blood levels of endorphins, and questionnaires.

Musculoskeletal System Responses

The mechanical pressure applied during massage is believed to help reduce tissue adhesion and musculoskeletal stiffness, which can be measured by the range of motion. The few studies investigating the effects of massage on range of motion did not show any significant differences, or they presented biases that made results difficult to interpret (Weerapong, Hume, and Kolt 2005). Ogai and colleagues (2008) recently noted reduced incidence of muscle stiffness when 10 min of kneading massage were included in the recovery phase after two intense exercise series on a cycling ergometer. However, stretching seems to be a simpler and more effective means to obtain this type of effect (Weerapong, Hume, and Kolt 2005).

Skin and Intramuscular Temperature

Massage induces a local increase in temperature, as well as hyperemia in the area being massaged. The temperature of the skin and the vastus lateralis muscle were shown to increase with massage, regardless of duration (from 5 to 15 min effleurage). However, the temperature increase remained relatively superficial and did not exceed a depth of 2.5 cm (Drust et al. 2003). Mechanical action on the subcutaneous skin capillaries and tissues could induce mastocytes to secrete vasodilating substances, such as histamine, sero-tonin, and acetylcholine, through mechanical or reflex mechanisms. Although this could not be proven, superficial vasodilation is likely to improve exchanges between cells and the blood (nutri-ent and oxygen supply, elimination of metabolic waste products and carbon dioxide) (Dufour et al. 1999).

Circulation Responses

Circulation responses include the following:

• *Venous circulation.* Sliding and static pressure improve venous return (Goats 1994b), as shown by Doppler measurement (Dufour et al. 1999). A slow rhythm, leaving 5 to 10 s between successive manipulations, is believed to provide an optimal effect. This effect may be due to a mechanical mechanism, whereby the pressure exerted causes the veins with antireflux valves to collapse, thus forcing venous return. A slow rhythm of massage could therefore allow veins to refill after being emptied by the manipulation. Massage tech-niques, such as manual lymphatic drainage, work along the same principles. Manipulations are directed along the superficial lymphatic network to help accelerate lymphatic return.

• *Arterial circulation.* Owing to the fact that the circulatory system is a closed circuit, massages are assumed to act on the arterial system indi-rectly, through their action on the venous system. Shoemaker and colleagues (1997) used pulsed ultrasound Doppler to quantify these effects. Massages were performed on small (forearm) and large (quadriceps) muscle groups using a range of techniques, including effleurage, kneading, and tapotement. No significant variations in blood flow or diameter of brachial or femoral arteries were measured before, during, or after massages. In contrast, low-intensity exercises involving prehension and extension of the knee almost tripled resting values. This led the authors to rec-ommend moderate exercise rather than massage to increase blood flow. More recently, Hinds and associates (2004) did not observe any differences between a group receiving massage and a con-trol group in terms of blood flow in the femoral artery, lactatemia, heart rate, or blood pressure. However, superficial circulation (figure 10.3) and skin temperature were significantly higher for the group receiving massage. The effect of massage on performance and recovery is thus question-

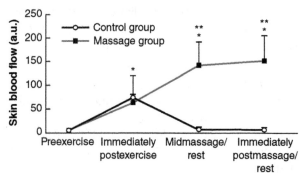

Figure 10.3 Cutaneous blood flow for the massage and control groups.

*Significantly different from baseline (resting values) ($P < 0.05$); **Sig-nificant difference between control group and massage group ($P < 0.05$).

Adapted, by permission, from T. Hinds et al., 2004, "Effects of massage on limb and skin blood flow after quadriceps exercise," *Medicine and Science in Sports and Exercise* 36: 1308-1313.

able, since neither skin temperature nor superficial circulation are associated with increased blood flow (Weerapong, Hume, and Kolt 2005).

• *Intermittent pneumatic compression (IPC).* Intermittent pneumatic compression devices were initially designed to treat patients suffering from venous insufficiency. The method was validated in a patient population (Gilbart et al. 1995; Harf-ouche et al. 2008; Mayrovitz 2007; Morris et al. 2003; Tochikubo, Ri, and Kura 2006) where it was shown to increase arterial blood flow and venous and lymphatic return, and to reduce edema. An increase in blood flow, similar to that induced by moderate exercise, was also observed in healthy subjects (Whitelaw et al. 2001).

Nervous System Responses

Nervous system responses to massage include a reduction of neuromuscular excitability, promo-tion of muscle relaxation, and the relieving of cramps and stiffness.

• *Neuromuscular excitability.* By stimulating sensory receptors, massage reduces neuromus-cular excitability and may also reduce muscle contractures and tension. This muscle relaxation was shown with roller-type devices (Naliboff and Tachiki 1991), but also with effleurage, sliding or static pressure, tapotement, and kneading. For the duration of the massage, these manipulations lead to a reduced H reflex (figure 10.4), which is an indicator of medullary reflex excitability (Gold-berg, Sullivan, and Seaborne 1992; Weerapong, Hume, and Kolt 2005).

The H reflex may be reduced by stimulating muscle, skin, and aponeuromechanoreceptors, as well as feedback loops for motoneuron excitability (Morelli, Chapman, and Sullivan 1999). This effect is generally localized and selective (figure 10.4) (Sullivan et al. 1991), but a contralateral effect may also be observed (Dufour et al. 1999).

• *Muscle cramps.* A cramp is a sustained and painful involuntary contraction affecting one or more muscles. The pain experienced is believed to be caused by stimulation of nociceptive muscle mechanoreceptors, and indirectly by the local ischemia and muscular acidosis induced by the contraction. Massage can be used to break out of the vicious cycle leading to persistent cramping. It may also contribute to rearranging muscle fibers, limiting the stimulation of nociceptive receptors, and reestablishing microcirculation (Weerapong, Hume, and Kolt 2005).

These mechanisms, however, remain hypothetical. First, massage does not appear to increase arterial circulation. Second, no published study investigates the effects of massage on the realignment of muscle fibers.

• *Pain.* Various explanations have been suggested as to how massage relieves low- to moderate-intensity pain (Dufour et al. 1999; Ernst 1998; Farr et al. 2002; Goats 1994b; Weerapong, Hume, and Kolt 2005; Willems, Hale, and Wilkinson 2009; Zainuddin et al. 2005). The first explana-

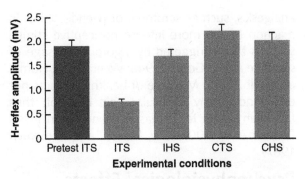

Figure 10.4 Mean H-reflex amplitudes and standard errors for the four experimental conditions (i.e., ipsilateral (I) and contralateral (C) massage of the triceps surae (TS) or hamstring muscles (HS)) and the pretest control condition. H reflexes were measured on the ITS (lower right extremity).

Based on Sullivan et al. 1991.

tion involves gate theory, or gate control (Melzack and Wall 1965): Sensory information triggered by touch is transmitted by rapid-conduction type $A\alpha$ and $A\beta$ nerve fibers. As long as signals from these fibers outnumber nociceptive, or pain, signals from the slower type $A\delta$ and C nerve fibers in the spinal cord, nociceptive signals will be inhibited (figure 10.5).

The second suggested explanation involves a signal descending from the midbrain to the spinal cord, which leads to secretion of endogenous

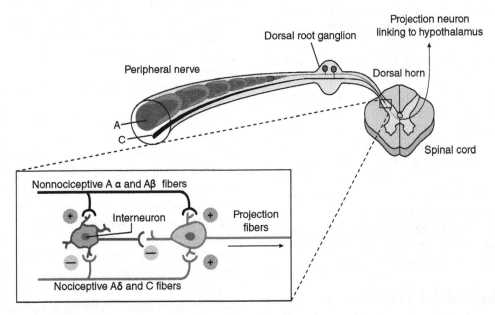

Figure 10.5 Schematic representation of gate control theory: Stimulation of the nonnociceptive $A\alpha$ and $A\beta$ fibers can be used to modulate the nociceptive signals from $A\delta$ and C afferent fibers at the dorsal horn of the spinal cord.

analgesics, such as serotonin or β-endorphins, in reaction to the more intense nociceptive stimuli such as those triggered by vigorous massage of sensitive areas (Goats 1994b; Weerapong, Hume, and Kolt 2005). Massage of healthy subjects not experiencing any particular pain has not, thus far, been linked to increased β-endorphin levels (Bender et al. 2007; Day, Mason, and Chesrown 1987).

Psychophysiological Effects

Several studies report a positive effect of massage on psychophysiological responses, as evaluated using questionnaires such as POMS (Profile of Mood States), STAI (State-Trait Anxiety Inventory) or, more specifically, using a perceived recovery scale. Thus, Weinberg and Jackson (1988) assessed the effects of 30 min of Swedish massage compared to physical activity, such as swimming, jogging, tennis, or racquetball, on 183 physical education students. Only massage and jogging led to significantly improved mood, including significantly reduced nervousness, confusion, fatigue, anxiety, anger, and discouragement (POMS test). Massage was the only technique associated with a reduced STAI test score. Similar results were reported for a study of boxers during training (Weerapong, Hume, and Kolt 2005).

Beyond the positive effect of massage on emotional well-being, some authors investigated athletes' perception of their own recovery. It is interesting to note that, to our knowledge, the majority of studies showed improved perceived recovery even though physiological (lactatemia, glycemia, heart rate) and muscle (maximal strength, time to exhaustion, vertical jump, specific tests) parameters did not necessarily show significant differences compared to control groups (Hemmings et al. 2000; Micklewright et al. 2005; Ogai et al. 2008; Weerapong, Hume, and Kolt 2005). According to Keir and Goats (1991), the psychological effect of a massage performed by an expert is based on the impression that the masseur has specifically located the sensitive areas. This placebo effect strongly contributes to the power of all manipulation techniques.

Massage and Recovery

Recovery responses to massage have been examined in terms of range of motion, arterial blood flow, metabolite elimination, force recovery, perceived recovery, and DOMS before, after, and between exercise sessions.

Before or Between Consecutive Exercise Session Effects

Massage is often provided before a competition, specifically because it is believed to increase blood flow. This increased blood flow is thought to improve performance by increasing oxygen and nutrient supply to cells, raising intramuscular temperature and the buffering effect of blood. To date, the data supporting these hypotheses remain weak. As previously mentioned, no significant increase in blood flow could be demonstrated during or just after massage. Comparing a group receiving massage 10 min before submaximal exercise (80% of maximal heart rate) to a control group revealed no differences for the variables measured during exercise ($\dot{V}O_2$, stroke volume, heart rate, blood pressure, cardiac output and arteriovenous oxygen differential, lactatemia) (Callaghan 1993). Massages were not shown to significantly affect stride frequency either when applied for 30 min on a population of sprinters (Weerapong, Hume, and Kolt 2005). However, the authors report that the highest stride frequency was measured for a member of the massaged group. It is therefore regrettable that this study did not report the stride length for all subjects.

When performed between two 5 km ergocycle exercises, massage alone does not allow better recovery compared to rest or active recovery (Monedero and Donne 2000). In this study, recovery was most effective when it combined massage and active recovery (figure 10.6).

In contrast, some authors indicated lower levels of fatigue when mechanical massage imitating a manual rolling massage was used (Portero, Canon, and Duforez 1996). Following a localized exhaustion test (30 maximal knee flexions and extensions at a rate of 180 s⁻¹), subjects who were massaged for 15 min had a better endurance time (66% of maximal voluntary force) and a lower reduction of maximal strength compared to subjects who simply rested for 15 min. The authors attribute this lower level of fatigue to better metabolite elimination.

Massage was also compared with resting during successive exercises, including specific boxing performance tests in a population of boxers (Hemmings et al. 2000). Although no difference was observed for parameters such as heart rate, gly-

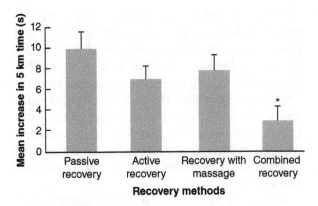

Figure 10.6 Mean increase in 5 km ergocycle trial time (s) ± SEM for different recovery methods: passive, active (15 min pedaling at 50% $\dot{V}O_2$max), massage, and combined recovery (3.5 min active recovery, 7.5 min massage and 3.5 min active recovery). n = 18. Results are expressed as the difference between time for 2nd trial and time for 1st trial.

*Significant difference between recovery methods (P < 0.05).

Based on Monedero and Donne 2000.

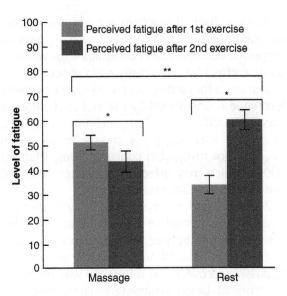

Figure 10.7 Perceived fatigue after two sets of exercises separated by a period of either rest or massage. Exercise involved 90 s isometric extension of the torso.

*Significant difference (P < 0.05); **Significant difference (P < 0.01).

Adapted, by permission, from H. Mori et al., 2004, "Effect of massage on blood flow and muscle fatigue following isometric lumbar exercise," *Medical Science Monitor* 10(5): CR173-CR178.

cemia, or performance during specific tests, the massaged group showed improved perceived recovery and increased lactatemia. The authors suggested that the increase in lactatemia may be linked to a greater investment by the boxers in this group due to their improved perceived recovery. Comparable results were observed during both isometric (figure 10.7) and dynamic ergometric bicycle exercises to exhaustion (Ogai et al. 2008; Robertson, Watt, and Galloway 2004). Although no differences in terms of lactatemia, heart rate, and average and maximal power were observed between groups (Robertson, Watt, and Galloway 2004), perceived recovery, total power (Ogai et al. 2008), and fatigue index (Robertson, Watt, and Galloway 2004) were all better in the massaged group. We can therefore hypothesize that massage increases tolerance to subsequent intense exercise through a placebo effect (Goats 1994b).

The results of Micklewright and colleagues (2005) comply with this hypothesis. Before performing a Wingate test, study participants either rested or received a 30 min massage. To specifically study the psychological effects, massage was applied only to the back, avoiding any muscles involved in the subsequent exercise. Although the POMS questionnaire did not show any difference between groups, the massaged group was able to maintain peak performance for longer, and its subjects developed significantly higher average

power (+2.5%) and total workload (+2%) (figure 10.8). The average power developed during a Wingate test has been correlated to performance in events such as a 25 m swim or 300 m run. Thus, a 2.5% gain is far from negligible. Although exercise was better tolerated after massage, the authors were not able to interpret the psychological effect because there were no differences in the POMS results between the two groups. The effect could thus depend on the individual athlete. Massage before or between exercises should be used with care. Although it can help prepare some

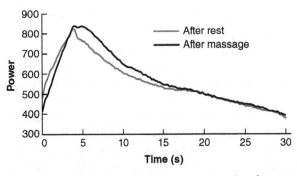

Figure 10.8 Power output on an ergocycle after 30 min rest or massage.

Reprinted, with permission, from D. Micklewright et al., 2005, "Mood state response to massage and subsequent exercise performance", *The Sport Psychologist* 19(3): 234-250.

athletes, individual reactions are not always predictable. It should therefore always be practiced in specific conditions and on athletes for whom a positive effect has previously been observed.

The need for caution when using this recovery technique is underlined by the fact that vigorous massage, either manual or mechanical, can lead to muscle microinjury (as shown by increased CK, LDH, or myoglobin concentrations) (Barnett 2006), which may affect performance (Cafarelli and Flint 1992; Callaghan 1993).

Zelikovski and associates (1993) evaluated the efficacy of intermittent pneumatic compression used for 20 min between two ergocycle exercises to exhaustion at 80% $\dot{V}O_2$max, and compared it to passive recovery. No difference was observed in terms of blood parameters (lactatemia, pH, bicarbonate, pyruvate, ammonia) or oxygen consumption. In contrast, pressotherapy treatment increased subjects' maintenance time by 45% during the second, higher heart rate, exercise. The mechanisms underlying this increased performance are uncertain, since there is no effect on metabolite elimination. The authors propose the following explanations: Exercise-related accumulation of fluid in the muscles and interstitial spaces increases the diffusion distance between capillaries and muscle cells, and changes membrane characteristics. These changes perturb muscle function. Thus, eliminating excess fluid by intermittent pneumatic compression could minimize the damaging effects of this perturbation. Finally, the feeling of well-being provided by the massaging action may also have a psychological effect, improving subjects' preparation for the second exercise.

Postexercise Effects

Massage is mainly used by athletes to reduce fatigue and, thus, the time needed to recover, particularly during competitions. The relaxing effects, both physical and psychological, may also contribute to the elimination of competition-related stress (Cafarelli and Flint 1992).

However, 30 min of massage provided after a semimarathon did not speed the recovery of maximal strength either 1 or 4 days after the event (Tiidus et al. 2004). In a review of the effects of massage on a population of athletes, Callaghan reports no benefit over 4 days of intensive cycling with a daily 161 km race (Callaghan 1993). During recovery, cyclists received either a massage or a placebo treatment. Enzymatic assays, performance during events, perceived recovery, and POMS questionnaires did not show any significant differences between the two groups. In another review, Weerapong underlines the lack of relevant results relating to the effects of massage on lactatemia. The few studies that do show reduced lactatemia combined massage with active recovery, which is known to accelerate lactate elimination by increasing blood flow (Weerapong, Hume, and Kolt 2005).

According to Barnett, massage applied after exercise neither minimizes the drop in maximal strength nor accelerates its recovery. On the other hand, it does have a beneficial effect on the athlete's perception of pain and recovery (Barnett 2006). This may not be an advantage, however, since underestimating the state of fatigue and returning to training early may increase the risk of injury.

DOMS Response

Muscle fever or DOMS (delayed-onset muscle soreness) appears between 24 and 72 h after intense and inhabitual exercise involving eccentric contraction. This is a significant problem for coaches and athletes, since it involves chronic pain associated with altered muscle function (lower strength and range of motion, increased stiffness and resting metabolism). DOMS is generally thought to be induced by mechanical constraints when stretching muscle fibers, causing microinjuries to sarcomeres, muscle membranes, and sarcoplasmic reticulum. This induces an inflammatory response and associated edema. At first, the migration of neutrophils and macrophages to the site of inflammation makes the injuries worse (and causes pain that peaks 48 h after exercise), but these cells also play a role in the process of cellular regeneration (Cheung, Hume, and Maxwell 2003; Gulick and Kimura 1996).

According to some studies on exercise-induced DOMS, massage may not affect range of motion, limit the drop in maximal strength, or accelerate its recovery (Hilbert, Sforzo, and Swensen 2003; Howatson and van Someren 2008; Zainuddin et al. 2005). Results from these studies showed that the evolution of maximal voluntary force was not significantly different for a group receiving massage (10 min arm massage 3 h after exercise) and a control group during the fortnight following eccentric biceps brachii exercises (Zainuddin et al. 2005) (figure 10.9). Massage did, however, reduce DOMS (–30%) and plasma creatine kinase concentrations at D + 4, as well as arm circumference at D + 3 and + 4. In the same way, 5 days of treatment with intermittent pneumatic compression for 20 min led to a significant reduction in arm circumference and improved range of motion at D + 2 and + 3 after a series of eccentric arm exercises. No effect on strength recovery was observed (Chleboun et al. 1995).

For Hilbert and colleagues (2003), the effect of massage is mainly psychological. Two hours after eccentric hamstring exercises, one group was massaged for 20 min while another received a simulation of massage. Shortly after exercise and up to 48 h later, subjects underwent tests for maximal strength and range of motion and answered POMS-type and perceived pain questionnaires. The authors also counted neutrophils in blood samples. The only difference observed for the massaged group was a reduction in perceived pain.

Other authors showed more striking effects using a protocol involving downhill running (Portero and Vernet 2001). Exercise was followed by 10 min of mechanical rolling massage of the rear thigh muscles of one leg for the following 6 days. The other leg served as a control. Edema, pain, and the drop in maximal strength at D + 2 were all lower in the treated thigh than in the untreated one (figure 10.10).

The effects of massage on DOMS have been extensively reviewed (Cheung, Hume, and Maxwell 2003; Ernst 1998; Weerapong, Hume, and Kolt 2005). Although results are heterogeneous, massage appears to significantly reduce DOMS when used starting 2 h after exercise. The mechanical pressure may improve blood and lymph microcirculation, thus reducing edema, local ischemia, and pain. Increased numbers of circulating neutrophils have also been observed compared to a nonmassaged group (Smith et al. 1994). Based on this, it was suggested that massage inhibits neutrophil migration to the site of muscle injury, thus limiting the inflammatory process. While reducing or disrupting the inflammatory process may seem beneficial in the short term, it must be remembered that it is part of the process involved in regeneration of damaged muscle fibers. In the long term, disruption of this process could actually be damaging for the athlete.

Before concluding, we must cite Moraska's study (2007) (figure 10.11) of three groups of athletes (n = 317) who were massaged by 95 student physiotherapists for 15 to 60 min after a 10 km race. The three groups were differentiated based on how experienced the student physiotherapists were, that is, how many hours of massage they had already performed (450, 700, or 950 h). The athletes were not aware of this distinction. For all three groups, perceived fatigue and pain were reduced immediately after treatment and over

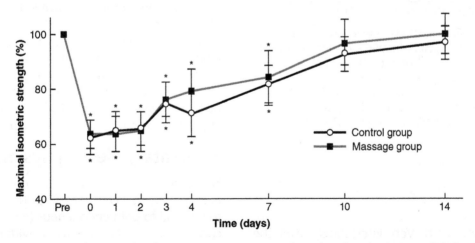

Figure 10.9 Progression of maximal voluntary isometric torque from baseline (pre), immediately after (0), and 1 to 14 days postexercise (eccentric exercise of biceps brachii) for the massage and control arms expressed as a percentage of baseline.

*Significant difference from baseline (pre-exercise value).

Reprinted, by permission, from Z. Zainuddin et al., 2005, "Effects of massage on delayed-onset muscle soreness, swelling, and recovery of muscle function," *Journal of Athletic Training* 40: 174-180.

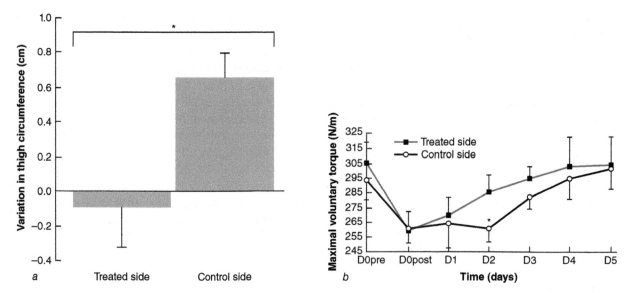

Figure 10.10 Variation in thigh circumference at D + 2 *(a)* and progression of maximal voluntary torque after downhill running *(b)*.

*Significant difference (P < 0.001).

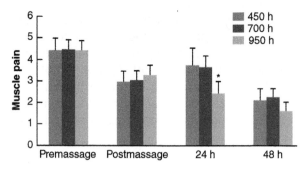

Figure 10.11 Race participants who received massages from student therapists with 950 h of didactic training reported significantly lower levels of muscle soreness over time compared with participants who received massage from therapists with 700 h or 450 h of training (P < 0.001). A total of 95 student therapists participated in the study: 31 with 950 h, 40 with 700 h, and 24 with 450 h of training.

*Significant difference (P < 0.001).

the following 48 h. Very interestingly, over the 48 h following the race, perceived pain was significantly lower in the group treated by the most experienced masseurs. This finding goes against the hypothesis that the benefit of massage is purely a placebo effect.

Electrostimulation

Electricity has been used as a therapeutic agent for centuries. Physicians in ancient Egypt used electric catfish to treat various painful disorders. Over the last 20 years, electrostimulation has taken off, particularly since the development of devices for use by the general public. Its many benefits in physiotherapy have led to a wider use of electrostimulation.

Because it is easy to use and because the recruitment mode is different from that of voluntary contraction, electrical stimulation is presented as an effective way to help athletes improve recovery, possibly by increasing arterial blood flow, improving venous return, and inducing hypoalgesia.

Elementary Electrophysiology

Nerve and muscle cells are the only cells in the human body that can be excited. This allows them to respond to both endogenous (during voluntary contraction) and exogenous (as with electrostimulation) electrical stimuli by generating electrical activity at the membrane known as an action potential.

When triggered in a nerve cell or its extension (axon), the action potential propagates rapidly to the neuromuscular junction, where it activates the muscle-cell membrane to induce contraction.

The intensity and duration of the electrical pulses required to trigger excitation are closely linked. They are characteristic of each type of nerve or muscle cell (see figure 10.12 for the recruitment thresholds of nerve fibers). Based on figure 10.12, we can see that a progressive increase in intensity or duration of electrical pulses has graded effects: first a sensory effect (excitation of the Aβ fibers), then a motor effect (excitation of α motoneurons), and finally pain (excitation of Aδ and C fibers) caused by the strongest stimulation. Because nerve cells are several hundred times more excitable than muscle cells, electrically induced muscle contraction in a healthy person will always be triggered indirectly by motoneurons.

During electrically induced motor excitation, the number of motor units recruited is directly proportional to the intensity and duration of the electrical pulses, termed spatial summation. The pulse frequency influences the degree of tetanic fusion of the motor units, termed temporal summation. It can also modulate the force produced, but does not affect the level of muscle activation. The intensities, durations, and frequencies of the electrical pulses generally used for healthy people vary from a few mA to a few tens of mA, from 100 μs to 1 ms and from 1 to 150 Hz, respectively.

During voluntary contractions, motor units are recruited according to Henneman's size principle (Milner-Brown, Stein, and Yemm 1973), with slow fibers being activated before rapid fibers. Since the motoneurons innervating Type II fibers are more excitable than those innervating Type I fibers, it was assumed for many years that electrostimulation reversed this order and preferentially stimulated rapid fibers. Today, it is thought that the order of motor-unit recruitment is not reversed, although it remains quite different from that observed during voluntary contractions. It appears to depend on other factors, such as the positioning and orientation of nerve fibers in the electrical field (Duchateau 1992; Knaflitz, Merletti, and De Luca 1990). Finally, motor-unit activation is also quite different for voluntary and electrically induced contractions. In the first case, trains of action potentials are controlled by the central nervous system, and they are specific to each motor unit (asynchronous activation). With electrical stimulation, all the recruited nerve fibers transmit identical trains of action potentials, as directed by the stimulator (synchronous activation).

The vast majority of devices available today use rectangular, biphasic symmetrical trains of pulses (figure 10.13). This allows effective stimulation without the risk of burning tissues. Stimulators generating interfering currents also exist. Their principle is illustrated in figure 10.14.

Two independent generators produce slightly different high frequency signals (a few kHz). When these signals cross each other in the tissue,

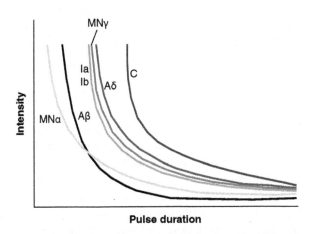

Figure 10.12   Recruitment thresholds for various nerve fibers. Aβ fibers: analgesic sensitive fibers; Aδ fibers: nociceptive sensitive fast fibers; C fibers: nociceptive sensitive slow fibers; MNα: α motoneuron; MNγ: γ motoneuron; Ia: primary ascending nerve fiber linked to intrafusal muscle fiber; Ib: ascending nerve fiber linked to Golgi tendon organs.

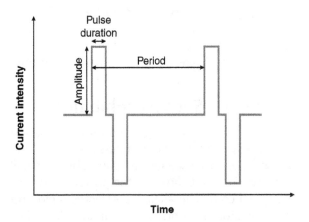

Figure 10.13   Example of current used in electrostimulation.

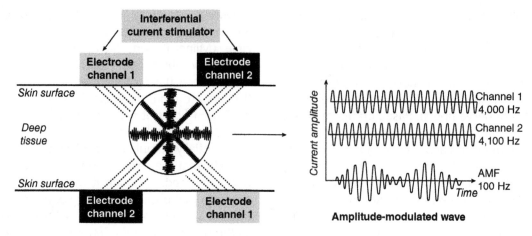

Figure 10.14 Diagram representing the principle of interference currents.

they generate a signal of modulated amplitude, with a frequency equal to the difference between the two source signals. The high frequency signals used in this technique penetrate more deeply into the tissues, and they are better tolerated (Goats 1990; Roques 2003). However, interfering current devices are available only in physiotherapy practices because they are quite difficult to use.

Physiological Effects of Electrostimulation

The choice of electrical pulses depends on the desired effect. The analgesic and trophic properties of electrostimulation are the most common recovery techniques.

TENS-type (transcutaneous electrical nerve stimulation) electrostimulation consists of stimulating specific sensory or afferent nerve fibers to relieve pain. Two action mechanisms are possible, depending on the parameters of the electrical current: segmental sensory inhibition and endorphin release.

Segmental Sensory Inhibition

Based on the principle of gate control, stimulation of large diameter $A\alpha$ and $A\beta$ fibers may inhibit the transmission of nociceptive messages by the small diameter $A\delta$ and C fibers in the posterior horn of the spinal cord. The analgesic effect is rapid and local, but stops as soon as stimulation is discontinued (Roques 2003). The current intensity is weak, causing only a prickling sensation (threshold of touch sensitivity, and the frequency is between 50

and 100 Hz). For localized pain, a small electrode (10–20 cm^2) is applied to the painful area, and another, larger electrode (50–100 cm^2) is applied transversally or at the level of the corresponding vertebra.

Endorphin Release

This is based on the principle that stimulating the afferent pain discriminating nerve fibers ($A\delta$ fibers) could increase β-endorphin production by the hypothalamus (Roques 2003). This raises the pain threshold and induces a generalized analgesic effect. Stimulation is applied via large electrodes placed on the spine, using very low frequencies (a few Hz) and high intensity. This produces unpleasant, almost painful sensations. β-endorphin secretion may persist for several hours after stimulation.

NMES-type (neuromuscular electrical stimulation) electrostimulation induces muscle contractions by triggering an action potential in the motor nerves. This can be used to induce multiple effects: Muscle activation is thought to increase blood flow in the same way as voluntary contractions. The pumping effect of the muscle, induced by rhythmic contractions, is believed to favor metabolite elimination and stimulate venous and lymphatic return. In addition, the electrical current may help limit edema by reducing microvascular permeability to plasma proteins. This could help maintain the osmotic gradient, which is altered by the inflammatory response, and it may limit the diffusion of liquids into the extracellular space (Holcomb 1997).

Blood Flow Response

Many authors have reported an increase in blood flow following electrostimulation. This effect appears logical, owing to muscle pumping and to the increased metabolic demands, but other mechanisms may also be involved, such as sympathetic inhibition of vasoconstriction and release of vasodilating substances by sensory neurons.

To test these hypotheses, Cramp and associates (2002) used a Doppler laser to measure the effect of three different current intensities (sensitivity threshold, below the motor threshold, and above motor threshold) using TENS stimulation (4 Hz, 200 µs) of the median nerve. They examined blood flow and skin temperature in the forearm, both at the site of stimulation and on the fingers, which are supplied by the same nerve.

Increased blood flow in the forearm was observed only when the current was strong enough to induce muscle contractions. Compared to the control group, skin temperature was not significantly altered by any of the conditions. In the fingers, no changes were observed under any of the electrostimulation conditions. These results do not support a role of the autonomous nervous system in the response to electrostimulation. For these authors, the mechanisms most likely to explain the increased blood flow are the muscle pumping effect and the increase in metabolic demands. Other authors reached the same con-clusions with comparable protocols on the triceps surae (Sherry et al. 2001) and trapezoid muscles (Sandberg, Sandberg, and Dahl 2007).

Miller and associates (2000) compared variations in blood flow, heart rate, and blood pressure induced by voluntary or electrostimulated contractions of similar levels on the triceps surae (figure 10.15). The only difference between these two contraction modes was vasodilation, which persisted for 15 s after the exercise, as indicated by measures of blood flow and vascular resistance in stimulated subjects. This result might be explained by an increased production of vasodilating metabolites (H^+, adenosine, phosphate) due to the enhanced recruitment of Type II fibers for electrostimulated contractions. These authors also conclude that the mechanisms behind the variations in blood flow were mainly due to muscle contraction rather than to the electrical current itself.

Pain Response

While it is accepted that the gate control mechanism is involved in pain modulation during electrostimulation, analgesia due to endorphin release in a healthy population is more controversial (Bender et al. 2007; Sluka and Walsh 2003).

For Hughes and colleagues (1984), analgesia induced by TENS in endorphin-release mode starts after 15 to 30 min of treatment and lasts several

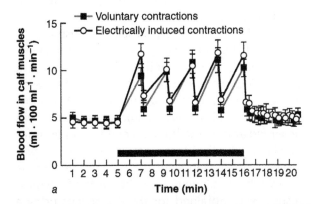
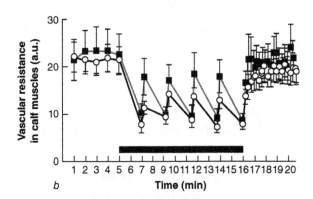

Figure 10.15 Effects of voluntary and electrically induced muscle contractions on blood flow (a) and vascular resistance (b) in the calf muscles. Baseline values are 1 min averages. The 10 min exercise period, indicated by the horizontal bar, was interrupted every 2 min to measure blood flow. Recovery values are 15 s averages. Values are means (± standard error of the measurement).

hours. They relate this analgesia to β-endorphin release since the effect was inhibited by injection of naloxone, the main antagonist of morphine receptors.

However, when pain was induced by electrical stimulation of nociceptive nerves (O'Brien et al. 1984), the results were contradictory. This study found no increase in β-endorphin concentrations or in pain threshold, and no effect of naloxone. Nevertheless, TENS stimulation in gate control or endorphin-inducing mode has been shown to increase the pain threshold for heat, cold, and mechanical pressure (Cheing and Hui-Chan 2003; Sluka and Walsh 2003; Tong, Lo, and Cheing 2007). In these studies, it was suggested that endorphin-inducing TENS increases encephalin, β-endorphin, and endorphin secretion, while gate control TENS increases dynorphin secretion. In addition, it is possible that the relatively high-intensity electrical stimulation may effectively distract subjects from nociceptive stimuli (Tong, Lo, and Cheing 2007).

Johnson and Tabasam studied the analgesic effect of gate control TENS on ischemia-induced pain (Johnson and Tabasam 2003). This pain is due to metabolite accumulation and is therefore more closely related to the pain caused by intense physical exercise than other types of pain. No significant differences between electrostimulation and placebo were detected, although the pain tended to be lower in the electrostimulated group.

Electrostimulation and Recovery

Electrostimulation is widely used as a recovery technique. It is therefore surprising to see that there are so few scientific articles published on this topic. The few studies we have found do, however, provide information on the effects of electrostimulation used between exercise series, and after exercise, compared to other recovery methods.

Between-Exercise Effects

Compared to passive recovery, electrostimulation between exercise series (5 series of 10 repetitions on a hack squat, at 80% of maximal voluntary force) resulted in a smaller reduction in power over the series, and in maximal strength at the end of the exercise (Maitre et al. 2001).

Lattier and associates (2004) compared the effects of 20 min recovery (passive, active; running at 50% of $\dot{V}O_2max$) or electrostimulation (after an uphill race; 18% slope) with progressive increases of speed until exhaustion. One hour after the end of the exercise, subjects performed a new uphill sprint exercise to exhaustion at 90% of $\dot{V}O_2max$. Maximal strength and sprint maintenance time did not reveal any significant differences among the three recovery techniques, although there was a trend for better times to exhaustion for the group that had received electrostimulation during recovery. The same recovery methods were tested in swimmers following a 200 m sprint crawl (Neric et al. 2009). In this study, active recovery and recovery using electrostimulation were equivalent, and were better than passive recovery.

Finally, recovery methods (active, passive, balneotherapy, and electrostimulation) were used on a football team during a 20-day training camp (Tessitore et al. 2007). Tests of anaerobic performance (squat jump, countermovement jump, rebounds, 10 m sprints) did not reveal any differences among the various procedures. However, perceived recovery for the active and electrostimulation groups was equivalent, and higher than for the passive and balneotherapy groups.

DOMS Response

Studies investigating the effects of electrostimulation on DOMS have not shown a significant effect on perceived pain, joint flexion, or restoration of functional capacities.

The effects of electrostimulation (user-determined high intensity, 8 Hz, 400 μs for 30 min) were compared to those of active (30 min running at 50% of $\dot{V}O_2max$) and passive recovery following a one-legged downhill race (7 km/h, 12% slope) (Martin et al. 2004). No significant difference was observed among the three recovery conditions for maximal voluntary force, activation level, or perceived pain 30 min after the end of the exercise or over the 4 following days (figure 10.16). Similar results were obtained for the quadriceps using a frequency of 125 Hz (Butterfield et al. 1997) and the arms using gate control (200 μs, 110 Hz) and endorphin-releasing TENS (200 μs, 4 Hz or 2 Hz) (Craig et al. 1996; Denegar et al. 1989), with no significant increase in β-endorphin levels.

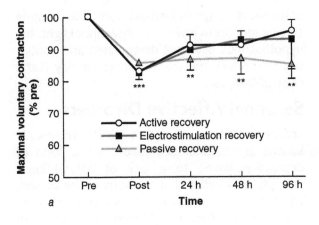

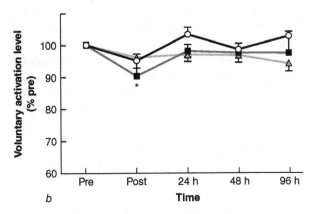

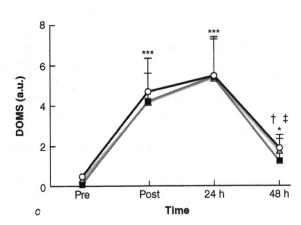

Recovery methods using TENS (90 Hz), cold application, or a combination of TENS and cold after eccentric exercises that mobilized the biceps brachii have also been studied (Denegar and Perrin 1992). Although these recovery methods reduced perceived pain, compared with a control group, this is likely to be a placebo effect. Indeed, a group receiving placebo treatment (TENS with no current applied) also reported a reduction in pain.

Finally, Vanderthommen and colleagues (2007) studied the effects of electrostimulation-based recovery (5 Hz, 250 μs, for 25 min) following 3 series of 30 eccentric quadriceps contractions. Although they did not detect any difference in terms of perceived pain and recovery of maximal strength compared to a control group, creatine-kinase levels were lower for the group receiving electrostimulation during recovery (figure 10.17). For these authors, the hyperemia induced by electrical stimulation may accelerate the elimination of cellular debris, thus reducing the inflammatory response.

Luminotherapy

Like all living organisms, humans depend on sunlight for life and well-being. Temperature, sleep, dreams, appetite, libido, physical fitness, and

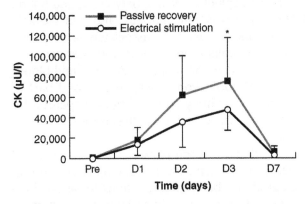

Figure 10.16 Recovery kinetics for isometric maximal voluntary contraction *(a)*, voluntary activation level *(b)*, and delayed-onset muscle soreness *(c)* when using active, electrostimulation, or passive recovery modes. Values are mean ± SE.

Significantly different from baseline (pre-exercise values): *$P < 0.05$; **$P < 0.01$; ***$P < 0.001$. †: Significantly different from values measured after 24 h ($P < 0.001$). ‡: Significantly different from values measured after 48 h ($P < 0.001$).

Adapted, by permission, from V. Martin et al., 2004, "Effects of recovery modes after knee extensor muscles eccentric contractions," *Medicine and Science in Sports and Exercise* 36: 1907-1915.

Figure 10.17 Serum creatine-kinase (CK) level (mean ± SD) for passive and electrostimulation-based recovery before exercise (pre-exercise), or 24 h (D1), 48 h (D2), 72 h (D3), and 168 h (D7) after exercise.

*Significant difference between the two conditions ($P \leq 0.05$).

Reprinted, by permission, from M. Vanderthommen et al., 2007, "Does neuromuscular electrical stimulation influence muscle recovery after maximal isokinetic exercise?" *Isokinetics and Exercise Science* 15: 143-149.

mood are all directly influenced by the intensity of natural light (see review by Cermakian and Boivin 2009).

Light is therefore the most powerful environmental synchronizer. It regulates the biological clock, or circadian rhythm, by determining the changes in a person's physical and intellectual capacities over a 24 h period.

Physiology of Light Exposure

The hypothalamus largely regulates the body's endocrine function. This area of the brain is directly linked to the retina. It interprets levels of light to control waking and sleeping states over a 24 h period.

More specifically, sunlight or bright light modifies the activity of the suprachiasmatic nucleus to promote the secretion of stimulatory arginine-vasopressin during wakefulness, and of a sleep-inducing peptide at night. During the day, light is detected by the photoreceptors in the retina. This triggers signals transmitted by the optical nerves toward the suprachiasmatic nucleus to stop melatonin production. Thus, during daytime, the suprachiasmatic nucleus promotes mental and physical activity by increasing body temperature and plasma adrenalin concentrations; at night, these phenomena are reversed.

Melatonin levels increase after nightfall to induce sleepiness, peaking between 2 and 5 a.m. The suprachiasmatic nucleus also synchronizes many bodily functions associated with falling asleep or waking up, such as body temperature, hormone secretion, urine production, and blood-pressure modulation. During the second part of the night, the body's internal clock prepares it for waking up by reversing the trends for these functions.

In addition to the phases of wakefulness and sleep, this system allows a large number of physiological and behavioral functions to be synchronized with the day–night cycle: motricity, hunger, thirst, sensory discrimination, thermoregulation, regulation of basal metabolism, hormonal secretions, receptor activity for many neurotransmitters, and so on.

On a clear spring or summer day, the number of lux (i.e., units of light) received varies between 10,000 and 100,000, while, in winter, it rarely exceeds 1,000 lux outdoors and 500 lux indoors. Several studies have shown that in winter, because of the shorter days and temperatures that promote staying indoors in low-intensity artificial light, the hypothalamus becomes deregulated and continues to produce melatonin during the day (Lahti et al. 2008).

Seasonal Affective Disorders

Seasonal affective disorders (SAD), commonly known as seasonal depression, are a form of depression linked to a lack of natural light. Depression is initiated at the start of the season, when daylight hours become shorter (most often in autumn and winter), and is only improved with the return of spring. Table 10.1 lists all the symptoms reported in scientific studies of seasonal depression.

Luminotherapy is the first-line therapeutic option for seasonal affective disorders (Lavoie 2002). Since this disorder is likely to be aggravated by overreaching or overtraining, this therapy is potentially beneficial for athletes, both in terms of optimizing recovery and improving performance. From this point of view, it is probably not pure chance that athletes tend to prefer sunnier climates for their winter preparation camps.

All the symptoms of seasonal affective disorder are listed in the screening questionnaire for overtraining provided by the French society for sports medicine (see www.sfms.asso.fr). In numerous sporting disciplines, winter is a period for serious training. Therefore, athletes may be at greater risk for overreaching leading to SAD. To develop more effective strategies to overcome these disorders, participants in athletics (athletes themselves, trainers, and medical personnel) need to understand their causes.

In the general French population, between 2% and 10% of people suffer from seasonal affective disorders (Attar-Levy 1998; Sartori and Poirrier 1996). Generally, the disorder appears for the first time between the ages of 20 and 25, but it has also been observed in children and adolescents. Prevalence in this population is between 2% and 5% (Sartori and Poirrier 1996). Women are four times more likely than men to be affected by this disorder (Sartori and Poirrier 1996).

Luminotherapy Treatment

Recent research tends to demonstrate the efficacy of luminotherapy in treating sleep, eating, and mood disorders.

Table 10.1 Symptoms Associated With Seasonal Depression

Nature of the problems	Associated symptoms
Psycho-affective	Anxiety Lowered dynamism and energy levels Relationship difficulties Reduced self-esteem Suicidal tendencies (rare) Irritability Social isolation Sadness
Alertness	Asthenia Concentration problems Slowed psychomotor reactions
Sexual	Lowered libido
Dietary and digestive	Irresistible desire to eat sugar-rich foods (79%) Hyperphagia (66–80%) Constipation Weight gain (2–5 kg, 75%)
Sleep	Disturbed sleep Hypersomnia (86–97%)
Other	Muscle cramps Aggravated pre-menstrual syndrome Headaches Lowered immune response

Data from Attar-Levy 1998; Cizza et al. 2005; and Lurie et al. 2006.

Sleep Disorders

Chronic fatigue syndrome is characterized by extreme lethargy and weariness, persisting even when the amount of sleep is adequate. It has been linked to a lack of sunlight in winter. At this time of year, because light intensity is low and daylight hours are short, the brain does not detect enough sunlight and continues to produce melatonin during the day, maintaining the body in a sort of standby mode when it should actually be awake. Asynchrony between the biological clock and sleeping and waking phases is the main cause of sleep disorders (Dumont and Beaulieu 2007). These can have effects on both physical and mental health, and even more so in athletes. Since light is the main synchronizer of circadian rhythm, it is easy to see why exposure to intense light can help treat circadian disorders (which are revealed by sleep disturbances). In athletes, exposure to intense light in the morning (on waking) also seems to favor and help maintain synchronized circadian rhythms.

Eating Disorders

Luminotherapy stimulates serotonin production, which plays a role in regulating the appetite (Adam and Mercer 2004). Furthermore, individual eating habits are controlled by hormones: Leptin induces a feeling of satiety, while ghrelin triggers hunger. Cizza and colleagues (2005) showed that in people suffering from winter SAD, the depressive state induced by a lack of light was associated with hyperphagia and increased fatty weight.

These eating disorders were not linked to changes in plasma leptin/ghrelin levels, suggesting that the body's sensitivity to these hormones is altered by the lack of light. Studies in animals showed that winter weight gain in many species was linked to a drop in leptin sensitivity, such that in winter, these animals feel less full and consequently eat more food (Adam and Mercer 2004). Several studies have shown encouraging results for the effects of luminotherapy as a treatment for bulimia and anorexia nervosa (Frank et al. 1998). As part of intense sport activity, luminotherapy may help athletes prevent changes to eating habits and maintain a diet compatible with intensive training and competition over the winter months.

Mood Disorders

Luminotherapy limits mood perturbations by improving global physical and mental balance and promoting restful sleep (Kohsaka et al. 1999). The production of serotonin, a natural antidepressant, can be stimulated by luminotherapy. This hints at potential benefits during competitions or strenuous training periods, which put the athlete under considerable psychological stress. Luminotherapy may be useful as part of athletes' mental recovery. Robazza and associates (2008) explain that mood and self-confidence may affect performance and athletes' commitment to their sports.

Phototherapy Applications

Although phototherapy is the first line of treatment for seasonal affective disorder, there is no consensus yet on the light intensity, spectrum, duration of exposure, and ideal timing for treatment. According to various studies, the duration of treatment depends on the intensity of the light source used.

In these studies, an intensity of 2,500 lux was generally used for 1 or 2 h, but more recent studies showed equivalent effects with 30 min of exposure at 10,000 lux (Lam and Levitt 1999). Because a reduced exposure time favors compliance with treatment, 30 min exposure at 10,000 lux has become the clinical norm.

However, the optimal timing for administering the treatment remains controversial. According to most studies, treatment on waking appears to be the most effective. The type of light seems to be less important than its intensity, although white light appears to be more effective than light using only one color in the visible spectrum (Hawkins 1992; Lam and Levitt 1999). It is not always necessary to look straight at the light, but the eyes must be kept open at all times. To reduce side effects, exposure duration can be reduced to 10 min at the start of treatment, and gradually increased to 30 min. Finally, phototherapy must be continued throughout the winter months, since the symptoms reappear rapidly when treatment is discontinued (Lahmeyer 1991).

Aromatherapy

Aromatherapy is based on the use of aromatic plant essences. The term commonly used for these aromatic essences is essential oils (EO).

In general, aromatherapy is used to treat various problems (coughs, headaches, sinusitis, asthma, digestive problems, insomnia, fatigue, sport injuries), as an antiseptic (against bacteria, viruses, fungi, and parasites), and even for general indoor hygiene. It is also thought to provide psychological benefits, particularly in the treatment of anxiety.

EOs contain high concentrations of active chemical compounds, and they can be dangerous. Several compounds are irritants or allergenic for the skin and mucous membranes. Others can be toxic at high doses or when used over prolonged periods (Lardry and Haberkorn 2007a, b).

Too few studies have been published on the topic of aromatherapy to allow us to make recommendations. Given the results obtained in a healthy population, we can assume that EOs may have many beneficial effects, either when used alone or in combination with other recovery methods.

As part of recovery, the effects of EOs may synergize with those of massage. Similarly, diffusing EOs in the air, in addition to having purifying and antiseptic effects, creates a calming atmosphere compatible with recovery. During training, if the conditions allow (indoor training area), EOs could be diffused to promote a positive effect on mood and stress, as well as alertness and reaction times (treatment of information in the brain).

Effects Attributed to Essential Oils

Numerous benefits are attributed to EOs. Examples include analgesia or pain relief (chamomile, pine, lavender, mint), energizing effects (mint, black pepper, rosemary, sandalwood, sage), calm (chamomile, myrrh, orange), anti-inflammation (lemon, incense, geranium), regulation of body functions (chamomile, lavender), and healing (eucalyptus, lavender, hyssop, rosemary). They may also act in synergy with the effects of other recovery methods to reduce physiological and psychological stress induced by training and competition (Lin and Hsu 2007, 2008). Tables 10.2 and 10.3 present the main EOs reputed to affect performance and recovery.

Table 10.2 Sports-Related Properties of the Main Essential Oils Used in Massage

Name	Relaxation/ stress relief	Circulation	Cramps	Pain	Inflammation	Muscle/joint
Yarrow				X	XX	XX
Basil			XX		XX	
Tropical basil	XX					XX
Bergamot					XX	XX
Birch			XX		XX	XX
Chamomile			XX		XX	
Camphor				XX		XX
Citronella					XX	XX
Lemon	XX				XX	XX
Cypress		XX				
Eriocephala			XX	X	XX	
Tarragon			XX	X	XX	
Eucalyptus				X	XX	XX
Gaultheria	XX		XX		XX	XX
Juniper	X					XX
Geranium bourbon	XX					XX
Ginger				XX		XX
Helichrysum		XX				XX
Laurel			XX	XX		XX
Lavender	XX		XX	XX	XX	XX
Mandarin	XX					
Marjoram	XX					XX
Peppermint	XX	XX		XX		XX
Mustard						XX
Nutmeg				XX		XX
Red myrtle	X	XX				
Niaouli		X				
Orange	X					
Bitter orange	XX					X
Scots pine						X
Pepper				XX		XX
Horseradish						XX
Ravensara	XX					
Rosemary 1,8 cineole		XX				XX
Balsam fir		X				
Mountain savory						X
Saro	XX					
Sassafras			XX			
Cedar						XX
Thyme (linalool)						X
Thyme (thymol)						XX
Ylang ylang	XX					X

XX = main indication; X = additional indication.

Based on Lardry 2007; Faucon 2009.

Table 10.3   Sports-Related Properties of the Main Essential Oils Used in Diffusion

Name	Relaxation	Pain	Fatigue	Inflammation
Angelica			X	
Bay		X		
Cardamom	XX			
Citronella	X			XX
Lemon			X	
Cumin		X		
Spruce			X	
Eucalyptus			XX	XX
Clove		XX	X	
Mandarin	XX			
Tea tree oil			XX	
Petasites	XX	X		
Scots pine			X	
Savory			XX	

XX = main indication; X = additional indication.

Based on Lardry 2007; Faucon 2009.

Modes of Action

EOs can be quite widely dispersed into the air by an electrical diffuser that uses cold nebulization. People in the environment then absorb EOs through their airways, where active molecules enter the bloodstream, and thus access all areas of the body. By stimulating the olfactory centers, EOs also act on the central nervous system.

EOs have been shown to have an effect on brain activity (Diego et al. 1998; Masago et al. 2000; Satoh and Sugawara 2003) and on the autonomous nervous system (Haze, Sakai, and Gozu 2002; Heuberger et al. 2001).

Psychological effects, including inducing calmness, reducing stress, improving mood and attention, have also been reported. These can be explained by the close relationship between how the brain treats olfactory and emotional types of information, linking personal experiences to smells (Cooke and Ernst 2000; Heuberger et al. 2001; Ilmberger et al. 2001; Weber and Heuberger 2008). EOs also have a definite placebo effect linked to personal expectations (Ilmberger et al. 2001).

EOs mixed with vegetable oils or neutral cream can be applied either as part of massage or as a poultice. As they penetrate the skin, EOs may be absorbed by the venous and arterial capillaries (Lardry and Haberkorn 2007b). Combined with the benefits of massage, and their possible effects on the central nervous system, EOs may help to reduce anxiety and stress, induce relaxation, and have a sedative or stimulating effect. They may also stimulate the immune system (figure 10.18) (Kuriyama et al. 2005).

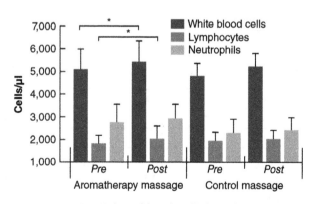

Figure 10.18   White blood cell, lymphocyte, and neutrophil concentrations before and after aromatherapy massage or control massage.

*Significant difference before and after treatment ($P < 0.05$).

Data from Kuriyama et al. 2005.

Performance and Recovery

It is important to note that most of the studies available on aromatherapy were carried out on animals or patients suffering from a disease. Almost no studies exist that investigate recovery and performance in healthy athletic subjects. Because of this, we will assume that results obtained in a general healthy population can be extended to athletes.

To our knowledge, only one study of EOs investigated their effects on recovery (Romine, Bush, and Geist 1999). After 2 min of walking at a moderate pace (approximately 90 steps/min), subjects recovered for 10 min, either in a neutral room or in a room where lavender EO was diffused. Lavender is believed to have sedative, relaxing, and soothing properties. Although the results were not statistically significant, the walkers exposed to lavender EO tended to have lower heart rate, systolic, diastolic, and arterial pressures than those from the control group. It is regrettable that the authors chose not to study a more tiring exercise.

Psychological Aspects

Several studies on students at exam time report that EOs act on mood by reducing stress and anxiety, as well as their physical symptoms (McCaffrey, Thomas, and Kinzelman 2009; Seo 2009).

For many authors, sporting performance is linked to psychological states such as mood (Cockerill, Nevill, and Lyons 1991; Hassmen, Koivula, and Hansson 1998; Norlander and Archer 2002; Seggar et al. 1997). Some even indicate that 70% to 85% of winning (or losing) athletes would be identifiable by their psychological profiles, mood, and personality tests (Raglin 2001). EOs may therefore be of great benefit for athletes.

Alertness and Cognitive Performance

EOs can help increase alertness and the speed with which information is treated in the brain, as shown by multitasking tests (Baron and Kalsher 1998). The relaxing effect of lavender EO was shown by increased EEG β waves, speed, and number of correct answers to mathematical questions. The stimulating effect of rosemary EO was shown by reduced anxiety, increased EEG α and β waves, and speed in answering mathematical questions (but not the number of correct answers) (Diego et al. 1998). Finally, sage EO has been associated with positive effects on anxiety, alertness, and cognitive performance in stressful situations induced by mathematics and memory tests, and a soothing action after tests (Kennedy et al. 2006).

Conditioning for Performance

As part of an experimental protocol, 76 children between 11 and 13 years of age were selected based on their lack of self-confidence. To start with, their scores in psychotechnical tests adapted to their age group were recorded (number of correct responses and time necessary to complete the test). Forty-eight hours later, the children were split into three groups. Those in the first group (control group) filled out a questionnaire in a room with no particular smell. Those in the other two groups completed psychotechnical tests once again, but this time with an olfactory stimulus (peppermint or strawberry). Before starting this second series of tests, the children were told that, unfortunately, since there were no more tests suitable for their age group, they would have to perform much more difficult tests designed for older children. They were instructed to just do their best. In truth, the tests were simpler than those performed on the first day. After another 48 h, all the children performed a further series of tests, this time similar to those completed on the first day. The children in the control group performed the tests in the same conditions as the first time. The two experimental groups were subdivided into four groups: two groups carried out the tests with the same olfactory stimulus as the previous time, and two groups with the other olfactory stimulus. Scores were compared between the first and last days, and the children who were exposed to the same smell showed a significant improvement. No difference was observed for the control or switched smell conditions. Results were thus improved by associating a smell with an experience of success, leading to an increase in self-confidence (Chu 2008).

Summary and Practical Recommendations

A review of the scientific literature shows that, unfortunately, no miracle method exists for improving athlete recovery. What benefits a patient population may not necessarily have the same effects on a healthy, athletic population. Although some methods are beneficial, others remain controversial, or are even refutable.

Massages do not appear to affect blood flow or range of motion, and the increase in temperature induced is only superficial. They are therefore unsuitable for warming up before exercising. They do not speed metabolite elimination between exercise series, but by improving perceived recovery, they might increase investment and tolerance for subsequent efforts. Massages appear to help relieve low- to moderate-intensity pain, and contribute to a feeling of well-being and to psycho-emotional responses. By reducing neuromuscular excitability, they promote the reduction of cramps and stiffness. Finally, some authors observed a reduction in DOMS. However, this may not be an advantage in the long run, since the hypothetical mechanisms underlying this effect may hinder the process of cell regeneration.

Intermittent pneumatic compression seems effective in reducing edema and improving venous return, lymphatic drainage, and perceived recovery.

When the parameters used are sufficient to induce muscle contraction, electrostimulation increases blood flow in the same way as voluntary contraction. Although increased metabolite clearance has not been clearly demonstrated, recovery using electrostimulation appears comparable to active recovery. The analgesic effects of TENS-type electrostimulation are the result of segmental sensory inhibition (possibly combined with a placebo effect). The possible contribution of endorphins to the effects of electrostimulation in healthy subjects remains highly controversial.

To date, electrical stimulation has not been proven to be effective as a treatment for DOMS in terms of perceived pain, improved range of motion, and restored functional capacity.

Luminotherapy and aromatherapy are methods for which almost no scientific data relating to recovery and athletic performance are available. Based on results obtained with healthy subjects, we can assume that these procedures may directly or indirectly benefit athletes through their effects on stress, alertness, motivation, and even how information is treated in the brain.

Optimizing Recovery in Elite Synchronized Swimming

Christophe Hausswirth, PhD, National Institute of Sport, Expertise and Performance (INSEP)

With: Karine Schaal, François Bieuzen, Christophe Cozzolino, and Yann Le Meur

In high-level synchronized swimming competitions, athletes must perform several routines successively. Between technical and free routines, preliminary rounds, and finals, the time span available to recover between each performance is often very short (less than 1 h). The swimmers performing in the duo competition, in particular, are usually confronted with the most demanding schedule. Selected as the strongest competitors within their team, they often combine duo, team, and sometimes solo events within a single meet. Optimizing recovery between each routine is therefore crucial to maintain a high level of performance throughout the competition. Given the high metabolic demand of each routine, requiring maximal aerobic work output and an important anaerobic contribution, it is important to identify which of the various recovery techniques available to swimmers are most efficient in promoting the return of physiological systems to a normal resting level before the next event. These techniques should therefore be prioritized to enable the athlete to return to a high level of performance.

We compared the effectiveness of six different recovery techniques used by the two elite synchronized swimmers representing France in the duo competition. On six separate days, they performed their competition routine twice, evaluated by their coach. The 70 min recovery period between routines included a different technique each day:

1. Active recovery: 15 min of swimming at 40% of $\dot{V}O_2max$, then 15 min of light technical synchronized swimming exercises.
2. Whole-body cryotherapy (WBC, Zimmer Medizin Systeme, GmbH, ULM, Germany): 3 min at −110 °C.
3. Contrast water therapy (CWT): 7 cycles alternating 1 min in cold water (9 °C), and 1 min in warm water (39 °C).
4. Steam room: 2 periods of 10 min, separated by a 1 min cold shower.
5. Passive recovery: 30 min lying in a horizontal position.
6. Massage: 30 min, lower and upper extremities.

For both swimmers, WBC was associated with the largest decrease in blood lactate. The second most effective technique was CWT for one swimmer, and active recovery for the other. The passive and massage recoveries were associated with the least efficient [lactate] recovery.

When asked to rate their perception of each recovery's effectiveness prior to beginning the second routine, both swimmers rated the passive recovery as the least efficient, while the active and the CWT always ranked among the top 3.

Measuring changes in heart-rate variability (HRV) in a given athlete can be used as an indicator of recovery from autonomic stress following exercise. For both swimmers, the WBC and active recoveries were associated with the strongest return of HRV toward baseline values before beginning the second routine, suggesting an efficient recovery of the autonomic nervous system with these two methods.

The swimmers' ability to return to a similar aerobic work output was affected by the recovery method. While both swimmers showed lower $\dot{V}O_2peak$ values when the second routine was performed after the steam room, massage, and passive recoveries, they both reached higher $\dot{V}O_2peak$ values after WBC, CWT, and active recoveries (figure 1).

For almost all recovery modes, blood glucose and lactate followed the same evolution, peaking at the end of each routine (with lower peaks at routine 2 than routine 1), returning to baseline levels (or slightly lower) during recovery and 15 min after routine 2 (figure 2). WBC was the only

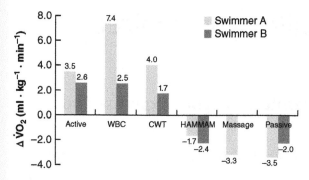

Figure 1 The difference in V̇O₂peak for each swimmer according to the recovery mode.

Reprinted from C. Hausswirth, *Optimizing recovery in elite synchronized swimming.*

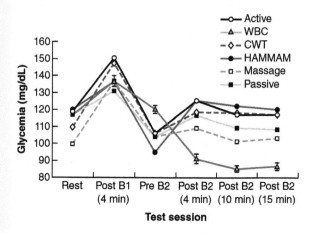

Figure 2 Evolution of glycemia (mg/dL) over the course of each test session. Mean values for both swimmers are presented.

Reprinted from C. Hausswirth, *Optimizing recovery in elite synchronized swimming.*

modality after which glucose actually decreased during routine 2 for both swimmers. This is due to the high rate of glucose release that accompanies the increased reliance on glucose as an energy source to fuel high-intensity exercise.

This could be explained by the thermogenic response to the intense cold treatment, boosting the overall metabolic rate in order to regulate internal temperature. Both swimmers indeed indicated the most negative thermal comfort ratings after WBC, followed by CWT. In this study, the swimmers did not take in any energy from food or fluids during the tests. Therefore, in a competition setting, adequate nutrition and hydration recovery strategies must be included. Even though in this specific context, both swimmers were able to increase V̇O₂peak following WBC, food intake may be especially important during recovery when this specific strategy is adopted, to offset the added energy expenditure induced by cold exposure.

WBC and CWT were the only two recovery techniques to consistently yield, for both swimmers, similar or lower ratings of muscle pain after routine 2. All other methods were associated with higher ratings. No trend appeared for the global rating of perceived exertion.

Even though some individual variations appeared in the response to these recovery methods, the trends that applied to both swimmers in the principal variables studied incited us to rank their overall effectiveness according to the following:

Physiological parameters (HRV, lactate recovery, δV̇O₂peak):

WBC > Active > CWT > Steam room > Massage > Passive

Perception/subjective parameters (perceived effectiveness and muscular pain ratings):

CWT > WBC > Active ≈ Massage > Steam room > Passive

From these results, it appears for these two highly and similarly trained athletes facing the specific metabolic demands and time constraints of synchronized swimming, the use of whole-body cryotherapy, contrast water immersion, and active recovery methods should be preferentially adopted over massage, passive, and steam room strategies.

Key Points

- After trialing different recovery methods, it is important to gauge their effectiveness and prioritize the recovery strategy.
- This should be done for both physical (objective) and mental (subjective) measures.
- Time is often limited in a competition setting, so the method that makes best use of the time available is a priority.

CHAPTER 11

Compression Garments

Antoine Couturier, PhD, and Rob Duffield, PhD

The ability to tolerate and recover from physical exertion during athletic activity is noted as an integral component of adaptation and performance enhancement (Barnett 2006). Media that assist postexercise recovery by improving skeletal blood flow in an attempt to improve muscle recovery are popular (Barnett 2006; Vaile et al. 2011). The training and competition demands of high-performance sport can contribute to chronic stress on the cardiovascular and musculoskeletal systems, consequently leading to deterioration of the venous system (Couzan 2006). Based on these demands, various manufacturers have developed compression garments for use by athletes to assist performance and to hasten postexercise recovery. Such technology has been widely adopted by athletes who expect substantial gains in terms of performance and quality of recovery in many sporting disciplines.

Physiological Responses to Compression

For areas located below the heart, arterial blood flow is ensured by contraction of the heart and hydrostatic pressure. Venous flow works against hydrostatic pressure when the body is in an upright position, so it is lower when standing than when lying down (Bringard et al. 2007). In addition, depending on the permeability of capillary walls and the pressure gradient between blood vessels and surrounding tissues, fluid may accumulate in the extravascular area, leading to edema (Bringard et al. 2007). In response to such physiological outcomes, several mechanisms can ensure the return of venous blood from the lower limbs to the heart. In the foot, the plantar venous plexus (Lejars' sole) is responsible for venous return, with each step sending a small amount of venous blood toward the calf, which acts like a pump for venous blood (see figure 11.1). When the muscles are relaxed, blood flows from the superficial veins through perforating veins, ending in deep veins. The antireflux valves in the perforating veins ensure that this process moves one way. When the calf muscles contract, venous blood is sent toward the deep popliteal and femoral network, from where it can be drawn toward the heart by the pumping action of the thoracic diaphragm (Bringard et al. 2007). Accordingly, compression therapy assists the compressive direction of venous flow. Medically speaking, it is used to help a venous system that is degraded by age and disease or a cardiovascular system in a weakened state (i.e., from disease or postoperative care) (Kahn et al. 2003).

From an athletic perspective, the demands on the musculoskeletal and cardiovascular systems during exercise are significant, and they require cardiodynamic adjustments to ensure sufficient cardiac output and ensuing venous return. According to Couzan (2006), most sports combine parameters favoring venous stasis such as the following:

- Imbalances between the heart's output and intake during endurance sports.

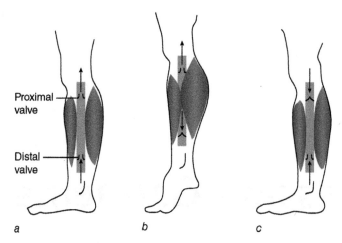

Proximal valve

Distal valve

a   b   c

Figure 11.1 Muscle action (pump effect) allowing venous return toward the heart. *(a)* Proximal and distal valves opened by venous blood flow returning to the heart. *(b)* Contracted calf muscles squeeze the intramuscular (twin veins) and intermuscular (popliteal vein) veins to move the blood toward the heart at high speed via the deep circulatory system. At the same time, reflux leads to closure of the valves downstream (distal) of the site of muscular vein compression. Venous blood flow is therefore directed only toward the heart. *(c)* When muscles relax, the inter- and intramuscular veins dilate, leading to a brief reflux that closes the (proximal) venous valves, preventing blood from flowing back. Muscle contraction empties the venous reservoir, thus reducing pressure in this compartment. This reduced venous pressure favors movement of blood from the arteries toward the veins, where pressure and volume are reduced. At the same time, blood is drawn from the superficial network through perforating veins and from the lower leg (foot).

- Sudden, extreme venous hyperpressure during impact sports.

- Blocked venous return during explosive or static efforts combined with blocked respiration or apnea.

- Slowed venous return caused by positions held for extended periods, associated with muscle contraction or specific compressing equipment.

- Hematoma and microthromboses caused by frequent and repetitive trauma, leading to vein dilation and parietal and valve damage.

In addition, endurance training increases venous compliance, predisposing athletes to varicose veins and venous insufficiency (Millet et al. 2006). Accordingly, it is evident that large adjustments are required during and following exercise to overcome the regulation of internal pressure and to maintain cardiovascular load.

The principle of compressive support is to artificially increase extravascular pressure and thus to assist the return of transmural pressure, or the difference between extra- and intravascular pressures, to values close to normal. External compressive support is expected to have multiple physiological effects, which may include the following (Couzan 2006; Perrey 2008):

- Dilation and venous stasis should be limited.

- Venous return and metabolite elimination should be improved by guiding superficial venous blood flow toward deep veins via perforating veins.

- Microcirculation should be improved.

- Excess fluid should pass more easily into venous circulation thanks to the increased interstitial pressure around capillaries.

- The muscular pump's efficiency should be improved.

- Longitudinal and anteroposterior oscillations due to shocks should be reduced. This will reduce alterations in neuromuscular transmission and optimize the recruitment pattern.

Compression garments may provide a varying range of superficial compression to the surface of the skin. For example, in France the Association Francaise de Normalisation classifies compression garments for venous insufficiency in four categories. This classification is based on the severity of the venous insufficiency to be treated: from light compression with about 10 to 15 mmHg at the ankle, which is recommended for slight venous insufficiency and heavy-feeling legs, up to extra-strong support, more than 36 mmHg at the ankle, which is prescribed for chronic, severe venous insufficiency and leg ulcers (Bringard et al. 2007). For classified vein diseases, decreasing pressure is applied from bottom to top, with maximum pressure at the ankle. Conversely, the pressure gradient is reversed, without cutting off blood flow, with the lower pressure on the foot and ankle and maximum pressure on the calf in athletes (Couzan 2006) (figure 11.2).

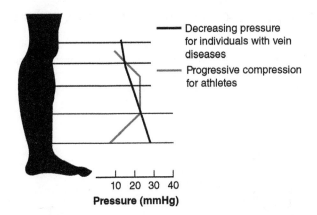

Figure 11.2 Pressure profiles for compression garments (mmHg).

Adapted, by permission, from S. Couzan, 2006, "Le sportif: un insuffisant veineux potentiel,"["The athlete: A venous insufficient potential"], *Cardio and Sport* 8: 7-20.

Effects During Exercise Performance

The use of compression garments to improve performance is popular among a wide range of athletes, and it is often promoted by respective manufacturers. Duffield and Portus (2007) compared the effects of three full-body compression garments (designed for use during sport) on cricket players' throwing ability (5 maximal and 6 precision throws, respectively) and performance during repeat sprints (20 m sprints over 30 min, interspersed with easy, moderate, and intense exercises every min). Results suggesting no significant differences were observed for the garments compared to a control condition. Fur-

ther, capillary blood lactate, pH, oxygenation, and perceived effort were also measured during the repeated-sprint test, without differences between the respective experimental conditions. Similar results have been reported for cycling exercise (Scanlan et al. 2008), simulated team sport circuits (Higgins, Naughton, and Burgess 2009; Houghton, Dawson, and Maloney 2009), and treadmill running (Ali, Caine, and Snow 2007). Specifically, Scanlan and colleagues (2008) used trained cyclists for an incremental test (measuring anaerobic threshold and $\dot{V}O_2$max), with or without compression garments (20 mmHg at the ankle, 17 mmHg at the calf, 15 mmHg at the thigh, and 9 mmHg at the gluteus maximus). Subjects cycled for 1 h at an exercise intensity close to the anaerobic threshold. Comparison with a control group revealed no significant differences for either mechanical (average power, peak power, total workload) or physiological (lactate, heart rate, $\dot{V}O_2$, muscle oxygenation) parameters.

Similarly, no beneficial effects were noted in a population of netball players during either power-specific tests (sprints or countermovement jumps, or CMJ) or match simulations (GPS measurement of speed and distances covered) (Higgins, Naughton, and Burgess 2009). Houghton and associates (2009) investigated the effects of compression shorts and T-shirts on thermoregulation during a field hockey match (4 × 15 min). They reported that no significant effect was evident for core temperature, sweating, heart rate, lactate, or perceived effort due to the garments compared to normal clothing. Finally, Ali and colleagues (2007) tested the effect of wearing support socks (20 mmHg at the ankle, 6 mmHg at the calf) during a 10 km race. Although differences were not statistically significant, heart rate and mean times tended to be lower for subjects wearing support socks (figure 11.3). This led the authors to hypothesize that venous return was slightly improved to explain performance trends. In addition, Ali and associates (2007) also previously used a sham compression garment involving elastic tights, which resulted in no physiological effects; although subjects did rate perceived muscle soreness as lower, suggesting some role of a placebo effect. Despite the lack of performance benefits noted from the aforementioned studies, other authors report beneficial effects of compression garment use. In a recent study, 21 athletes performed an incremental test to determine $\dot{V}O_2$max

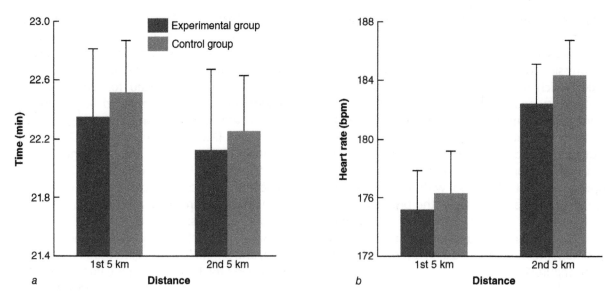

Figure 11.3   Effect of support compression on continuous running trials at 5 and 10 km. *(a)* Mean time; *(b)* mean heart rate.

Adapted, by permission, from A. Ali, M.P. Caine, and B.G. Snow, 2007, "Graduated compression stockings: Physiological and perceptual responses during and after exercise," *Journal of Sports Sciences* 25: 413-419.

on a treadmill with or without compression socks. Ten days later, the same athletes repeated the test, reversing the experimental conditions (i.e., without or with support socks). Exercise time, total workload, and performance at aerobic and anaerobic thresholds were all significantly higher for athletes wearing support socks (Kemmler et al. 2009). Conversely, Goh and colleagues (2011) recently reported no effect of lower body compression garments on run to exhaustion times at velocity at $\dot{V}O_2$max in either hot (32 °C) or cold (10 °C) environmental temperatures.

Although compression garments were initially designed to improve venous return, their most interesting effects on performance may be mechanical and neuromuscular, rather than cardiodynamic in nature. By comparing the effects of compression shorts, elastic tights, or normal shorts on oxygen consumption at different intensities, Bringard and associates (2006) reported that, in the absence of differences in $\dot{V}O_2$max, the energy cost at 12 km/h was lower in athletes wearing compression shorts, and that the $\dot{V}O_2$ slow component was 26% or 36% lower compared to those wearing elastic tights or normal shorts, respectively (figure 11.4). The improvement in energy cost was attributed to better muscle coordination and greater propulsive force,

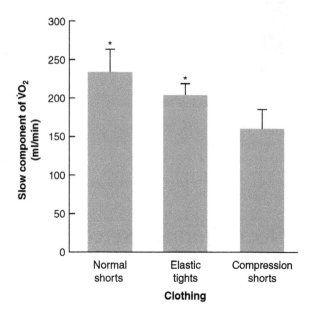

Figure 11.4   Mean (± SE) oxygen uptake amplitude, calculated as the difference between minutes 2 and 15 during constant heavy running exercises (80% $\dot{V}O_2$max) for three different clothing conditions: normal shorts, elastic tights, and compression shorts.

*Significantly different (P < 0.05) from compression shorts condition (n = 6).

Adapted, by permission, from A. Bringard, S. Perrey, and N. Belluye, 2006, "Aerobic energy cost and sensation responses during submaximal running exercise—positive effects of wearing compression tights," *International Journal of Sports Medicine* 27: 373-378.

due to reduced muscle oscillation and optimized neuromuscular transmission. In a review of the literature on elastic garment support, these same authors report similar results, including higher maintenance of jumping power, greater strength and power development, reduced muscle oscillation during repeated vertical jumps, and reduced electromyographic activity for the flexor muscles of the leg during sprint exercises (Bringard et al. 2007). Previous research has outlined that maximal efforts involving repeated peak power may be improved by the use of compression clothing (Kraemer, Bush et al. 1996), although such findings are not uniform to all studies (Doan et al. 2003).

Regardless, these benefits are proposed to result from improvements in the stretch-shortening cycle due to the superficial elastic support surrounding the muscle and tendon. In this case, the benefit is due to the additional external "spring" effect on the muscle during contraction, rather than to muscle compression (Kraemer, Bush et al. 1996; Doan et al. 2003). In summary, there is a mixture of evidence as to the ergogenic benefits of compression clothing during exercise. While some benefits may be present for explosive speed and power movements, such outcomes are not likely due to any physiological alteration in venous return. Conversely, few studies highlight compression garment benefits for performance or physiological responses during prolonged exercise performance. As such, the benefits to improve performance may be minimal.

Compression and Recovery

While performance outcomes are often the central focus for any athlete or sport scientist, compression garments may have greater advantages to improve recovery following an exercise bout (Duffield, Cannon, and King 2010). However, to date, the evidence remains equivocal, since the two known studies comparing compressive support to other recovery methods report varying conclusions (Gill, Beaven, and Cook 2006; Montgomery, Pyne, Hopkins et al. 2008). During a 3-day basketball tournament, 29 players were assigned into three recovery groups: carbohydrate intake and stretching, cold-water immersion, and compression stockings (mean pressure 18 mmHg). Comparison of pre- and posttournament tests (sprints, agility test, vertical jumps)

suggested that cold-water immersion provides the best recovery, followed by carbohydrate intake combined with stretching, and finally compression stockings (Montgomery, Pyne, Hopkins et al. 2008). However, Gill and associates (2006) compared the recovery (up to 3 days postmatch) of 23 rugby players using contrasting temperature immersion, compression garments, active recovery, and passive rest after a competitive match. Passive rest was reported to be less effective than the other three methods, which were all comparable in the rate of recovery of postmatch creatine kinase (CK) release. Accordingly, despite a lack of performance markers, Gill and colleagues (2006) concluded that compression therapy (as well as immersion and active recoveries) were effective at enhancing the rate of recovery of CK expression following exercise resulting in microdamage.

Recent studies on the effects of compression garments to assist recovery following exercise also report mixed findings. The effects on recovery of compression garments worn during exercise and up to 24 h after exercise were investigated in several studies. The benefit of wearing lower body compression garments (Duffield, Cannon, and King 2010) and full-body garments (Duffield et al. 2008) were assessed in training conditions (repeated sprints for 20 m, plyometric jumps, scrum machine, match simulations) at 24 h intervals in a population of team-sport athletes. No differences in performance (figure 11.5), maximal strength, lactate accumulation, pH, heart rate, or markers of muscle injury were observed compared to a control group, during or after recovery. In contrast, perceived pain and fatigue were significantly lower when the rugby players wore compression garments compared to normal clothing.

In another study involving cricket players in a cold environment (15 °C) (see previous section, Effects During Exercise Performance, for a description of the protocol), tests carried out after 24 h recovery revealed similar lack of differences, apart from higher skin temperatures, lower levels of markers for muscle injury, and perceived pain. In this case, wearing compression garments during the recovery phase may have assisted blunt markers of muscle damage and improved perceived recovery of soreness (Duffield and Portus 2007). These researchers also highlight a potential role for compression garments as thermal insulators in cold conditions, or when exercises are

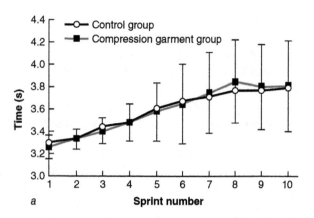 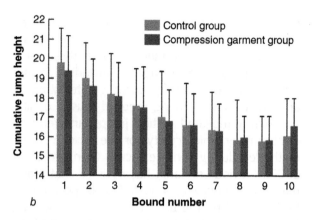

Figure 11.5 Comparison of compression garment and control conditions for *(a)* 20 m sprint time (mean ± SD) and *(b)* total distance (mean ± SD) for 10 plyometric jumps performed during a 10 min exercise protocol.

Adapted from *Journal of Science and Medicine in Sport,* Vol. 13(1), R. Duffield, J. Cannon, and M. King, "The effects of compression garments on recovery of muscle performance following high-intensity sprint and plyometric exercise," pgs. 136-40, copyright 2010, with permission from Elsevier.

interspersed with long resting periods. However, Goh and colleagues (2011) found that compression garments in both hot and cold conditions resulted in higher skin temperatures, without differences in core temperature or ensuing effects on treadmill running performance. Accordingly, while the information on the effects of compression garments on recovery is still mixed, some benefits may be present. Although few studies outline improved recovery of performance, some evidence for reduced markers of metabolism and damage indicate potential benefits. Further, perceptual recovery following competition or training may also be assisted by use of postexercise compression.

Anaerobic Metabolite Clearance

The effect of compression garments on lactate accumulation as a marker of anaerobic clearance also remains equivocal. Berry and McMurray (1987) investigated the use of compression socks, initially designed for patients with venous insufficiency (18 mmHg at the ankle, 8 mmHg at the calf), on healthy subjects during exhausting cycling exercise. Lactate accumulation was systematically lower during recovery when compression socks had been worn during both exercise and recovery, as opposed to solely during exercise. However, there was no difference between subjects who wore compression socks or those who didn't on ensuing recovery. This suggests that some of the influence of compressive clothing existed both during and following exercise. Further, because

plasma volumes remained unchanged, the authors concluded that compression socks reduced lactate diffusion out of the muscle after exercise. In other words, lactate is retained in the intracellular matrix, and the inverted pressure gradient provided by the support may assist lactate efflux. Conversely, the same authors tested a similar protocol using elastic sport garments during running exercises on a treadmill (Berry et al. 1990). Lactate accumulation, hematocrit, oxygen consumption, and heart rate measured before and during exercise and during recovery revealed no differences among the three conditions. In this case, it would seem that the pressure exerted by the commercially available compression garments was insufficient to obtain significant results. This appears to be confirmed by recent studies reporting physiological recovery following exercise (Duffield et al. 2008, Duffield and Portus 2007).

Delayed-Onset Muscle Soreness (DOMS) Response

The studies specifically analyzing the effect of compression garments on muscle stiffness include one or more exercises with an eccentric component that is intense enough to cause muscle microinjury (Maton et al. 2006). Subjects used either compression garments or other recovery methods to investigate the benefits of compressive therapy to blunt exercise-induced DOMS response. However, as compression garments were not worn during exercise, no information was provided on how effective they are in preventing DOMS. Specifically, following 30 min downhill walking (25% slope) carrying a rucksack (12% of body weight) 8 subjects wore compression socks on one leg 5 h per day for 3 days, while the other leg served as a control. Although the exercise was perceived as low intensity by subjects, it was sufficient to induce muscle injury, as determined by decreased voluntary force, voluntary activation, and peak twitch force (Perrey et al. 2008). Moreover, wearing compression socks reduced pain at 72 h postexercise (figure 11.6) and restored neuromus-

cular function at a faster rate than without the garments (Perrey et al. 2008).

Similarly, beneficial effects were also reported following intense exercises involving lower (5 series of 20 step jumps) (Davies, Thompson, and Cooper 2009) and upper limbs (2 series of 50 maximal eccentric biceps contractions, including 1 maximal concentric contraction every 4 repetitions) (Kraemer et al. 2001). While wearing compression clothing did not improve recovery of muscle contractile function, levels of muscle injury markers, perceived pain and swelling, and joint flexion were all lowered. Kraemer and colleagues (2001) suggest that compression clothing may provide external pressure, limiting movement and preventing excessive damage to the contractile fibers and essentially assisting to reduce the inflammatory response. However, current evidence to support this finding remains equivocal (Maton et al. 2006; Duffield, Cannon, and King 2010). Furthermore, perceived pain has been reported to be lower following high-intensity exercise when compression garments have been worn (Duffield, Cannon, and King 2010; Duffield and Portus 2007), with trends for reductions in the inflammatory response also reported in one (Duffield and Portus 2007). However Ali and associates (2007) had previously reported that elastic tights may also impart similar perceptual benefits. Regardless, in line with previous propositions, compression garments may reduce the constraints involved in eccentric contraction by providing muscle support (Kraemer et al. 2001). Accordingly, the reduction in the direct effects of load bearing skeletal contraction may help reduce the ensuing accumulation of DOMS responses. As such, it may be of use for recovery for compression garments to be worn during exercise as well.

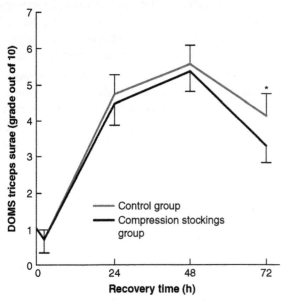

Figure 11.6 Rating of perceived delayed-onset muscle soreness (DOMS) following an exercise protocol to compare graduated compression stockings and control conditions (mean ± SEM).

*Significant difference between the two conditions (P < 0.05).

Adapted, by permission, from S. Perrey et al., 2008, Graduated compression stockings and delayed onset muscle soreness (ISEA Conference, Biarritz). In *The Engineering of Sport* 7, vol. 1, edited by M. Estivalet and P. Brisson (Paris: Springer), 547-554.

PRACTICAL APPLICATION

Compression garments must be adapted to individual needs and activity (exercise or recovery). A progressive pressure profile, rather than a degressive one, is preferable.

- Regarding performance, the supporting role of compression garments may improve the energy cost. However, their effects on the cardiovascular system appear to be minimal.

- Regarding recovery, compression garments have been shown to enhance perceived recovery. In comparison with other methods, they appear to be better than passive recovery.

Summary

Based on the present array of results reported in the scientific literature, it is difficult to find unequivocal findings regarding the performance and physiological benefits of sport compression garments for athletic recovery. Compression garments designed for a healthy athletic population must be adapted not just for individual needs, but also to the activity performed (exercise or recovery). In addition, compression garments with progressive, rather than degressive, pressure profiles may be more suitable for athletic populations. Most medical compression garments have a degressive pressure profile, and as such, they may be inappropriate for athletic populations. However, most sport compression wear has reduced pressure gradients and may not provide sufficient compression to induce physiological adjustments.

However, the following conclusions may be possible:

- In terms of performance, effects on the cardiovascular system are minimal, although the supportive role of compression clothing may improve the energy cost and efficacy of the muscle's stretch-release cycle.

- In terms of recovery, medical support (applying higher pressure than support designed for sporting activity) may have a negative effect by preventing metabolite diffusion out of the muscle. While athlete perception of recovery may be improved with compression garment use, skeletal-muscle function does not appear to be improved by compression garments.

- When compared with other recovery methods, compression is less effective or comparable to immersion, active recovery, or consumption of carbohydrate combined with stretching, but it may be better than passive recovery.

Recovery Between Races During Canoe Sprint World Championships

François Bieuzen, PhD, National Institute of Sport, Expertise and Performance (INSEP)

Every year in July or August, the Canoe Sprint World Championships occur. The world championships in 2011 were important for two reasons. First of all, more than 85% of Olympic quotas for London 2012 were allocated from the results of these championships. Second, the format for male contenders has changed. The International Canoe Federation removed the 500 m race and replaced it with a 200 m race for the 2012 Olympic Games. As a consequence, the 200 m, formerly a recreational distance, has now become a major event for most countries, inducing changes in warm-up and recovery strategies because of the specific demands of this discipline. Indeed, the duration of a 1000 m run for an elite athlete in K1 (kayak with one athlete) is generally between 3:30 and 3:40 min, whereas a 200 m run lasts only about 35 s.

Consequently, the French national team decided to specialize one of their athletes on the 200 m men's K1. Previously, this athlete was a specialist 1000 m paddler, but he regularly participated in 200 m races. In April 2011, the decision was made to change his training structure and programs and also to adjust his recovery to the new requirements of the discipline. Several major changes have thus taken place.

In the past, his recovery strategy involved an active recovery session followed by massage. The active recovery took place immediately after the run and lasted between 5 and 20 min. The intensity was low and close to the aerobic threshold. Thirty to 40 min later, an experienced physical therapist took charge of the athlete for a massage lasting 1 h.

At present, his recovery strategy is organized and adapted to the demands of a 200 m run. The new strategy is more focused on the effect of all-out exercise lasting 30 to 40 s, such as the increase of metabolic by-products. Table 1 shows a comparison of the recovery strategies following a 1000 m event and a 200 m event.

The main objective of the present strategy is to accelerate the clearance of the metabolic by-products from the blood, by means of stimulation with an electromyostimulator device combined with a forearm compression garment, and to quickly restore the energy levels with appropriate fluid containing CHO. Active recovery (whose effectiveness has been widely demonstrated) has been dropped to avoid delaying the resynthesis of muscle glycogen. In hot environmental conditions, a cooling jacket is also extensively used. Figure 1 shows an image of the athlete using these recovery methods.

Key Points
- It is important to adjust the recovery methods to suit specific events within the sport.
- This requires adopting and experimenting with new recovery techniques and equipment.
- Different strategies may need to be prioritized.

Table 1 Comparison of Recovery Strategies Following 1000 m and 200 m Events

	Timing	Recovery strategy
1000 m (before April 2011)	0 min to 5–20 min	Active recovery
	30–40 min to 90–100 min	Massage (upper and lower body)
	0–60 min	Hydration (Water + CHO + Na$^+$)
200 m (from April 2011 on)	0–20 min	Electromyostimulation with Veinoplus sport device
	20–60 min	Massage (upper and lower body)
	0–60 min	Hydration (water + CHO + Na$^+$)
	0–next run	Cooling jacket (Cryovest)
	0–next run	Custom-made forearm compression garment

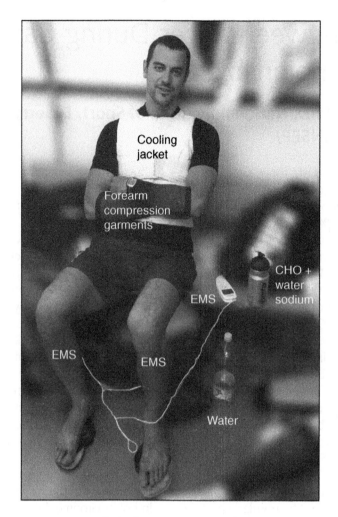

Figure 1 Athletes may use many recovery strategies.

Photo of French national kayak team member Maxime Beaumont, courtesy of François Bieuzen.

CHAPTER 12

Local Thermal Applications

Sylvain Dorel, PhD

With: Cédric Lucas and Lorenzo Martinez Pacheco

Local thermal applications, particularly cold applications, are generally used for the clinical treatment of soft-tissue injuries. Although ice remains one of the most widely used means of application, some other options exist today. In addition to their known positive effects as part of therapy, it is reasonable to suppose that these techniques may have positive effects if used as part of a prevention strategy. Application of local cryotherapy could help improve recovery by reducing the symptoms linked to the appearance of delayed-onset muscle soreness (DOMS), restoring the functional capacity of the muscles used more rapidly, and reducing the risk of injuries linked to muscle microlesions. Therefore, this chapter attempts to show that these techniques may be transferrable to the more general context of recovery in top-level performance in sport. We will present the different local thermal application techniques for muscles (and tendons). We will describe the current knowledge about the physiological point of view, which may or may not justify their use, and the latest scientific information for each technique. We will also illustrate the procedures themselves with some practical examples applying each of the techniques, and will provide a certain number of recommendations for their appropriate use.

Physiological Responses to Local Thermal Applications

Cold and heat applied locally to the muscles lead to almost exactly opposite physiological responses. The main local effects of cooling are a decrease in metabolic activity, capillary vasoconstriction (followed as a rebound effect by vasodilation when the source of cold is removed), an anti-inflammatory effect characterized by a lower level of edema, a drop in nerve conduction (analgesic effect: reduced pain sensation), and a reduction in the compliance of the muscle (i.e., increased stiffness). Local heat application leads to an increase in metabolic activity and vasodilation (increase in the supply of oxygen, nutrients, and antibodies), a drop in activity of the neuromuscular spindles, and reduced sensitivity when stretching. This chapter also covers the specificity of local applications compared to other cold-application techniques (particularly immersion). However, it is recommended that the reader refer to the appropriate section for full coverage of the various techniques.

Currently, the methods for local cold application are becoming more diverse. Some of the main ones are as follows:

- Ice, applied directly or through a wet cloth or in an ice pack. This technique is widely used.
- Cryogenic gel packs (or thermal cushions) and instantaneous cold packs containing water and a particular type of salt.
- Gas (CO_2 microcrystals) or pulsed cold air as a spray, aerosol, or with a more elaborate compressor-based device.
- Cooling vests containing ice or cryogenic gel packs. This technique is more specifically used as part of recovery or when repeated efforts are to be performed in conditions of significant thermal stress (high ambient temperature, increase in core temperature, and so on).

We will see that some of these methods are frequently associated or combined with other techniques (particularly compression, massage, or alternate use with heat). In these cases, it is only possible to assess the cumulated effects, as opposed to those resulting from cold alone.

In contrast to cold, local heat application is much less used in practice, and the scientific data relating its effects remain limited. The source of heat can be created by means of a damp cloth, a heat pad (or hot compress), or by substances that have been heated, such as clay, algae, or paraffin.

All these techniques induce mainly local physiological effects. However, some of them also lead to core adaptations (particularly in terms of core temperature or heart rate). To better understand how the most commonly used methods provide benefits in the context of recovery, the current section studies whether (or not) they would influence the following physiological and functional variables: skin temperature, muscle temperature (or core temperature), muscle soreness or pain, and processes linked to the appearance of local inflammation. Finally, we will then move on to the functional response (up to the performance in sport).

Effects on Skin and Muscle Temperature

Local applications induce significant physiological responses through action on the (deep) intramuscular temperature in the zone treated. Indeed, although skin temperature is an indicator of the thermal effect, it does not always reflect the influence of the application at the deeper level within the muscle. A target skin temperature of around 10 °C to 15 °C is generally accepted as necessary in order to induce significant physiological effects of the cold at the muscle level (MacAuley 2001).

Direct Ice Application

The amplitude of the response is mainly affected by the method, duration, and frequency of applications. Direct application of ice to the skin through a damp cloth appears to be one of the most effective and least dangerous methods for the skin (in terms of the risk of burns) (MacAuley 2001). The use of a cloth containing ice cubes or crushed ice (i.e., humid cold) results in a 12 °C drop in temperature in 15 min (Belitsky, Odam, and Hubley-Kozey 1987). In the same study, with a plastic bag containing ice (drier cold) or a cryogenic gel pack, the decrease is only of 9.9 °C or 7.3 °C, respectively. In line with these observations, using a dry interface between the skin and the source of cold, Dykstra and colleagues (2009) showed that a small bag filled with ice and water leads to a slightly greater reduction in intramuscular temperature than a small bag with ice cubes only, and to a much greater reduction than a crushed ice pack (both during and after the application phase).

The drop in skin temperature is relatively important in the first 10 min, but it slows down over the next 10 min. Current recommendations therefore indicate an optimal application time of 15 to 20 min. This is for two main reasons: This duration allows a sufficient drop in skin (and muscle) temperature and prevents the negative effects of the application, particularly those linked to nerve lesions within the skin, which may be caused by much longer local cold applications. To amplify or maintain the drop in intramuscular temperature, it is possible to repeat 10 to 15 min applications at regular intervals. This allows the superficial temperature to return to acceptable levels, while continuing to act on the deep temperature over a longer period (MacAuley 2001).

The size of the contact area between skin and the ice pack and the quantity of ice both influence cooling effect. However, in a recent study, the amount of ice was shown to have a greater effect on the magnitude of the drop in temperature (Janwantanakul 2009). This author therefore suggests that ice packs weighing at least 0.6 kg provide the best results for reducing skin temperature and, consequently, intramuscular temperature.

An important parameter that is more difficult to take into account is the insulating effect of the subcutaneous adipose tissue (Hocutt et al. 1982; MacAuley 2001). Studies have, nevertheless, clearly shown significantly greater drops in intramuscular temperature, after 20 min application of a crushed ice pack, in populations with a thinner adipose layer (<8 mm vs. >20 mm) (Myrer, Draper, and Durrant 1994; Myrer et al. 2001). Classically, the deep intramuscular temperature (3 cm below the adipose tissue) continues to drop for 10 to 20 min after the application has been stopped. Interestingly, this additional decrease is significantly faster in subjects with a thinner adipose layer. Thus, whether taken at a depth of 1 or 3 cm, the muscle temperature remained significantly lower 30 min after treatment for this population (–9.04 °C and –7.35 °C as a depth of 1 or 3 cm, respectively) compared to subjects with a larger subcutaneous fat reserve (–5.48 °C and –4.52 °C) (Myrer et al. 2001). Because of these observations, despite the lack of consensus on all the practical aspects of the application, it seems important to take into account this fat mass in the part of the body being tested when deciding on the protocol to reduce the muscle temperature.

Ice Massage

One alternative technique consists of massaging with ice cubes or an ice pack rather than simply applying the pack to the skin. This method increases the contact between the ice and the skin and, therefore, the conductivity. It reduces local blood flow, thus limiting reheating (MacAuley 2001; Waylonis 1967). This method could therefore be of interest in preventing acute trauma (such as a joint sprain) and for recovery from exercises that cause localized muscle damages, such as heavy eccentric exercises inducing significant microlesions (especially if this is unusual for the athlete). In addition, it has been shown that compression associated with cold, even though this is quite similar to the basic ice application regarding the final temperature reached, nevertheless leads to a much more rapid drop in intramuscular temperature (–5 °C in 18 min with compression vs. –5 °C in 28 min without compression) followed by maintenance of this temperature (Zemke et al. 1998).

Cold-Water Immersion

Another means to locally apply cold to some parts of the body (e.g., leg or forearm) is to immerse the specific zone in cold water. Compared to application of a pack of crushed ice, immersion in cold water at 10 °C for 20 min leads to a slightly lower drop in intramuscular temperature after this treatment: in the range of –5 °C to –7 °C, although the difference is not very significant (Myrer, Measom, and Fellingham 1998). However, these authors have shown that coming back to the initial temperature after cooling is quicker with crushed ice (reheating starts 5 min after the end of the application) than with immersion (an additional drop of 1.8 °C occurs over the 30 min following immersion). Full details about the immersion technique are presented in chapter 14.

In several cases, we have seen that the drop in intramuscular temperature continues for several minutes after the end of the application. However, it appears important to maintain relative inactivity of the part of the body treated in this posttreatment period if this effect is to be obtained. Indeed, even very moderate exercise (e.g., walking at 5.5 km/h) after a 20 min application of ice on the triceps surae causes an almost complete return to normal intramuscular temperature after 10 min (Myrer, Measom, and Fellingham 2000) (figure 12.1). Because of time constraints, this period of inactivity is not always easy to adhere to. Therefore, an alternative is to use ice packs, which can be attached to the area to be treated ("to-go ice bag") to allow athletes to continue their daily activity. The expected advantage of this approach is that it extends the duration of application and also promotes the positive effects of the application through the compression induced by the ice pack. Despite permitting a drop in skin temperature, this technique is ineffective in reducing the deep intramuscular temperature of the triceps surae in athletes who continue walking during the application (Bender et al. 2005). This technique could, however, be of benefit if the application involved an almost inactive muscle group (e.g., upper body muscles) during the recovery phase.

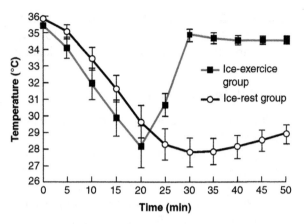

Figure 12.1   Progression of intramuscular temperatures during ice application (20 min) and over the 30 min following the application for two groups: one observing a rest period (ice and rest), the other resuming moderate walking activity over the 10 min following the application (ice and exercise).

Reprinted, by permission, from J.W. Myrer, G.J. Measom, and G.W. Fellingham, 2000, "Exercise after cryotherapy greatly enhances intramuscular rewarming," *Journal of Athletic Training* 35: 412-416.

Hot–Cold Contrast

Finally, the method alternating applications of cold and heat (known as *hot–cold contrast method*) does not have the same effects. In contrast to the one-shot application of ice for 20 min, alternating 20 min applications of a hot gel pack (5 min at 75 °C) and an ice pack (5 min) does not have a real effect on intramuscular temperature (Myrer et al. 1997) (figure 12.2). The efficacy of this technique

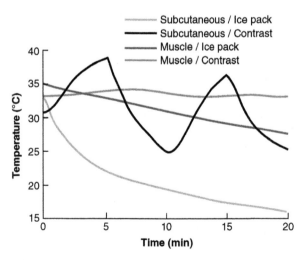

Figure 12.2   Progression of skin and intramuscular temperatures over 20 min treatment with ice or a method contrasting heat and cold.

Adapted, by permission, from J.W. Myrer et al., 1997, "Cold- and hot-pack contrast therapy: subcutaneous and intramuscular temperature change," *Journal of Athletic Training* 32: 238-241.

can hence be questioned, since the desired physiological effects require a significant fluctuation in deep temperature. On this topic, techniques using contrasting temperatures are more commonly based on immersion (see chapter 14 for more on immersion recovery). Contrasting water immersion is a procedure for which the benefits, based on the vaso-pumping phenomenon induced by successive vasoconstriction and vasodilation of blood vessels, have been experimentally demonstrated.

Cryotherapy With Pulsed Cold Air

A new technique for local cold application has emerged in recent years: cryotherapy using cold air or a refrigerated gas (also known as *neurocryostimulation*). This method consists of pulsing cold air or gas (generally CO_2) between –30 °C and –80 °C, at a high intensity and pressure, in dry conditions, directly onto the skin over the muscle to be treated. This convection method cools the tissues to a greater extent than more common techniques. A study by Mourot and colleagues (2007) confirms that this procedure leads to a very significant and rapid reduction in skin temperature. For example, a 2 min application of cold gas (CO_2 at –78 °C and 75 bars) causes a 26 °C decrease in skin temperature of the hand (down to 7 °C), while an ice pack application (15 min) induces a temperature drop of only 19 °C. After cooling, the early rise in temperature is faster after projection of cold gas (+ 6 °C) compared to ice (+ 3 °C). However, temperatures rise to an identical 21 °C at 17 min in both cases (2 min after application of ice and 15 min after application of cold gas). Thereafter, progression is similar.

Given the recent emergence of this technique, little scientific and medical evidence of its utility is available; therefore, its effect on deep temperature remains to be confirmed. However, it is interesting to note that applying it to the back of the hand, as reported by previous authors, leads to a significant drop in the skin temperature measured in the palm of the same hand. The extent of the drop is similar to that obtained with immersion in cold water. Although the adipose layer is not thick in this area, it is reasonable to assume that pulsed cold air can cool deep tissues. However, since the cooling time at skin level is quite short (because of the low duration of application and the speed of reheating), it remains to be established whether pulsed cold air allows a (sufficiently durable) drop in intramuscular temperature.

The heat shock induced by rapid variations in temperature is one of the main effects sought when using this type of technique. This shock also leads to central reactions: increase in systolic and diastolic blood pressure (indicating extensive vasoconstriction), combined with quick ortho-sympathetic activity, followed by an increase in cardiac parasympathetic activity on return to normal temperature (Mourot, Cluzeau, and Regnard 2007). This effect is interesting because the reactivation of the parasympathetic nervous system after exercise (which could be examined through study of heart-rate variability) is currently considered to have a positive influence on athletes' recovery capacity. Although the experimental evidence for this effect of pulsed cold air on postexercise parasympathetic activity remains to be established, other studies using local cold applications just after exercise (e.g., immersion of the face in cold water) have recently confirmed that it induces an accelerated reactivation of the parasympathetic system (Al Haddad, Laursen, Ahmaidi, et al. 2010).

Effects on Muscle Pain, Soreness, and Inflammation

Some exercises produce significant local mechanical constraints that may lead to muscle pains over the following hours and days. This delayed-onset muscle soreness (DOMS) is symptomatic of a certain number of physiological changes in the muscle cells and interstitial fluid: inflammatory processes, edema formation, muscles-cell lysis, and so on. We will now address the question of how local thermal applications can positively influence various nervous (pain sensations) and muscular and blood criteria, which all represent indirect indicators of the extent of the DOMS phenomenon.

Local Cold Application and Recovery of Muscle Damage and Soreness

It has been clearly shown that vasoconstriction and reduced metabolic activity lead to reduced tissue swelling, inflammation, immediate pain, and extent of the injury (Enwemeka et al. 2002). Thus, the experimental proof exists: Local, rapid, and prolonged application of cold after muscle injury reduces cellular destruction by leucocytes and improves the flow of nutrients in the tissue (Schaser et al. 2006). In the case of acute treatment of musculoskeletal trauma, cold is a valid means to improve cell survival after local hypoxia

induced by the inflammatory process and formation of edema (Merrick et al. 1999; Swenson, Sward, and Karlsson 1996).

Although we have seen that local cryotherapy, used at appropriate intervals or for relevant durations of application, has an effect on skin and muscle temperatures, its actual influence on DOMS remains to be shown. Currently there is no irrefutable scientific evidence demonstrating a real benefit of cold application in the prevention and treatment of DOMS beyond an analgesic effect, whatever the duration of application (Cheung, Hume, and Maxwell 2003). Unfortunately, no consensus has been reached on this topic. Some authors have studied treatment with ice (or combining ice and exercise) after exhausting concentric/eccentric exercise involving elbow flexion, and have shown that it provides no benefit in terms of perceived pain and blood creatine kinase (CK) concentration (a marker historically used as an indicator of muscle damage, which is today judged overdependent on numerous central variables) (Isabell et al. 1992).

In parallel, no difference in perceived muscle soreness was observed, either immediately, 20 min, 24 h, or 48 h after exhausting exercise followed by a 15 min massage with ice (Gulick et al. 1996; Yackzan, Adams, and Francis 1984). In their review, Cheung and colleagues (2003) indicate that the level of experience does not appear to have a major effect on this absence of results. This partially contradicts the belief that highly trained athletes are more resistant to DOMS induced by eccentric exercise and, thus, have a greater positive response to cold application.

Consequently, given the positive effects of cold as a treatment for acute injury or trauma, this suggests that muscle soreness is associated with a different or less extensive inflammatory process. In this way, we can reasonably hypothesize that local cold application may have a positive effect as a recovery process from exercises that are locally traumatizing or from those that induce significant muscle injury, an extensive inflammatory reaction, and formation of edema (e.g., heavy eccentric or plyometric exercises). One of the most complete studies on this topic is that of Howatson and associates (2005), which investigated the influence of a 15 min ice massage after traumatizing (inducing muscle damages) eccentric exercise of the forearm flexors. No beneficial effects of the treatment were observed for any variable measured over the following hours and days (perceived soreness; figure 12.3),

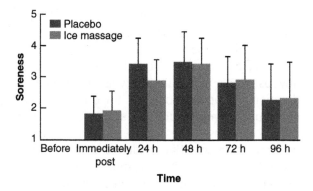

Figure 12.3 Perceived soreness immediately, 24 h, 48 h, 72 h, and 96 h after performing eccentric exercises with the forearm flexors without treatment (placebo) or following ice massage. Rated on the Talag Scale, a subjective visual analogue scale rating from 1 (no soreness) to 7 (unbearably painful).

Adapted, by permission, from G. Howatson, D. Gaze, and K.A. van Someren, 2005, "The efficacy of ice massage in the treatment of exercise induced muscle damage," *Scandinavian Journal of Medicine and Science in Sports* 15: 416-422.

circulating CK concentration, data for isokinetic strength at several speeds, or range of motion—all parameters corresponding to signs or symptoms of DOMS. These authors concluded that the beneficial effects of cold in the treatment of acute injury are not easily transferable to the field of muscle recovery, even in the case of recovery from traumatizing eccentric exercise. However, in this study, only a single application of ice was used.

Intracellular pH

Measurement of pH is an interesting way to prove that intense muscular exercise may affect homeostasis. An intramuscular increase in H^+ concentration due to intense muscular activity (particularly related to significant glycolytic metabolism) leads to a reduced intracellular pH. This phenomenon is generally considered to be responsible for the drop in performance, or delayed muscle recuperation after exercise. In animals, cold application has been shown to significantly increase intracellular pH (Johnson et al. 1993). But can these results be transferred to recovery after traumatizing exercise in humans? A recent study by Yanagisawa, Niitsu, and colleagues (2003) examined the effects of local cold application after traumatizing eccentric exercise of the plantar flexors on inflammatory response indicators, intramuscular pH, and muscle soreness. Although they did not detect a significant effect of cold on this last variable throughout the recovery period, the results indicate that

cold application has a positive effect on pH (i.e., a higher pH measured 30 and 60 min after exercise) and on edema development (i.e., reduction in swelling 48 h after exercise). Thus, this partially confirms what we stated previously: Procedures involving local cold application appear to be beneficial in the specific case of very traumatic localized exercise, which may lead to muscle injury and a significant inflammatory response.

Circulating Biomarkers

Contradictory results have been obtained for circulating biomarkers. Nemet and associates (2009) recently examined the effect of applying an ice pack (for 2 × 15 min) on pro- and anti-inflammatory responses (cytokines) and on levels of anabolic and catabolic mediators after exercise involving repeated sprints. They show a faster reduction of the pro-inflammatory mediator IL-1β, no effect on IL-6 (confirming results from previous studies), and an even faster reduction in the anti-inflammatory mediator IL-1ra compared to a control group not receiving cold treatment. In addition, cold application leads to a much faster reduction in levels of anabolic hormones (IGF-I and IGFBP-3) linked to a higher increase in a catabolic agent (IGFBP-1) during the recovery phase after the training session.

Their results can be summarized as follows: Local cryotherapy applied just after interval-training sprint exercises reduces pro- and anti-inflammatory cytokines and anabolic hormone levels (thus reducing the anabolic effect of the previous training). Therefore, these results corroborate studies that showed cold to sometimes have a deleterious effect on sporting performance (see next section), and others indicating no actual effect on muscle injury and DOMS existed. The effects of chronic cold applications (i.e., several times per week over several weeks) on muscle damage or pain, whether induced by repeated traumatizing exercise or not, have not yet been clearly established.

This is particularly linked to the fact that studies in humans are scarce. In a population of endurance-trained rats, it has been shown that immersion in cold water after each exercise led to more extensive muscle injury at a faster rate (Fu, Cen, and Eston 1997). In fact, the authors showed that repeated cold applications during protocol involving daily exercise leads to reduced perceived pain and edema formation, thus masking the usual physiological signals alerting of muscle injury. In this case, cryotherapy could be dangerous in the medium term, allowing repetition of exercises that are overly traumatic for the muscles and, consequently, would not improve muscle-fiber regeneration. This raises the question of appropriate dosage, particularly regarding the time between cold application and a return to exercise as part of training. Should training be resumed immediately after cold application, or even the next day if the exercise is very likely to induce muscle damages?

Chronic Application and Muscle Regeneration

In an untrained population, cryotherapy has been shown to have a deleterious effect on the adaptive processes induced by endurance and strength training over 6 weeks (Yamane et al. 2006). The effects observed were, in particular, reduced muscle-fiber regeneration, reduced hypertrophia, and lowered increase of artery diameter. These results are limited because they concerned only untrained subjects and exercises that do not induce DOMS. However, they remain of interest, specifically since they put forward the interference of cold-related effects (in chronic application) with the usual muscle-regeneration processes, suggesting that there is a risk of delaying rather than favoring improved performance in the medium term.

Muscle Function and Performance Effects

In practical terms, with regard to recovery in sport, cold is applied with the aim of improving the return to normal motor performance, that is, reducing the time necessary to restore the functional capacities of the muscles involved. This is particularly relevant when it is used during or after a phase of multiple competitions.

Acute Postapplication Performance

The very short-term functional benefits of local cold application are a subject of heated debate. Most studies show a trend for negative effects on performance immediately after treatment. Thus, Fischer and associates (2009) recently showed performance during a shuttle sprint test and vertical jump to be reduced immediately and 20 min after a 10 min application of ice to the hamstrings. These results confirm those of Cross and colleagues (1996), who also showed an increase

in times of a shuttle sprint test performed after immersion of the legs in cold water for 20 min. Maximal quadriceps strength, whether tested in concentric or eccentric conditions, was also reduced immediately after application of ice (Ruiz et al. 1993). However, this negative effect seemed to be maintained only for the eccentric condition 20 and 40 min later. Most of these authors explain their short-term negative results by the usual arguments relating to the deleterious effects of cold on muscle contractility and nerve-conduction velocity.

Thus, it seems difficult to avoid causing deterioration in muscle function in the period immediately after application of ice. We can therefore question how appropriate it is to use ice applications as an aid to sporting performance, particularly during competitive events. Consequently, most authors advise athletes to take care when using ice (often employed to reduce pain or as therapy, i.e., after an injury) in the context of a competition (team sport, combat sport, and so on) because of its non-negligible effects on muscle function when competitive activity is rapidly started again (within the 20 min following treatment). However, other authors (Kimura, Thompson, and Gulick 1997) did not show a negative influence of prolonged cold application on peak eccentric torque exerted by the plantar flexors, at any velocity (i.e., 30 and 100 °/s). These authors highlight the fact that the effect of cold can significantly vary as a function of muscle-fiber typology, and that the magnitude of cooling may also play an important role in muscle-force production. This could partly explain the differences between these results and those of Ruiz and colleagues (1993), cited previously. No differences in muscle tension are generally observed when the temperature is greater than 18 °C to 20 °C.

The potential negative effects of cold on maximal strength may be counterbalanced by accelerating posttreatment reheating by performing a warm-up exercise before resuming activity. Indeed, it has been shown that the reduction in jumping, shuttle sprint, and 40 m sprint performances could be compensated by warming up the muscles for a few minutes after the application (Richendollar, Darby, and Brown 2006). This method was used in the study by Kimura and associates (1997), and it may explain the absence of negative effects on maximal eccentric strength observed after cold application. So, what about the contrasting hot–cold bath or ice therapy, which could almost certainly have the same effects?

Some studies do not indicate a negative effect of cold application on other indicators of subsequent muscle function (i.e., other than maximal strength). Tremblay and colleagues (2001) report that quadriceps proprioception (i.e., perception of the leg's position and force produced) are not significantly altered by applying crushed ice to the thigh for 20 min. This partially confirms some previous results on agility (Evans et al. 1995), but does not corroborate the negative results reported by Wassinger and associates (2007) after application of ice to the shoulder, either with regard to shoulder proprioception or throwing accuracy over 30 s. However, this study was limited because performance was tested during an unusual task in untrained subjects.

In addition, it was shown that a 20 min treatment with ice could improve recovery of maximal shoulder-strength capacity in baseball players, although the results were more relevant when the application was followed by low-intensity active recovery for 20 min (Yanagisawa, Miyanaga, et al. 2003). This study is a singular example showing that this type of mixed recovery (ice plus active recovery) could also be associated with a significant reduction in muscle soreness (although, as a whole, the level of DOMS measured during this experiment was consistently low).

Repetition Performance

The results presented by Yanagisawa, Miyanaga, and colleagues (2003) slightly qualify the conclusion stated previously that a rapid resumption of the activity just after cold treatment is necessar-

ily incompatible with a good repeat performance. However, even when associated with a gradual return to activity (by warming up), after application of cold, the level of performance will return to its initial state in the best case: This is not a benefit when compared with passive recovery. So, should local cold application be abandoned between repeated exercises? The answer is not necessarily clear cut, to the extent that some recent studies have shown a trend for an improvement in repeat performances in specific situations. Thus, local immersion of the forearms in cold water for 20 min improves repeat climbing performance in trained athletes compared to passive recovery (Heyman et al. 2009) (table 12.1). Given the shallow depth of water (70 cm), the authors attribute the positive results to the temperature (which drops significantly) rather than to an effect of hydrostatic pressure. This means that these results can be extrapolated to local cold applications using ice. The hypotheses advanced to explain the observed benefits remain the same as those generally used in relation to effects induced by cold: Local vasoconstriction reduces acute inflammation and perceived muscle pain (which can be considerable in the forearm) and also reduces the subsequent loss in strength.

Given these effects, it is possible that cold may have a beneficial effect on muscle endurance. It is interesting to note that the scientific proof in this field is much more obvious. Based on the 10 studies reviewed on the topic by Kimura and colleagues (1997), an improvement is shown in functional muscle endurance following application of cold, whatever the cooling method used and the type of muscle contraction tested. The positive

Table 12.1 Effect of Four Recovery Methods on the Repetition of a Maximal Performance in Trained Climbers

Recovery modalities	DURATION OF CLIMBING (S)		NUMBER OF MOVEMENTS	
	C1	C2	C1	C2
Passive recovery	573.0 ± 307.3	415.5 ± 180.1 ‡	151.1 ± 108.0	114.9 ± 67.5 †
Active recovery	471.6 ± 199.4	455.9 ± 173.3	123.5 ± 72.1	119.1 ± 56.4
Electrostimulation	562.2 ± 358.5	445.1 ± 221.7 †	156.7 ± 136.7	121.6 ± 80.6 †
Cold-water immersion	551.0 ± 312.0	549.2 ± 250.7	147.0 ± 31.5	150.0 ± 91.9

Performance (time and number of movements) was maintained with active recovery and cryotherapy applied to the forearm muscles (20 min), but decreased with electromyostimulation and passive recovery.

† Significant difference between initial performance (C1) and repeat performance (C2) ($P < 0.01$).

‡ Significant difference ($P < 0.001$).

Adapted, by permission, from E. Heyman et al., 2009, "Effects of four recovery methods on repeated maximal rock climbing performance," *Medicine and Science in Sports and Exercise* 41: 1303-1310.

effects are related in particular to a lower perceived pain or exertion during the effort, a lower decrease of strength, an increase in viscosity, and a drop in the products of metabolism, as well as a slower increase in temperature during endurance exercises (Kimura, Thompson, and Gulick 1997).

In the context of strength training, Verducci (2000) studied the effect of repeated applications of ice packs to the shoulder, arm, and elbow (3 min) on different performance indicators during a session of resistance and strength training exercises (22 reps at 74% of 1RM every 8 min until exhaustion). This cryotherapy treatment was found to have a positive effect on speed and power produced and, consequently, on the total work output generated throughout the session (+ 14.5%). As with the contrasting temperatures method, the 3 min applications used in this

Environmental Temperature and Cooling Vests

Cooling can be useful in the specific case of repeated exercises performed in conditions that are likely to lead to a significant increase in muscle or core temperature. A hot atmosphere (as experienced during the 2008 Olympic Games in Beijing, with an average temperature around 30 °C) induces several acute physiological responses that may have a detrimental effect on performance or its repetition. Among these adaptations are the following:

- Significant increase in loss of fluids (dropping plasma volume)
- Significant cutaneous capillary vasodilation
- Reduced blood flow in active muscles
- Reduced stroke volume
- Increased heart rate
- Increased oxygen consumption
- Increased glycolytic metabolism and, hence, glycogen depletion

The effects on recovery are linked, in particular, to the drop in lactate catabolism in the liver and to the reduced local blood flow in the active muscles. In parallel, the thermal stress affects cerebral treatment processes: Central nervous fatigue and altered mental performance can easily interfere with the athlete's cognitive responses. This leads to reduced attention and decision-making capacities, and reduced speed and precision in the task.

In response to the negative influence of hot climatic conditions on sporting performance, linked to the increase in body heat during exercise, it was suggested (in the 1980s) that this stress could be reduced by cooling the body before the event or exercise (precooling). This idea was further developed throughout four Olympic Games in the last 20 years, all of which were performed in high-temperature conditions (Atlanta, 1996; Sydney, 2000; Athens, 2004; and Beijing, 2008). Many methods were tested to artificially reduce body temperature just prior to exercise, the most common of these being exposure to cold air and cold-water immersion. However, from a logistical point of view and because of the duration of exposure, it seems difficult in most cases to use these precooling methods, particularly in the context of top-level sporting competitions (Quod, Martin, and Laursen 2006). Thus, cooling vests (or jackets) were developed. These seem to adequately cool the athlete's body and improve performance (Arngrimsson et al. 2004; Cotter et al. 2001; Hasegawa et al. 2005; Uckert and Joch 2007).

Classically, cooling vests, with or without sleeves, have special compartments into which ice packs can be inserted. The whole setup generally weighs between 2 and 4 kg, depending on the model. In the context of athletic activity, these vests containing ice packs have been shown to be both practical and accessible. The use of this type of product has rapidly grown in athletics in the past 10 years. In parallel, although the current scientific data are somewhat debatable, the majority of them underlined a significant advantage of this precooling technique for optimization of athletic performance.

study do not cause a drop in muscle temperature. However, it is interesting to note that in this specific situation, intermittent cryotherapy (in contrast with a unique prolonged application) allows better preservation of the explosive exercise and raises the threshold for manifestation of fatigue (i.e., task failure, in this case). This positive effect seems to be mainly related to a lower increase in local (or even core) temperature during sessions performed in a warm atmosphere (i.e., gym). This adaptation is certainly close to the positive effects of precooling techniques used prior to performance carried out in a hot ambience (see next section).

Endurance Exercise Effects

The works on the effects of cooling vests are recent. They mainly concern performance during endurance exercises. Cotter and colleagues (2001) showed that, in challenging conditions (34 °C, 60% relative humidity), wearing a cooling vest (for 45 min and with exposure to an environment at 3 °C) prior to performance had a positive effect during a pedaling exercise (35 min of duration with 15 min in time-trial condition). These positive effects were revealed by significantly lowered temperatures (core and skin), heart rate, and indicators of perceived exertion before and during exercise compared to a control group. A functional benefit was also observed, since the average power developed during the last 15 min

of the time trial was greater (+ 17.5%) for the precooling condition. These encouraged results present interesting perspectives, and they have been confirmed by other studies investigating the same type of exercises. Hasegawa and associates (2005) studied the effect of wearing a cooling vest and hydration during pedaling exercises for 1 h at 60% of $\dot{V}O_2$max (32 °C, 70–80% humidity) on the subsequent performance during a time to exhaustion trial at 80%. Hydration, like the cooling vest, had a positive influence on performance. Combining the two procedures allowed an even more significant improvement of the time to exhaustion (figure 12.4).

Other studies investigating running endurance performance showed similar results. Thus, Arngrimsson and associates (2004) described positive effects of wearing a cooling vest during warm-ups on rectal, core, and skin temperatures; on heart rate; and on indicators of perceived thermal discomfort, at the start of a 5 km race (32 °C, 50% relative humidity; see table 12.2). Cooling athletes during warm-up thus reduces cardiovascular and thermal stress, particularly at the start of the subsequent effort, which results in a better speed during the second part of the event. Through this, it enables a significant improvement in total performance, as evidenced by a gain of 13 s.

However, in this study, the speed at the start of the event was predetermined (for 1.6 km). So, it is therefore legitimate to wonder whether the

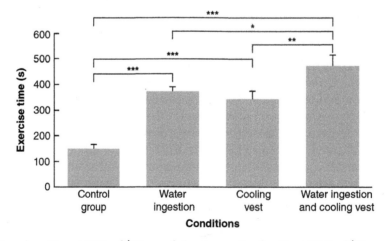

Figure 12.4   Times to exhaustion at 80% of $\dot{V}O_2$max following cycling for 1 h at 60% of $\dot{V}O_2$max in various conditions (control group, with water intake, with cooling vest, and with both water intake and cooling vest).

* Significant differences between the two conditions (P < 0.05).

** Significant differences between the two conditions (P < 0.01).

*** Significant differences between the two conditions (P < 0.001).

Adapted, by permission, from H. Hasegawa et al., 2005, "Wearing a cooling jacket during exercise reduces thermal strain and improves endurance exercise performance in a warm environment," *Journal of Strength and Conditioning Research* 19: 122-128.

Table 12.2 Effects of Wearing a Cooling Vest on Cardiovascular Variables, Temperatures, and Perceived Exertion and Thermal Discomfort, Recorded at the End of Warm-Up Preceding 5 km Maximal Running

Variables	CONDITION	
	Control (n – 7)	Vest (n – 17)
Esophageal temperature	37.4 ± 0.4 *	37.1 ± 0.5
Rectal temperature	38.2 ± 0.4 *	38.0 ± 0.4
Mean body temperature	35.6 ± 0.6 *	33.8 ± 1.2
Mean skin temperature	37.9 ± 0.4 *	37.5 ± 0.4
$\dot{V}O_2$ l/min	0.51 ± 0.08	0.52 ± 0.14
Heart rate, beats/min	122 ± 22 *	111 ± 20
Respiratory exchange ratio	1.04 ± 0.05	1.04 ± 0.04
Rating of perceived exertion	8.8 ± 2.0 **	7.7 ± 2.0
Thermal discomfort	3.1 ± 0.7 *	2.5 ± 0.8

* Significant difference between conditions (P < 0.05).

** Difference between conditions (P = 0.071).

Adapted, by permission, from S.A. Arngrimsson, 2004, "Cooling vest worn during active warm-up improves 5-km run performance in the heat," *Journal of Applied Physiology* 96: 1867-1874.

same precooling protocol would have led athletes to optimize a freely defined pacing strategy at the start of the event. A faster start with the cooling vest could have led either to further enhancement of the global performance or to an inappropriate strategy, leading to an increased total time. To test this second hypothesis, a recent study by Duffield, Green, and colleagues (2010) assessed the benefits of precooling (by immersion of the lower limbs in cold water prior to warm-up, and cooling using ice packs during warm-up) on performance during a time-trial exercise involving 40 min of cycling in

a warm atmosphere (33 °C, 50% relative humidity) performed with free pacing strategy. Interestingly, the same type of positive results was found as in the study described previously: an increase in average power produced, specifically over the last 10 min of the exercise (figure 12.5). Thus, the main physiological adaptations are related to the reduction in temperature.

Even more interesting results were obtained by Webster and associates (2005) on a time to exhaustion in running at 95% of $\dot{V}O_2$max (37 °C, 50% relative humidity). We believe that

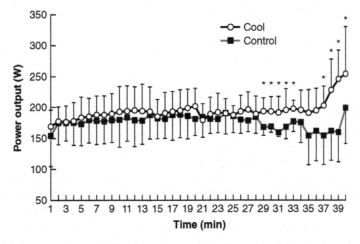

Figure 12.5 Time course of the power output (W) during a 40 min self-paced cycling time trial for precooling (Cool) and control conditions.

*Significant difference between conditions (P < 0.05).

Reprinted, by permission, from R. Duffield et al., 2010, "Precooling can prevent the reduction of self-paced exercise intensity in the heat," *Medicine and Science in Sports and Exercise* 42: 577-584.

characteristics such as the vest's weight, particularly its capacity to dissipate heat, should be taken into account when assessing results from this type of study. In line with this suggestion, the most significant improvement in performance in this study (+ 49 s) was obtained using a vest with the best dissipation characteristics. This last point has led us to list some practical recommendations in the following sections.

All these studies tend to confirm that precooling with cooling vests reduces thermal and cardiovascular stress induced by endurance performances carried out in hot, humid atmospheres. Indeed, in the context of a high ambient temperature, cooling vests appear to be a compensatory solution, reducing body temperature by reducing skin temperature and, thus, favoring thermoregulation by convection (dissipation of heat through blood flow) (Uckert and Joch 2007). The benefits of using this type of cooling jacket when preparing for a performance in regular temperature conditions (around 20 °C and 30% humidity) are much more difficult to discern, although a trend for improvement has been observed (Hornery et al. 2005).

Short-Duration Maximal-Exercise Effects

It is more difficult to study effects on shorter duration maximal exercises involving mainly anaerobic metabolism (ranging from 1 and 2 s, up to 60 s), and the benefits of precooling are more controversial. Most studies indicate that applying cold prior to the effort using a cooling vest does not have a positive effect. Cheung and Robinson (2004) showed no significant variation in peak or average power or in perceived exertion during repeated sprints. Duffield and colleagues (2003) observed no improvement in the performance of a group of field hockey players during four sprint series (15 × 5 s sprinting, separated by 55 s recovery, at 30 °C and 60% relative humidity) when a cooling vest was worn 5 min before the start and during the recovery periods between series. Only the skin temperature and indicators of perceived thermal discomfort were significantly lower. The results described by Castle and associates (2006) are not very encouraging either, since once again, wearing a cooling vest for 20 min before repeated sprints for 40 min (at 33.7 °C and with 51.6% humidity) does not affect peak or average power. It must be noted that in this study, the temperature of the vest was quite high (around 10 °C),

thus reducing its capacity to dissipate heat. However, in the same study, local application of ice packs to the thighs significantly improved the power developed on an ergocycle (+ 4%). This concurs with some of the previously cited studies (including Verducci 2000), which report a positive effect of the use of local ice applications during the different recovery phases of repeated strength-training exercises when applied to the muscle groups involved.

The influence of precooling using a cooling vest on a unique, brief all-out performance is even less obvious. Deleterious effects have been found following the use of this type of vest prior to maximal pedaling exercises for 45 s (Wingate test, at 33 °C and 60% humidity). These effects are even more marked in tests without a warm-up period (Sleivert et al. 2001).

Current Uses and Modes of Application

This section aims to identify the practical aspects of use for each technique (application conditions, recommendations, and risks) and, to a certain extent, to assist athletes and trainers in their choice of the methods presented. We will not present techniques for which the main application is more clearly linked to therapy than to muscle recovery (ice-cube massage, cryogenic spray, and so on). See chapter 14 for more information about means of application of cold or heat related to immersion (Jacuzzi, ice bath, cryotherapy chamber), and chapter 10 for the methods related to general physiotherapy.

Classic Local Cryotherapy

Classical local cryotherapy involves applying a cold object to the skin to induce a drop in skin and muscle temperatures.

Conditions and Practical Use

Local cold application can involve a pack or pouch containing ice cubes or crushed ice, which may be bathed in water, depending on what is available. Of all cooling aids, ice is the most economical and easiest to find, whether on-site at a training facility, in a care center, directly on-site at a competition, or even at a hotel. The physiotherapist (physical therapist), like the athletes themselves, can plan to carry an ice pack or freezer bags as part of a kit. The option of ice-cube bags, (bags filled with water and then frozen) is interesting, provided that the hotel room is equipped with a refrigerator. Since ice can melt rapidly in some conditions (in particular in a warm climate), meaning that the drop in temperature may not be sufficient to achieve the desired effect, ice cubes have to be added during the session.

Application of ice can be either static (by simple application) or dynamic (by massage over a larger area). Ice must be applied to a clean area, which should be free from cuts or scrapes if possible. It is advisable to always place a cloth between the ice pack and the skin, whether dry or wet. Since some people have fragile skin that cannot endure wet cold, due care should be used when applying ice (e.g., by placing a dry cloth next to the skin). Whatever the case, applying an ice pack directly to the skin increases the risk of burns. Ice must not be applied for extended periods, since its efficacy on the muscle is reduced. The application must last at least 15 min (and should not exceed 30 min). Finally, the most effective solution concerning the volume of the ice pack, as we have seen, is a small bag with ice in water weighing at least 600 g.

For dynamic applications, there is no need for a cloth. Massage with an ice pack involves a larger area and induces a much faster drop in temperature; it is therefore not necessary to exceed 20 min of exposure. Ice-pack application techniques, whether static or dynamic, are to be used only after the athlete returns to a calm physical and mental state. This differs from therapeutic applications, which should be applied immediately after acute musculoskeletal injury (such as a sprain). The application can be repeated with a minimum interval of 20 min and several times throughout the day, although a single application can also be effective. This technique is particularly recommended when very traumatizing muscular exercise (of an eccentric type) is performed, since it helps reduce edema formation. However, we remind readers that the advantage of reducing the natural inflammatory response remains controversial for the scientific and medical communities, both in terms of recovery and of preservation of the athlete's physical integrity.

Alternative techniques used to apply cold locally include the following:

- Cryogenic gel packs, for which the main disadvantage is the logistic difficulty of repeating the application rapidly, since it is necessary to let the pack cool before re-use.

- Instant cold packs, which can be used only once. These cold packs are more useful for therapeutic applications, just after an acute trauma.

- Massage using ice cubes, which may be a more analytical indication, both for the muscle and tendon, or even the ligament. In combination with deep transversal massage, for therapy, the ice cube allows rapid superficial analgesia; this is therefore more commonly used for therapy than for recovery. It is appropriate to use a simple ice cube to replace the ice pack when muscles, or muscle groups, of low anatomic volume (e.g., those in the forearm) are to be treated. For instance, this technique was used in the world badminton championships in 2009 (leg massage with ice pack, in the distal to proximal direction, i.e., from the foot to the hip).

Risks, Disadvantages, and Contraindications

The major risk when applying cold is of causing skin burns. Athletes must heed their sensations and must not try to exceed their capacity to withstand the pain caused. As part of local, static applications, special care must be taken when applying cold to zones where nerves pass under the skin, particularly at the knees and elbows, given the risk of paralysis *a frigore*. The main contraindications to the use of ice packs are the same as those for all cryotherapy methods, and they must be diagnosed by a doctor. The patient must avoid the use of ice when it is contraindicated. When an unusual clinical disorder appears following application of ice, it would be wise to immediately stop the procedure. The following contraindications may be encountered:

- *Cryoglobulinemia.* This disease is characterized by the presence of proteins in the plasma (cryoglobulins such as IgG or IgM immunoglobulins), which precipitate and solidify when the plasma temperature drops. Clinically, this can cause purpura (skin hemorrhage characterized by small red patches corresponding to a leakage of red blood cells out of the capillaries), Raynaud's phenomenon, mucosal hemorrhages, or even arthritis. These proteins can be detected by laboratory tests.

- *Raynaud's phenomenon.* This is a localized vascular spasm of one or more fingers in response to cold or an emotional stress, leading to cold sensations, burning pains, sensitivity disorders, or intermittent changes in finger coloration.

- *Skin sensitivity disorders.* Because of the risk of skin burns, precautions must be taken when applying cold to a zone where sensitivity is reduced or absent. Based on current evidence, although this technique is interesting when performances that may cause muscle injury are repeated as part of competition (climbing tournament, baseball game, and so on), it does not seem advisable to use it chronically during intense training phases. Indeed, the main risk in the medium term is of masking the usual physiological signals of muscle injuries and continuing to repeat exercises that may cause excessive muscle trauma. This could have a deleterious effect on the adaptive processes linked to training, particularly those related to myofibril regeneration and muscle hypertrophy.

Cold-Gas and Cold-Air Cryotherapy

Gas-based cryotherapy involves spraying liquid nitrogen or CO_2 to induce a rapid cold shock. Cold-air cryotherapy uses a machine to cool ambient air before spraying it onto the area to be treated. A temperature of -30 °C would probably be necessary to cause cold shock. Compared to the more commonly used local cryotherapy techniques, gas-based and pulsed cold-air cryotherapy accentuate the following effects:

- Anti-inflammatory
- Vasomotor: Immediate reflex vasoconstriction is followed by faster cutaneous vasodilation than with normal cryotherapy, minimizing microcirculatory effects. By associating these phenomena with the gas pressure, this technique provides edema-draining massage by acting on capillary stasis.
- Muscle relaxant: reduction in muscle tone
- Analgesic: slowing nerve conduction

Conditions and Practical Use

Gas-based therapy is applied to a predefined area. Any cuts must be protected and the skin must be dry. Equipping the generator with a temperature probe ensures safe use by maintaining the skin at a temperature between 2 °C and 10 °C. Before starting the session, the treatment must be explained to the athlete. The operator starts by spraying the gas in a slow, regular, continuous sweeping movement over the target area at a distance of 10 to 20 cm, avoiding fixing a point during spraying so the intensity of cold and pressure does not cause rapid tissue injury. If the generator does not have a temperature probe, the operator must permanently check for white crystals fixing themselves to the skin. If this type of skin whitening appears, the skin must be rapidly rubbed to limit any possible tissue injury. However, it is normal for the body's hair to become white during the application. The duration of spraying is quite short: The skin temperature should be maintained between 2 °C and 5 °C for a total duration of 30 to 90 s (depending on the tissues to be treated) if CO_2 or liquid nitrogen are used, and between 1 and 6 min with pulsed cold air. If the subject cannot withstand the treatment, the operator should increase the distance between the nozzle and the skin, reduce the duration of treatment, or discontinue treatment. While this application mainly aims to allow an early return to effort after joint trauma, during the reathletization phase, or as part of physiotherapy, its use can almost certainly be extended to the treatment of muscle soreness (DOMS) to improve the recovery from very traumatic exercise (with high mechanical constraints). The application can be repeated 2 or 3 times the first day and once per day over the next 4 or 5 days. In addition, even if complementary experimental proof is needed, this technique appears to promote the neurovegetative process induced by the thermal shock, favoring reactivation of the parasympathetic nervous system after exercise.

Risks, Disadvantages, and Contraindications

The main risk of gas-based cryotherapy remains skin burns, much more so than with normal cryotherapy. The treatment should be adapted during the session in line with the subject's reactions. The areas where nerves are present under the skin should be treated with care. In addition, it is difficult to use gas-based cryotherapy on large areas, given the amount of gas required. It is therefore more appropriate to use it for small areas. From a logistical point of view, the gas-propelling machinery requires management of supplies, which increases both the cost and the amount of space required to store equipment. Less space is required for cryotherapy machines that use cold pulsed air, which do not require canisters of gas. In addition, regular maintenance is required, particularly for defrosting. Machines can be transported from one room to another, but it may be complicated, even impossible, to move them greater distances, such as to a competition. Because of this, these machines are mainly found in specialized centers, which are also better equipped to cover the costs associated with this type of equipment. Another disadvantage of these techniques is the dearth of studies comparing the different cryotherapy techniques, with most manufacturers performing their own research. This lack, as we have seen, is even more obvious when muscle recovery is considered. Finally, the contraindications are the same as for the more common cryotherapy techniques.

Local Heat Application

Thermotherapy is the application of heat to the skin as part of therapy, heating the body to temperatures above physiological levels. The heat propagates by conduction (which facilitates heat exchange by direct contact). For a thermal agent to be considered hot, its temperature should be between 34 °C and 36 °C (at least), up to a maximum of 58 °C. This upper limit is fixed based on the skin's sensitivity.

The main expected effects of thermotherapy are as follows:

- Improved cellular nutrition and oxygenation
- Improved body defenses due to bactericidal action
- Antisoreness (DOMS) effect
- Analgesic effect
- Antispasmodic effect
- Improved cellular recovery
- Optimized lymphatic drainage
- Enhanced tissue repair processes

Local heat application leads to an increase in metabolic activity and vasodilation (causing an increase in the supply of oxygen, nutrients, and antibodies), and a drop in activity of the neuromuscular spindles and in sensitivity to stretching. Although we haven't insisted on this technique in the paragraph dealing with the physiological bases, physiotherapists agree that its properties make thermotherapy an interesting method to deal with stiffness.

Conditions and Practical Use

Heat can be applied using various supports: algae, paraffin, gel packs (hot packs), or towels soaked in hot water.

Algae are semiliquid physical agents formed from a mixture of mineral water, sea water (or salted lake water), and substances, either organic or inorganic, resulting from biological and geological processes. They are used in local treatments as part of recovery.

Paraffin is a solid, opalescent, odorless hydrocarbon. It is less dense than water and melts easily. It is a by-product of the production of oil-based lubricants from petrol. Paraffin has a lower density than algae (around 30%), which reduces its capacity to store heat.

Hot packs are packs of chemical gel that can be heated rapidly in a microwave, making them practical to use.

Finally, a hot towel can be applied. This method is easy to implement and is the most economical solution when applying heat. However, the benefit of the rise in temperature is of limited duration.

For a local application, the temperature must be between 30 °C and 45 °C. The upper temperature limit depends on the subject's sensitivity and tolerance. The application can last between 20 and 25 min and should be used at least 2 h after the physical effort. A single application appears to be sufficient.

The main conditions to be respected are as follows:

- Check for the absence of cuts, inflammation, infection, and burns.
- Ensure that the recipient's skin is dry.
- Never apply the treatment directly to the skin (interleaving with a cloth prevents the risk of burns).
- Ensure that the subject remains inactive throughout treatment.
- Avoid covering the subject's head.

Risks, Disadvantages, and Contraindications

Whatever the method of heat application chosen, the main risk is of skin burns, but there is also a risk of ischemia when the weight of the application exceeds 600 g. When algae or paraffin are applied as a compress, the main disadvantage is that it is impossible to rapidly reheat compresses after use. Indeed, reactivation of the thermal effect of these compresses requires quite a long soak in hot water.

The main disadvantage of thermotherapy is that it can be difficult to raise the temperature to an optimal and homogeneous level. Reheating heat sources in a microwave must be performed in several short bursts (30 s), with intermittent stirring of the gel, to homogenize the temperature and to ensure that it is not too high. Different manufacturers agree on a heating time between 1.5 and 2 min.

Compresses or hot packs allow a patient to remain dry; this is not the case when using towels soaked in hot water. This must be anticipated. In addition, the fact that the application must be very reduced and localized can constitute a disadvantage if an effect on the whole body is desired. In this case, immersion procedures may be more appropriate.

The main contraindications of thermotherapy are infections, low blood pressure, active hemorrhage, acute inflammation, active skin problems (e.g., fungal infection), and altered sensitivity.

Alternating Hot and Cold

The isolated physiological effects of cold and heat are already known. Another method consists of alternating the two in what is known as *contrasting therapy* (or hot–cold method). This is expected to have numerous effects: stimulation of blood and lymph flow as a result of vasoconstriction/vasodilation (increase in blood flow), and reduced edema, pain, and muscle soreness. All these effects improve recovery. However, as we have seen previously, the supposed pumping action has not been definitively proven, particularly because this technique does not lead to a sufficient decrease in intramuscular temperature (Hing et al. 2008; Myrer, Draper, and Durrant 1994). The effect on edema has also been questioned by the same authors. Even the review by Cochrane (2004) on the effects of contrasting therapy on recovery finds a lack of clear proof of the benefits of this technique, and points to the need for further study to determine the real physiological effects. Finally, the contrasting temperature technique is more commonly used with the immersion technique (see chapter 14). Nevertheless, contrasting hot and cold techniques are widely used in athletics. We will therefore provide details on the main practical aspects here.

Conditions and Practical Use

When it is to be applied to the extremities, this technique requires the use of two containers, one with hot water (between 38 °C and 44 °C) and the other with cold water (between 10 °C and 20 °C). The extremities are dipped alternately in these containers. The heat- and cold-application methods presented previously can all be used as part of alternating hot and cold treatments. Any method involving contrasting temperatures should be applied at least 1.5 h after exercise. It is also advisable to avoid resuming exercise during the hour following the application: The peripheral vessels must maintain adequate elasticity to allow their contraction and dilation.

It is best to start with a hot bath, lasting approximately 7 min, followed by cold application for 1 or 2 min, followed by cycles of 4 min of hot and 1 min of cold. The total duration can be up to 30 min. According to authors and physiotherapists, treatment can be terminated with either hot or cold application. The methods and application times are, however, very varied. No clear trend emerges from the comparison of different studies. Nevertheless, to achieve physiological effects at the intramuscular level, the temperature of the hot sequence must be at least 40 °C.

Finally, from a more subjective point of view, athletes treated using this method after training or a competition confirm an impression of less muscle and joint stiffness and indicate a sense of lightness. Although this subjective criterion has not been extensively addressed, it should not be ignored; it is often advanced to explain the positive effects of this technique.

Risks, Disadvantages, and Contraindications

The major risk, once again, is of inducing superficial burns. The contraindications are the same as for both cold and hot applications, to which defects in arterial circulation, algoneurodystrophy, the initial processes related to vascular spasms (Raynaud's disease, intermittent lameness, and so on), hypertonia, and acute inflammation of the peripheral joints, ligaments, and muscles must be added.

Cooling Vests

Although numerous refreshing products are available to athletes (cold shower, cold room, acclimatization pool, iced fog ventilator, cooling vest, portable air-conditioned tent, and so on) today, several practical questions must be considered. These concern, among others, difficulties linked to transport, storage, cost, and whether or not the technique is easy for athletes to use. Trainers and athletes must therefore be particularly attentive to these logistical aspects in order to avoid unpleasant surprises when they travel for competitions. In this context, cooling vests have emerged as a very interesting technique: They are easy to transport and can be used in many circumstances, and they do not require any particular logistics apart from the need for a freezer to maintain them at an appropriate temperature.

Based on the scientific data indicated previously, the benefits of wearing cooling vests appear to be limited to endurance exercises. Thus, the recommendations presented here are for endurance performances in hot climates with a greater or lesser degree of relative humidity.

Conditions and Practical Use

Because the technique is very recent, it is important to ensure that the product used conforms to a certain number of technical criteria. The vest must be light (weigh less than 3 kg) and adapted to the athlete's body. The target temperature of the vest should be between 1 °C and 5 °C if a real effect on skin temperature and, if possible, body temperature, is to be achieved. Thus, the current challenge for research and innovation in this field is to create vests that combine both practical criteria and a maximal capacity to dissipate heat. Several models are available on the market, offering different cooling techniques and thermal properties (Quod, Martin, and Laursen 2006). During the Olympic Games in Beijing in 2008, one of the most popular cooling vests was the Arctic Heat brand. However, to our knowledge, manufacturers currently take little account of the characteristics related to time and heat dispersion.

Because of this, a new type of vest called Cryovest, was designed for the French delegation participating in the Beijing Olympics in 2008.

Castagna and colleagues (2010) compared the efficacy of the Arctic Heat vest and the Cryovest based on a pedaling exercise in a hot atmosphere (30 °C and 80% relative humidity) for 30 min at 25% of maximal aerobic power (MAP), followed by 15 min at 60% of MAP and 20 min recovery. The Arctic Heat model (1,900 g) consists of four horizontal bands (containing crystallized gel) on the front and back of the vest, covering a total surface of 0.1039 m^2. The Cryovest model (1,920 g) is made up of eight pockets (four on the front and four on the back), each containing cold packs measuring 15 $\times$ 15 cm (FirstIce, United States), and covers a surface of 0.1800 m^2.

In terms of technical quality, the Cryovest presents a larger cooling surface in contact with the skin (+ 75%) compared to the Arctic Heat. In addition, using the Cryovest favors a more stable skin temperature (around 20 °C) throughout the 60 min exposure. This encourages heat dissipation rather than creating great temperature contrasts of shorter duration that would lead to significant vasoconstriction, thus dramatically reducing the exchanges between the body and the environment.

During exercise, this results in significantly lower rectal temperatures when using a Cryovest. It has been clearly established that exercise tolerance is poorer in conditions that create a greater thermal stress for the subject. The heart rate and oxygen consumption (i.e., yield) were significantly lower during exercise performed with the Cryovest, and they dropped rapidly during the recovery period. Similarly, a lower level of dehydration was observed with the Cryovest than with the Arctic Heat. Finally, all of these criteria are significantly different from a control condition wearing a standard T-shirt.

This study therefore confirms that the technical characteristics of cooling vests should be taken into account, since they play a role in the kinetics of thermal exchange with the skin. These differences significantly influence exercise times for events performed in a warm atmosphere, and they also play a determinant role in the subsequent recovery phase. Given the characteristics of the exercise studied, which appears globally similar to a warm-up, the gain observed here could be exploited during competitions by advising athletes to wear this type of vest in the period preceding the competitive effort.

Risks, Disadvantages, and Contraindications

No real contraindications or risks are associated with the use this technique. Nevertheless, some precautions should be mentioned. In real terms, the cooling jacket must be used in the precompetition or pretraining phase for 20 to 60 min. It is worn during the warm-up phase to minimize the activity of the thermoregulation processes directly linked to warming up in a hot atmosphere. At the same time, it should not affect warm-up, which, in this type of extreme environment, must be adapted so as not to induce too great an increase in heart rate and core body temperature before the main event. Similarly, despite the improved thermal comfort due to wearing the vest, hydration must be closely monitored because it has remained a factor of performance in this type of condition (see chapters 7 and 8 for nutrition and hydration strategies).

As is the case for other performance-enhancing options (e.g., response to hypoxia), adaptive responses to this body-cooling technique may result in extensive interindividual variability. Thus, it is strongly advised to test the precooling vests in situations that are close to those of the competition so as to anticipate adaptations that may be more or less desirable. Finally, in our opinion, athletes must carefully choose the model of cooling vest that they wish to use, based on the specific constraints of their sport (duration, repeated effort, and so on), the expected environmental conditions, and the technical characteristics of the different models available on the market.

PRACTICAL APPLICATION

The following table outlines the recommendations and technique effectiveness for reaching a target skin-temperature reduction.

Technique	Application time	Temperature decrease	Effectiveness/notes
Direct ice application: wet and cold*	15 min	~12 °C	Apply for 15–20 min at 10 to 15 min intervals. (Choose the longer time points for subjects with increased fat body mass.) Ice pack should weigh at least 0.6 kg.
Direct ice application: dry and cold**	15 min	~7 °C to 10 °C	
Ice massage	18 min/28 min†	~7 °C to 10 °C ~5 °C (intramuscular)	May be preferred technique for joint sprain and localized muscle damage.
Ice immersion	20 min	< or equivalent to direct ice application	Come back to the initial temperature after the application is longer.
Hot–cold contrast	20 min (5 min/5 min)	Large oscillation (skin) No effect (intramuscular)	While no temperature reduction, see discussion of immersion recovery (in chapter 14) for other applications.
Pulsed cold air	2 min max (repeated)	Up to 26 °C (skin of the hand)	Repeated short applications because the early rise in temperature after the application is faster. Limited experimental evidence.
All			Posttreatment inactivity is required for treatment effects.

*Damp cloth containing ice cubes or crushed ice. **Ice only or a cryogenic gel pack. † With/without compression.

Summary

Cold application should last almost 20 min and should not be followed by activity of the muscles concerned during the next 20 to 30 min. Ice massage and repeated applications are interesting additional methods to improve the cooling effect. The cold application before or between two all-out explosive exercises should be avoided if time after application is less than 30 min: It is unbeneficial for muscle performance.

Despite the fact that local application of contrasting heat–cold technique is widely practiced by athletes, the interest remains debatable. Therefore, the immersion method is preferable.

Repeated local cryotherapy has a beneficial effect on muscle endurance and limits the decrease of strength in repeated exercises. This process can be recommended in specific contexts for which perceived muscle pain is very important: the repetition of maximal performance resulting in a very high level of local solicitation (e.g., forearm treatment for climbing performance) or carried out with a non-negligible heat stress (e.g., indoor strength training).

New methods appear in the field of local cryotherapy. Pulsed cold air is interesting because it seems to improve recovery from very traumatic exercises (associated with a high level of delayed-onset muscle soreness) and to promote (by the thermal shock) the reactivation of the parasympathetic nervous system after exercise.

The use of cooling jackets is an interesting and beneficial post- and precooling technique. It induces a better maintenance of endurance performance carried out in a hot environment.

Air-Pulsed Cryotherapy

Gaël Guilhem, PhD,[1] **François Hug, PhD,**[2] **Antoine Couturier, PhD,**[1]
Jean-Robert Filliard, PhD,[1] **Christian Dibie, MD,**[1] **Sylvain Dorel, PhD**[1,2]

[1] National Institute of Sport, Expertise and Performance (INSEP)
[2] University of Nantes

Exercise-induced muscle damage (EIMD) is characterized by cytoskeletal alterations that originate primarily from unaccustomed exercise. Such muscle perturbations can be the consequence of training resumption, beginning of strength training, changes in training contents, or increase of duration or intensity of exercise. As EIMD persists after the sore phase resulting from muscle damage, muscular deficits are often underestimated when soreness disappears. Such process represents a non-negligible risk of trauma for musculoarticular complex. The study of innovative methods aiming at improving muscle recovery appears relevant from both a clinical and sport performance point of view. Cold application (cryotherapy) has been suggested to minimize the symptoms of EIMD.

Thus, the present study aimed (i) to determine the time course of muscle performance recovery of mechanical (force) and physiological (soreness, edema) indexes after a about of strenuous eccentric exercise and (ii) to quantify the effects of localized air-pulsed cryotherapy intervention (−30°C) on these parameters.

Two healthy trained males executed 3 sets of 20 maximal isokinetic eccentric contractions of the elbow flexors of the dominant arm (the one used for throwing a ball) at $120°·s^{-1}$ in order to evoke EIMD. Maximal isometric torque (MVC), delayed-onset muscle soreness (DOMS), and muscle edema of the biceps brachii (measured by the means of the transverse relaxation time, T_2) were assessed the day before the eccentric exercise (PRE) and at 5 subsequent occasions (1, 2, 3, 7, and 14 days postexercise). After the exercise, participant 1 received a cold air–pulsed treatment (CRYO), while participant 2 passively recovered (PAS). The CRYO intervention was applied just after and at 1, 2, and 3 days postexercise:

1. −30°C pulsed-air applications generated by a Cryo 6 skin-cooling system (Zimmer Medizin Systems, Neu-Ulm, Germany)

2. Maximal available air flow power (intensity = 9)

3. Vertical and horizontal motions, keeping a constant 5 cm distance between the tube nozzle and the skin

4. 4 min applications with 1 min rest between each application to avoid any burns due to extreme cold

5. 5 min of rest after the last application

Both participants exhibited a decrease in MVC and an increase in DOMS and T2 levels after the eccentric exercise of the elbow flexor muscles. The force-generating capacity was similarly affected for both participants throughout the recovery time, although the CRYO participant presented a slightly lower MVC reduction at 1 and 2 days postexercise. This was concomitant to a lower perceived soreness for the CRYO participant at 2 and 3 days, while the increase in DOMS was lower for the PAS participant 1 day after eccentric exercise. Muscle soreness results from inflammation and fluid accumulation (edema), which are able to stimulate nociceptive and metabosensitive type III and IV muscle afferents that trigger DOMS. The lower rate in perceived soreness could be attributable to a lower muscle edema for the CRYO participant throughout the exercise recovery in comparison to PAS. Muscle edema level subsequently showed a similar return to baseline at 7 days for both participants.

Besides its analgesic effect, the application of cold has been shown to reduce edema formation, decrease cell metabolism, and consequently allow the uninjured cells neighboring the site of initial trauma to survive the period of hypoxia induced by edema. Based on these data, cryotherapy has been suggested to minimize the symptoms of EIMD. In this context, many forms of localized (e.g., ice, cold-water immersion) or global (e.g., whole-body cryostimulation) cooling have been used, showing contradictory findings. Our results showed that cryotherapy reduced delayed muscle edema, which could consequently decrease DOMS, while force-generating capacity recovery was not improved by cooling.

In conclusion, although cooling is widely used in clinical practice, the present repeated air-pulsed cryotherapy treatment did not demonstrate clear significant benefits to the muscle function recovery (i.e., maximal strength capacity). The observed reduction in the development of muscle edema may have affected the perceived soreness but failed to improve strength recovery after a severe monoarticular eccentric exercise. Finally, considering the potential high interindividual variability in strength loss and amount of damage induced by this eccentric exercise, the influence of cryotherapy could depend on the exercise modality and the duration of cooling intervention.

Key Points

- −30 °C air-pulsed application, even repeated, does not improve muscle performance after muscle damage.

- Cooling may prevent the development of delayed muscle edema without having any impact on strength recovery.

- Damage indexes depend on exercise modality and show interindividual variability that must be considered when evaluating the effectiveness of a recovery method.

CHAPTER 13

Variations in Thermal Ambience

Christophe Hausswirth, PhD

With: François Bieuzen, PhD, Marielle Volondat, Dr. Hubert Tisal,
Dr. Jacques Guéneron, and Jean-Robert Filliard, PhD

Exercise-related stress is often increased due to environmental conditions, particularly those relating to temperature change. Every sporting activity has an ideal ambient temperature. Any deviation from this reference temperature will have a negative effect on performance.

Indeed, physical activity in a warm or cold atmosphere means that the body and the mechanisms involved in temperature regulation have to work harder. Although very effective, these thermoregulatory mechanisms may not be able to cope with extreme conditions. They do, however, allow the body to adapt during chronic exposure. Artificial heating or cooling of ambient temperature is an expanding technique, both to prepare athletes for competitions in difficult conditions and to improve the body's recovery capacity. This works even better when athletes train in an environment that exposes them to these conditions on a daily basis.

Physiological Responses to Air Temperature

To maintain vital functions, the core temperature in warm-blooded animals, like humans, must be virtually constant (Candas and Bothorel 1989).

This core temperature is the body's reference temperature, and it reflects a balance between heat gain and loss. To maintain vital functions, chemical energy must be continuously supplied. Although temperature fluctuations occur daily (or even hourly), these remain low, in the range of 1 °C (34 °F). In contrast, core temperature may rise above the normal range (from 36.1 °C to 37.8 °C, or from 97 °F to 100 °F) during muscular exercise (Maughan and Shirreffs 2004) or in specific environmental conditions. Athletes practicing in extreme conditions (hot and humid) run a very high risk that their core temperature might rise abnormally and dangerously above these levels.

Heat Exchange

Cellular activity and biochemical transformations needed to maintain vital functions require a constant supply of energy. According to the first principle of thermodynamics, this energy may be transformed and stored in different forms. Thus, in muscle, 25% is converted into mechanical energy (allowing movement and, by extension, physical activity) and 75% produces heat, or thermal energy. A large proportion of metabolic energy is therefore dissipated as heat, which must then be evacuated to maintain a stable core body temperature. For this, heat gain due to energy metabolism (or to the environment when the external temperature is high) must be balanced against heat loss due to exchanges with the surrounding medium (Candas and Bothorel 1989). Heat exchanges between the body and the outside environment occur through the skin, at the body's surface.

Heat can be exchanged with the air in four ways: conduction, convection, radiation, and evaporation. The blood carries heat produced by the body toward the periphery (the skin). The skin provides the interface for heat transfers between deep organs, which must be protected, metabolically active tissues, and the surrounding air. All four means of heat exchange participate in these transfers. During physical exercise in normal environmental conditions, sweating is the main means of thermoregulation. It is centrally regulated, and the rate of sweat secretion is proportional to the amount of heat that must be dissipated to ensure a relatively constant core-body temperature. Heat production through muscle activity constitutes the main thermoregulation mechanism during physical exercise in a hot environment (Nielsen 1996). High relative air humidity can, however, adversely affect this mechanism.

Thus, heat exchanges are tightly controlled in the body. Resting core temperature is approximately 37 °C (99 °F), but may exceed 40 °C (104 °F) during exercise. Extreme muscle temperatures around 42 °C (107.6 °F) indicate that the body is storing, rather than dissipating, significant amounts of heat. This is because the heat produced by exercise-related metabolic reactions cannot be evacuated fast enough. We now know that a temperature increase of 1 °C or 2 °C (1.8 °F or 3.6 °F) promotes good muscle function (heat reduces tissue viscosity, improves tendon elasticity, raises the speed of nerve conduction, and modifies muscle enzyme activity), but temperatures above 40 °C (104 °F) affect the nervous system, and thus perturb the various thermoregulatory mechanisms.

The body has heat receptors located in the hypothalamus, which controls and registers core temperature like a thermostat. Thus, at the slightest deviation from the reference temperature (e.g., during physical exercise), the thermoregulatory centers are immediately informed. They then activate various regulatory mechanisms to read just the core temperature. Two types of receptors are involved in this process: central receptors located in the hypothalamus (where they register the temperature of the blood irrigating the brain) and peripheral receptors located in the skin (registering temperature variations in this organ). Data registered by peripheral receptors are communicated to the hypothalamus and the cortex.

If the temperature exceeds the reference value (for example, during exercise), several physiological responses will be implemented. The hypothalamus starts by stimulating the sweat glands, which will secrete sweat and promote heat loss through evaporation. As the temperature rises, so does the rate of sweating. When the blood or skin heats up, the hypothalamus sends instructions to the smooth muscles in the walls of blood vessels, causing peripheral vasodilation. Increased blood flow in the skin allows blood, heated in the core, to be cooled. This favors heat transfer to the periphery and heat loss through radiation and convection, guarding against the risk of dehydration that might occur if evaporation was the sole cooling mechanism.

In a very hot environment, humidity affects the amount of heat that can be eliminated by sweat evaporation. When the atmosphere is both hot and humid (close to 100%), sweat evaporation becomes virtually impossible (Candas, Libert, and Vogt 1983; Nielsen 1996), and the core temperature tends to rise rapidly and dangerously. High humidity means that a large number of water molecules are present in the air. The concentration gradient of water vapor between the skin's surface and the environment is therefore reduced, lowering the air's ability to accept further water molecules and limiting sweat evaporation. Any factor interfering with the process of heat elimination thus leads to an increase in body temperature and to discomfort. The comfort level for a given air temperature is therefore affected by relative humidity (table 13.1) and air movement.

The table shows, for example, that with negligible air movement, an actual air temperature of 29.4 °C corresponds to a perceived temperature of 33.9 °C combined with 70% humidity, 36.1 °C at 80% humidity, and 38.9 °C at 90% humidity.

Cold Stress

At 20 °C, the body temperature of a person resting and wearing light clothes requires no regulation. Cold stress occurs in any environment causing heat loss that constitutes a threat to

Table 13.1 Influence of Relative Humidity (%) on Athletes' Perception of Air Temperature

Relative humidity	21.1 (70)	23.9 (75)	26.7 (80)	29.4 (85)	32.2 (90)	35.0 (95)	37.8 (100)	40.6 (105)	43.3 (110)	46.1 (115)	48.9 (120)
0	17.8	20.6	22.8	25.6	28.3	30.6	32.8	35.0	37.2	39.4	41.7
10	18.3	21.1	23.9	26.7	29.4	32.2	35.0	37.8	40.6	43.9	46.7
20	18.9	22.2	25.0	27.8	30.6	33.9	37.2	40.6	45.0	48.9	54.4
30	19.4	22.8	25.6	28.9	32.2	35.6	40.0	45.0	50.6	57.2	64.4
40	20.0	23.3	26.1	30.0	33.9	38.3	43.9	50.6	58.3	66.1	
50	20.6	23.9	27.2	31.1	35.6	41.7	48.9	57.2	65.6		
60	21.1	24.4	27.8	32.2	37.8	45.6	55.6	65.0			
70	21.1	25.0	29.4	33.9	41.1	51.1	62.2				
80	21.7	25.6	30.0	36.1	45.0	57.8					
90	21.7	26.1	31.1	38.9	50.0						
100	22.2	26.7	32.8	42.2							

*Air temperature in °C (°F): From left to right, no shading indicates a comfortable temperature and lack of danger; gray shading indicates a low risk with a possibility of heat-induced cramps; light gray shading indicates high risk, probable heat exhaustion, and possible heatstroke; and dark gray shading indicates very high risk and probable heatstroke.

homeostasis. A drop in skin or blood temperature leads to a hypothalamic response, activating the mechanisms to counteract cold, and thus increasing the body's heat production. The first reactions are shivering, nonshivering thermogenesis, and peripheral vasoconstriction.

Shivering is a reflex response to cold, which translates as a succession of involuntary muscle contractions. It increases resting metabolic heat production four- to fivefold. Nonshivering thermogenesis is the result of stimulation of the metabolism by the sympathetic nervous system. Resting metabolic rate is controlled by the thyroid hormones, and an increase in metabolic rate leads to increased endogenous heat production. Sympathetic stimulation also leads to peripheral vasoconstriction by acting on smooth muscles located in the walls of the arterioles near the surface of the skin. Their contraction reduces the size of the vessels and thus the blood flow to the skin, thereby preventing excessive heat loss. As the skin's temperature drops, the metabolic rate of skin cells is also reduced, thus limiting the need for peripheral oxygen.

In contrast to the response to high external temperatures, the body's thermoregulatory process to counteract cold is quite limited. Once this limit has been reached, a significant amount of heat is lost if appropriate protective clothing is not worn. In a cold environment, heat exchange by conduction, convection, and radiation can lead to a loss of calories greater than what is produced by endogenous systems. It is difficult to define the precise conditions leading to this excessive heat loss and hypothermia. Indeed, numerous factors, both intrinsic (adipose tissue, circulatory system, hormonal system, level of training, state of fatigue) and extrinsic (type of clothing, external temperature, wind, humidity, altitude) are involved in the thermal equilibrium. These affect the gradient between loss and gain of heat. The general principle is that the greater the temperature differences between the skin and the environment, the greater the loss of heat. But, as we have just explained, the rate of heat loss also depends on anatomic and environmental factors. Wind, for example, is a cooling factor, increasing heat loss by conduction and convection. Similarly, thermal stress is increased by high humidity and more intense cold. The same actual temperature can feel much more uncomfortable when the air is very humid or in windy conditions.

Whole-Body Cryotherapy

The first very-low-temperature cold rooms appeared in Japan in 1989, when Yamauchi used a cryogenic chamber to treat rheumatism. The indications for whole-body cryotherapy (WBC, also described as whole-body cryostimulation) were subsequently extended to various inflammatory conditions—arthritis and multiple sclerosis (Fricke 1989), rheumatoid arthritis (Metzger et al. 2000)—and to skin disorders such as psoriasis (Fricke 1989). WBC was then offered to treat pain and prevent posttraumatic edema, limiting exposure to 2 or 3 min (Zagrobelny 2003).

In the sporting realm, WBC has been used at temperatures close to –100 °C with the aim of limiting the spread of muscle lesions (Swenson, Sward, and Karlsson 1996). It has also been offered as a prophylactic treatment to reduce the risk of muscle lesions during intense training periods. However, how well WBC restores biological constants following intensive training remains to be shown.

Although we still lack scientific hindsight on the link among recovery, sport, and cryostimulation, we will provide a summary here of the effects of WBC on some relevant parameters. This will help us identify possible applications in the field of recovery in sport. Upcoming studies should help to assess its effects during the postexercise period. However, data already available in the scientific literature show benefits on some inflammatory parameters, possible improvement of antioxidant status, and improvements in mood and mild depression. Complementary studies focused on physical activity should indicate whether athletes recover better when using this treatment.

Inflammatory Marker Responses

Some authors have measured various markers of inflammation in subjects exposed to very low temperatures. Banfi, Melegati, and colleagues (2008) showed that treating top-level rugby players with WBC for 1 week led to reduced rates of pro-inflammatory cytokines (IL-2 and IL-8) and increased levels of anti-inflammatory cytokines (IL-10). This is the only study in which the results can genuinely be related to recovery after intense muscular exercise. According to the authors, WBC should improve muscle recovery, although they were not able to measure to what extent.

In this 5-day study, 10 top-level rugby players were placed in cryogenic chambers at –60 °C for 30 s, then at –110 °C for 120 s. In addition, subjects followed their normal 3 h daily training without changing their workload. While no significant difference was measured in terms of immunoglobulin or C-reactive protein (CRP) levels (two markers of acute inflammation), the authors did show that creatine kinase (CK) and PGE2 prostaglandin concentrations were significantly reduced after 5 days of WBC (figure 13.1). No control group was included in this study. The authors explain that the drop in creatine kinase is likely to result from the stimulation of noradrenalin (NA) secretion during exposure to cold, an effect demonstrated in Rønsen and colleagues' 2004 study. No assay of NA was performed in this study.

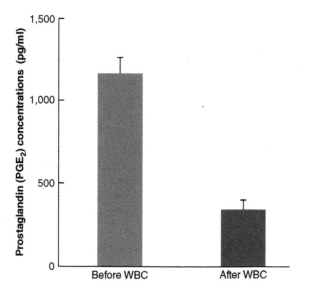 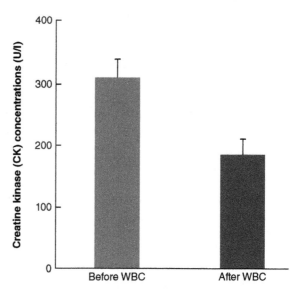

Figure 13.1   Variation in serum concentrations of muscle injury markers (prostaglandin and creatine kinase). The left panel shows a significant reduction in prostaglandin PGE2 ($P < 0.0001$) after WBC. The right panel shows a significant reduction in creatine kinase ($P < 0.01$) after WBC.

Reprinted from *Journal of Thermal Biology*, Vol. 34, G. Banfi et al., "Effects of whole body cryotherapy on serum mediators of inflammation and serum muscle enzymes in athletes," pgs. 55-59, copyright 2008, with permission from Elsevier.

Banfi, Melegati, and colleagues (2008) observed a reduction in PGE2 associated with reduced CK levels. PGE2 is synthesized at the site of inflammation, where it acts as a vasodilator in synergy with other mediators, such as histamine and bradykinin. These mediators cause increased vascular permeability, leading to edema. Their reduction, after 5 days of WBC, seems to be a good indicator of improved muscle recovery. However, the absence of a control group in the study is a flaw that makes it impossible to reliably conclude on the efficacy of WBC in recovery. This research does, however, offer some suggestions as to the parameters likely (or not) to favor improved recovery.

Hormonal Responses

Most studies of WBC have focused on the kinetics of biochemical markers and how various hormones evolve in response to exposure. It is widely believed that changes (or lack thereof) to endocrine parameters are relevant topics for research on improved athlete recovery. In this context, a recent study by Smolander and colleagues (2009) compared WBC exposure (–110 °C for 2 min) to cold-water immersion (0–2 °C for 20 s). The two groups were subjected to either treatment weekly over a total of 12 weeks. Various hormones—growth hormone (GH), prolactin, and the thyroid hormones (TSH, T3, T4)—were analyzed. The authors concluded that there were no significant variations in hormone levels for the WBC group. Prolonged exposure to cold seemed to have no effect on the concentrations of these hormones.

Based on this lack of effect of WBC on hormone levels, we can conclude that this procedure conforms to sporting ethics. These results are supported by those of another recent study by Banfi, Krajewska, and associates (2008), who indicated that for a group of 10 athletes, none of the hematological parameters (e.g., red blood cells, white blood cells, hematocrit, hemoglobin, platelets, and so on) was affected by five 2 min exposures over a week. In an earlier study, Leppäluoto and associates (2008) showed that exposure to WBC (3 times per week for 12 weeks) induced a significant increase in plasma noradrenalin (NA) levels (figure 13.2). The authors explain that the increases in NA levels recorded over the 12 weeks could play a role in relieving perceived pain, an effect seen in other studies using traumatizing exercise. However, no scale of pain perception

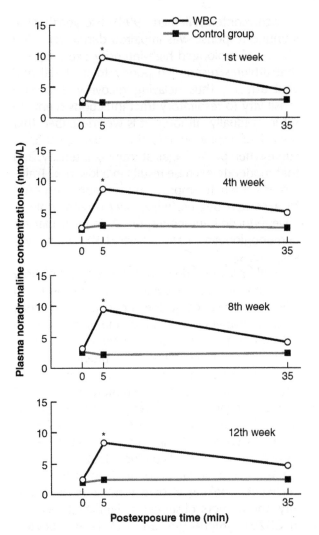

Figure 13.2 Variations in plasma noradrenalin concentrations after 1, 4, 8, and 12 weeks of whole-body cryotherapy (WBC).

* Significant difference from initial value (P < 0.01).

Adapted, by permission, from J. Leppäluoto et al., 2008, "Effects of long-term whole-body cold exposures on plasma concentrations of ACTH, beta-endorphin, cortisol, catecholamines and cytokines in healthy females," *The Scandinavian Journal of Clinical & Laboratory Investigation* 68(2): 145-153.

was offered in this purely descriptive study, which involved straightforward cryostimulation by WBC exposure.

Immune Responses

For a number of years, the immune system has been of particular interest to sport physiologists. The incidence of sore throats in very fit athletes initially helped doctors to detect overtraining syndrome. These intuitions were confirmed more recently by some very well-run American and British studies.

In this context, Nieman (1994) observed that the immune response was impaired during repeated phases of prolonged high-intensity exercise, and that athletes responded poorly to bacterial and viral attacks, thus delaying recovery. Excessive sensitivity to respiratory tract infections seems to set in gradually, although it is well described that the risk of respiratory infection follows a J-shaped curve when plotted against training intensity, and that moderate exercise results in a low risk. Training is known to improve the immune response to a certain degree, while overworked athletes have reduced immune responses, in particular for immunoglobulins, natural killer (NK) lymphocyte subgroups.

While no study deals with the kinetics of how the immune system evolves after exposure to WBC, mainly because the procedure is so new, models involving swimming and immersion in cold water have been used for the last few years in Nordic countries. These have provided indications on how the immune system is affected. This practice, which was developed more on a cultural than on a scientific basis, has always been empirically linked to improved resistance to infections.

In this context, Janský and associates (1996) carried out a 6 wk study on the effects of immerging 10 patients in water at 14 °C for 1 h 3 times per week. Looking at several markers of immunity, the authors observed a significant increase in CD25+ lymphocytes and CD14+ monocytes. Interleukin-6 (IL-6), a factor stimulating T-lymphocyte production, was also shown to increase, but not significantly. Although it lacked a control group, this preliminary study indicated a possible stimulation of the immune system by limited (less than 1 h) exposure to cold. These results were reinforced by a study comparing populations swimming regularly in cold water or not (Dugué and Leppänen 1999). Plasma levels of IL-6, monocytes, and leukocytes were all higher in cold-water swimmers. The authors concluded that the immune systems of cold-water swimmers better controlled the inflammatory response and that repeated exposure to cold (either by immersion or other methods) could explain the improvement in defense against infections. It could therefore be suggested that repeated exposure in cold rooms (i.e., WBC) stimulates the immune system and reduces susceptibility to infections in acclimatized people. New studies on WBC should shed light on these hypotheses and offer insights into the relationships among immunity, cold, and athletic recovery.

Respiratory Function Responses

The incidence of exposure to cold ambient temperatures on respiratory function has been the focus of several studies. The body is known to react to cold by stimulating the sympathetic nervous system, thus inducing bronchodilation (Marieb 1999). Bandopadhyay and Selvamarthy (2003) studied respiratory function in 10 subjects exposed to Arctic cold for 9 weeks. The results showed that the forced expiratory volume per second (FEV1) was significantly reduced in the first few days. It then recovered its initial level after 4 weeks of exposure. At the end of the 9 weeks, the authors observed a significant improvement in FEV1, but this is not maintained over time. These respiratory consequences of exposure to cold were recently studied by Smolander and associates (2009), who subjected 25 nonsmokers to WBC. The subjects underwent three 2 min WBC sessions per week for 12 weeks. Peak flow (PF) and FEV1 were measured 2 and 30 min after each session. No change in PF or FEV1 (measured 2 min after the sessions) was recorded over the 3 months of the study. On the other hand, for measurements performed 30 min after sessions, PF and FEV1 values were significantly reduced by the end of the first month. The authors explained that the sympathetic effect, a reflex to the cold, seemed to return to basal levels after 30 min, when the parasympathetic system becomes more active. The authors concluded that WBC should be used with care in people with respiratory problems.

Antioxidant Status Responses

Physical exercise is characterized by an increase in oxygen consumption, and consuming high levels of oxygen is associated with increased free-radical production (Jenkins 1988; Sen 1995). Modulation of oxygenated free-radical production plays a clear role in muscle recovery after exercise (Gauché et al. 2006). High-intensity exercise and exercises involving many eccentric movements are a true stress, producing metabolic by-products with significant effects on cellular structures. Oxygen-derived free-radical (ODFR) species—involved in oxidative stress—differ in terms of structure, but all are extremely reactive compounds that, once produced, oxidize various cellular components. This oxidation can lead to cellular dysfunction and, among other things, to inflammatory disorders. WBC has sometimes been used to reduce oxidative stress.

The first studies, carried out by Siems and Brenke (1992), showed that acute exposure to WBC between 1 and 5 min caused oxidative stress in experienced swimmers. One hour after exposure to cold, the concentration of intraerythrocyte oxidized glutathione, a marker of oxidative stress, had increased more significantly in subjects exposed to WBC than in a control group. This was combined with reduced concentrations of uric acid, a true scavenger of reactive oxygen species (Ames et al. 1982). Because of this, the authors suggested that the global increase in anti-oxidant protection resulted, in the long term, from repeated exposure to mild oxidative stress. In addition, during cooling and stimulation of the body, mitochondria exposed to low temperatures produce 10 times more superoxide anions as a result of increased lipid peroxidation (Bartosz 2003).

A study by Dugué and colleagues (2005) showed increased total plasma antioxidant (TPA) capacity after three weekly cold-room sessions for 12 weeks. These results contradict the authors' initial hypothesis that values would be significantly reduced, and explain the improved protection. The study was unable to confirm any hypothetical increase in antioxidant protection due to repeated exposure to cold. Indeed, most studies investigating cryostimulation and its influence on ODFR and lipid peroxidation focused on treatment of rheumatoid arthritis (Metzger et al. 2000; Yamauchi 1989).

A single study investigated the effects of a WBC session (at −130 °C) on the ratio of pro-oxidants to antioxidants (Lubkowska et al. 2008). Plasma total oxidant status (TOS) was significantly lower 30 min after exposure to WBC for 3 min (figures 13.3 and 13.4). The next day, the TOS level was still significantly lower than the basal level before WBC exposure. In addition, total antioxidant status (TAS) values were significantly lower 30 min after exposure to cold, but they did not differ from basal values the next day. However, in the case of athletes, cryostimulation is combined with physical exercises as part of regular training. It is therefore difficult to know to what extent lipid peroxidation is the result of training or cryostimulation (Bloomer et al. 2006; Swenson, Sward, and Karlsson 1996).

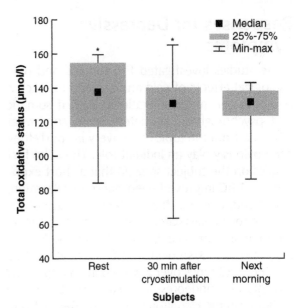

Figure 13.3 Variations in plasma total oxidant status in resting subjects, 30 min after cryostimulation and the next morning.

* Significant difference between 30 min after cryostimulation and next morning (P < 0.05).

Adapted from *Journal of Thermal Biology*, Vol. 33, A. Lubkowska et al., "Acute effect of a single whole-body cryostimulation or prooxidant-antioxidant balance in blood of healthy, young men," pgs. 464-467, copyright 2008, with permission from Elsevier.

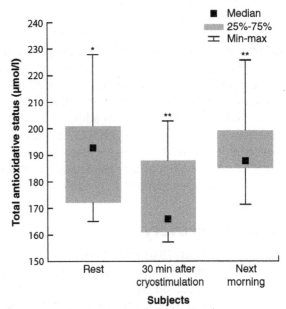

Figure 13.4 Variation in plasma total antioxidant status in resting subjects, 30 min after cryostimulation and the next morning.

* Significant difference between resting and 30 min after cryostimulation (P < 0.05).

** Significant difference between 30 min after cryostimulation and next morning (P < 0.01).

Adapted from *Journal of Thermal Biology*, Vol. 33, A. Lubkowska et al., "Acute effect of a single whole-body cryostimulation or prooxidant-antioxidant balance in blood of healthy, young men," pgs. 464-467, copyright 2008, with permission from Elsevier.

Responses for Depressive Symptoms

Some studies investigated the somatic and psychological effects of cold and seemed to reach a consensus on mood alterations. Even if somatic and psychological parameters seem somewhat removed from the topic of recovery in sport, they can obviously play an indirect role. Thus, the first studies on the subject showed that a short exposure to WBC improved sleep, sense of relaxation, and mood, and that these effects can persist for hours or even days (Gregorowitcz and Zagrobelny 1998). In a more recent study, Rymaszewska and associates (2003) studied the effects of WBC (−150 °C, 160 s, 10 times over 2 weeks) in 23 depressed patients on antidepressant treatment. Using the 21 items on the Hamilton Depression Rating Scale (HRDS), the authors concluded that WBC exposure had a positive effect on HRDS scores, and thus helped alleviate symptoms of depression. Given these results, the authors published another study that was similar to their previous work, but which included a control group of 34 patients (Rymaszewska, Ramsey, and Chladzinska-Keijna 2008). After 3 weeks, the HRDS scores for the 26 patients suffering from depression were reduced by 34.6% in the WBC group, against only 2.9% in the control group. One neurobiological hypothesis states that depression results from a deregulation of the hypothalamic-pituitary-adrenal (HPA) axis. The authors relate the improved mood regulation and HRDS scores to this axis. In addition, it seems that WBC also has positive effects on patients' biological rhythms. All these results could provide some help for the temporary psychological problems frequently encountered by athletes during training.

Athlete Recovery

A preliminary study (Barbiche 2006) evaluated the effects of WBC on muscle stiffness reported by high-level athletes during muscle reinforcement physiotherapy (physical therapy) for the knee, following surgery on the anterior cruciate ligaments. The 17 subjects were exposed to WBC for 3 min per day for 3 weeks. The results revealed a trend indicating reduced muscle pain reported by patients exposed to WBC compared to the control group ($P = 0.07$). However, complementary studies will be necessary to confirm these results on a larger sample cohort and to explain this positive effect of WBC on muscle stiffness and recovery.

Dry-Heat Sauna

Saunas have been in use for 2,000 years, and they are still widely used today, particularly in Finland, where no fewer than 2 million saunas exist for a population of 5.2 million.

Finnish high-level athletes have played an important role in promoting the use of saunas for recovery. At the Los Angeles Olympic Games in 1932 and the Berlin Games in 1936, saunas were brought directly from Finland and installed in the Olympic villages. This led to significant media coverage and many questions about the application of saunas for top-level athletes.

The use of dry heat is now widespread; it is used to alleviate hypertension, anxiety, and irritability. It also has an effect on the incidence of colds (Einenkel 1977; Ernst et al. 1990), asthma, and other bronchoconstrictive disorders (Preisler et al. 1990). Ambient temperatures higher than the skin's temperature stimulate all the thermoregulatory pathways and mechanisms. Heat can be recovered by the body through radiation and convection.

In the scientific literature, the term *sauna* designates a wide range of conditions: During sauna sessions, the optimal exposure temperature is between 80 °C and 90 °C, and the relative humidity varies between 15% and 30% (Kukkonen-Harjula and Kauppinen 2006). Other authors include a much wider range of humidity levels, between 3% and 50% (Paolone et al. 1980; Shoenfeld et al. 1976). In practice, sauna users modify the humidity levels themselves by pouring water over hot stones. Depending on the study, the duration of exposure can also vary considerably, ranging between 5 and 20 min. For Kauppinen and Vuori (1986), exposure should be limited to 10 min. With regard to how it is used, a sauna session is often repeated three times and followed by immersion in cold water (5–15 °C) with the head above water, a cold shower, or simply by exposure to room temperature (23–24 °C). In some countries (like Finland), these techniques are repeated regularly throughout the week, sometimes even daily, and cold water is always used between exposures to heat. Finally, temperatures in Finnish saunas often exceed 110 °C (Kukkonen-Harjula and Kauppinen 2006).

During exposure to sauna, body temperature increases in line with the increase in ambient temperature. This rise in body temperature induces physiological changes at the cardiovascular, pulmonary, and neuromuscular levels, as well as changes to the inflammatory, hormonal, and immune systems. Changes to cardiovascular activity primarily result from sweating and the associated peripheral vasodilation and movement of blood toward the periphery (Paolone et al. 1980). Figure 13.5 illustrates how sauna temperature can be optimized as a function of humidity level. For temperatures between 80 °C and 100 °C, relative humidity should be between 10% and 20% (Leppäluoto 1988).

Extreme temperatures, used in moderation, have been shown to be safe for people in good health and patients suffering from appropriately managed cardiovascular complications. Heating the body is supposed to be beneficial for athletic recovery, to treat muscle pain, and as part of rehabilitation after injury (Brukner and Khan 2001). Nevertheless, very few of the numerous publications examining the effects of sauna use deal with its effect on recovery after physical exercise. The increase in core-body temperature during sauna use has a negative effect on subsequent athletic performance in warm conditions (34 °C) (Simmons, Mündel, and Jones 2008).

Many factors such as varying experimental conditions (temperature, humidity level, duration,

and technique of cooling associated with sauna use), subjects' familiarity with saunas, nutrition, variations in hormonal secretions over the next 24 h, the physical exercise preceding exposure, and individual characteristics (age, gender, health, percent adipose tissues) must be taken into consideration when defining the effects of sauna use (Kukkonen-Harjula and Kauppinen 1988) and its influence on thermoregulatory responses (Yokota, Bathalon, and Berglund 2008). All these factors should be used to define the precise context in which athletes may use this tool to optimize recovery.

Cardiovascular System Responses

Exposing a large area of the body's surface to heat in a sauna can be a source of stress for the heart, causing ectopic heartbeat, hypotension, heat-induced fainting, and tachycardia. In extreme conditions, it could even cause death (Turner 1980).

The rise in temperature causes blood flow to be reorganized. Indeed, when skin temperature exceeds 41 °C, heart rate increases by 6.6 L/min, blood flow to the viscera increases by 0.6 L/min, kidney circulation drops by 0.4 L/min, and muscle blood flow drops by 0.2 L/min (Rowell, Brengelmann, and Murray 1969). These physiological modifications are induced by sweating, which is needed to maintain body temperature. Because sweating is intimately linked to vasodilation of skin capillaries, skin blood flow increases by 20% to 40%, without altering the heart's ejection volume. On the other hand, cardiac output is increased by 70% or more when sauna temperature exceeds 90 °C (Kauppinen 1989). Water loss associated with maximal output through the skin reduces blood pressure. This is then compensated by increasing the heart rate (Kauppinen and Vuori 1986). Indeed, resting heart rate (i.e., 70 bpm) increases by 60%, 90%, and 130% in saunas at temperatures of 80 °C or 90 °C (dry heat) and 80 °C (humid heat), respectively (Kukkonen-Harjula et al. 1989) (table 13.2). This is combined with an increased consumption of oxygen (Hasan, Karvonen, and Piironen 1966) and increased cardiac output due to failure of vasomotor control and sudden skin vasodilation (i.e., reduction in peripheral resistance) (Greenleaf 1989).

Following 7 min of exercise on a rowing ergometer, Taggart and colleagues (1972) monitored athletes by electrocardiogram (ECG) over a 5 min sauna session (90–100 °C). They observed reduced T-wave amplitude and a reduction in

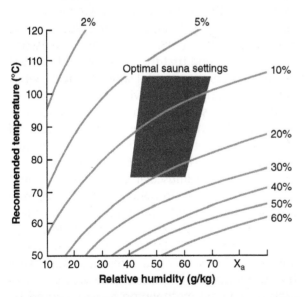

Figure 13.5 Relationship between recommended temperature and relative humidity for exposure to high air temperatures (between 80 °C and 100 °C). Recommended humidity is between 40 and 70 g of water vapor per kg, equating to a relative humidity between 10% and 20%.

Adapted, by permission, from J. Leppäluoto, 1988, "Human thermoregulation in sauna," *Annals of Clinical Research* 20: 240-243.

Table 13.2 Effect of the Duration of Sauna Exposure on Oral Temperature and the Subjective Perception of Heat Stress

Heat condition/time	Length of exposure (min) (range)	Oral temperature (°C)	Rating of heat stress (0–10)
80 °C DRY HEAT			
Before bathing		36.5 ± 0.2	3.3 ± 0.3
End of first sauna exposure	22 ± 5 (19–32)	37.3 ± 0.4*	3.8 ± 0.4
End of second sauna exposure	16 ± 6 (4–24)	37.5 ± 0.3*	
100 °C DRY HEAT			
Before bathing		36.2 ± 0.4	3.7 ± 0.5
End of first sauna exposure	15 ± 5 (9–21)	37.6 ± 0.5**	4.9 ± 1.4
End of second sauna exposure	11 ± 2 (8–14)	38.2 ± 0.5**	
80 °C DRY AND HUMID HEAT			
Before bathing		36.3 ± 0.3[a]	
End of first sauna exposure	22 ± 6 (17–33)		3.6 ± 0.17
End of second sauna exposure	10 ± 1 (8–12)	39.5 ± 0.7**	8.4 ± 1.5

Values are indicated as average ± standard deviation.

* Significant difference between sauna exposure and basal values, i.e., before sauna (P < 0.01).

** Significant difference between sauna exposure and basal values, i.e., before sauna (P < 0.001).

[a]Average values (three missing values).

Springer *European Journal of Applied Physiology and Occupational Physiology*, Vol. 58(5), 1989, pg. 545, "Haemodynamic and hormonal responses to heat exposure in a Finnish sauna bath," K. Kukkonen-Harjula et al., pages 543-550, table 1. © *European Journal of Applied Physiology*. With kind permission from Springer Science and Business Media.

the S-T segment. These results were statistically different from those obtained in a control group recovering for 5 min in normal environmental conditions. The hemodynamic changes observed by ECG may be the result of myocardial ischemia caused by reorganized blood flow (Taggart, Parkinson, and Carruthers 1972). In addition, a study of 10 subjects (average age 44) using ECG during a 10 min sauna session at 70 °C to 74 °C and 3% to 6% humidity showed that sauna use after exercise is potentially dangerous for healthy people (Paolone et al. 1980). Nevertheless, according to other studies, the risks of myocardial infarction (Romo 1976), heart attack (Suhonen 1983), and sudden death (Vuori, Urponen, and Peltonen 1978) are lower in a sauna than in the course of normal daily activities.

Neuromuscular System Responses

Exposure to heat results in increased transmission of nervous influx and proprioception, as well as improved reaction times (Burke, Holt, and Rasmussen 2001). Heat therapy leads to increased muscle elasticity and joint flexibility, and it is likely to reduce muscle spasms (see review by Wilcock, Cronin, and Hing 2006). In addition, heat increases circulation within joint capsulae and reduces synovial fluid viscosity (Kauppinen and Vuori 1986). However, Prentice (1982) and Sawyer and associates (2003) found that the effect of heat on joint flexibility is only apparent if subjects stretch while in the high temperature environment.

Some studies of neuromuscular modifications examined the effects of sauna exposure followed immediately by immersion in cold water. Results from a primary study on 16 subjects showed a reduction in neuromuscular activation, measured using a surface electromyogram (EMG) 20 min after exposure to extreme temperatures (10 min at 75–85 °C and 30–40% humidity, followed by a cold bath at 16 °C) (De Vries et al. 1968). These authors excluded a vagal effect and suggested that the reduction in muscle tone was linked to reduced gamma motoneuron activity caused by the increase in temperature.

Indeed, the slight acidosis induced by sauna use is related to a depletion in intracellular potassium that leads to a reduction in neuromuscular function and, consequently, a reduction in muscular contractile force. In light of this, it seems that sauna exposure does affect muscular performance. In addition, immersion in a cold bath (16 °C) after this exposure does not appear to restore basal muscle strength.

Sauna use is also likely to play a role in reducing pain and, therefore, improving recovery. According to Kukkonen-Harjula and Kauppinen (2006), sauna use relieves the pain induced by musculoskeletal disorders. These authors suggest that the combination of intense heat and cooling induces an analgesic effect by increasing β-endorphin secretion (Kukkonen-Harjula and Kauppinen 1988). Indeed, a study carried out on subjects suffering from fibromyalgia who were not accustomed to sauna use revealed a better tolerance to pain after 6 weeks of treatment (intense heat/cooling) than after warm-water immersion (Nadler, Weingand, and Kruse 2004). These authors also show a significant effect of heat on reducing perceived muscle pain when using continuous exposure (8 h/day for 2 to 5 days).

Hormonal System Responses

Heat exposure through sauna use induces hemodynamic and endocrine changes similar to those encountered during physical exercise. Sauna exposure (for example, at 80 °C) increases noradrenalin concentrations in the blood and urine (Kauppinen and Vuori 1986) to such an extent that, in young subjects (<14 years), the concentration can triple (relative to basal level), reaching a level equivalent to that seen during maximum-intensity physical exercise (Kukkonen-Harjula et al. 1989). Nevertheless, in contrast with what is seen during stress or physical activity (Christensen and Schultz-Larsen 1994), most studies show no increase in plasma-adrenaline concentrations following sauna exposure (Taggart, Parkinson, and Carruthers 1972). When an increase in noradrenalin concentration was detected, it was associated with increased plasma adrenaline concentration (Taggart, Parkinson, and Carruthers 1972; Tatár et al. 1986). This could be due either to the fact that subjects were not used to this stress or to a bias created by some analysis methods (Kukkonen-Harjula and Kauppinen 1988).

The effects of sauna use on circulation and hormone secretion result mainly from activation of the sympathetic nervous system and from the action of the hypothalamo-hypophyseal axis (Kauppinen and Vuori 1986). These secretions are generally a response to stress that result in energy mobilization, as well as efforts to maintain body temperature (to counteract the effects of sweating). In addition, water is retained by the kidneys, thanks to reduced glomerular filtration. Despite active retention by the kidneys, water and sodium loss induce a reduction in plasma volume, which leads to hemoconcentration, activation of the renin-angiotensin-aldosterone system (Kukkonen-Harjula and Kauppinen 1988) and increased arginine-vasopressin levels (Kukkonen-Harjula and Kauppinen 1988; Leppäluoto, Tapanainen, and Knip 1987).

Thus, plasma angiotensin II, aldosterone, prolactin, and vasopressin concentrations are all significantly increased. Endocrine variations are brief and do not have a permanent effect (Kukkonen-Harjula and Kauppinen 1988). For example, a 15 min sauna exposure to a temperature of 72 °C at 30% humidity significantly increases plasma concentrations of growth hormone (GH) from 9 to 36 μg/L, but concentrations return to their initial value after 30 min (Leppäluoto, Tapanainen, and Knip 1987). According to this study, the increased secretion of GH during sauna use is partly linked to an increase in GHRH (growth-hormone-releasing hormone).

With regard to anti-inflammatory effects, the plasma prostaglandin E2 concentration (PGE2, assimilated with hormonal function) does not appear to be significantly affected by exposure to dry heat (0% humidity) from 80 °C to 100 °C (Kukkonen-Harjula et al. 1989). A study of seven subjects by Jezová and colleagues (1989) measured levels of adrenocorticotropic hormone (ACTH), which is secreted by the anterior lobe of the pituitary gland, and of β-endorphins following sauna exposure. These authors show that exposure to 85 °C to 90 °C at 7% relative humidity for 30 min significantly increased plasma concentrations of these two hormones, from 15 to 42 pg/ml and from 60 to 130 pg/ml, respectively. Nevertheless, Leppäluoto and associates (1975) were unable to show an increase in ACTH in subjects who were used to sauna exposure.

The results from Jezová and colleagues (1985) showed that after 15 min at room temperature after sauna exposure, only the β-endorphin concentration was significantly higher. This could explain the analgesic effect of this recovery mode after exercise (figure 13.6). In addition, the same authors showed that the plasma cortisol concentration—which tended to increase during sauna use lasting 30 min—increased even more rapidly over the 15 min of recovery and became significantly higher: from 12 ng/ml (i.e., basal level) to 19 ng/ml on average. This increase is likely to act on glucose metabolism by increasing glucose synthesis in the liver, thus promoting energetic recovery by glycogen repletion. According to Kauppinen and Vuori (1986), the effects of sauna use on cortisol and ACTH concentrations are variable, and can be explained by differing experimental conditions and a possible effect of habituation to sauna exposure (Kukkonen-Harjula and Kauppinen 1988). In addition, because cortisol varies greatly with circadian rhythm (see chapter 9), the results of some of the studies must be treated with caution.

Finally, the psychological effects of sauna use have been studied on many occasions. Sauna exposure is often primarily used to accelerate a psychological return to calm in athletes. Putkonen and Elomaa (1976) carried out a study on five athletic 18-year-old men who were exposed three times for 10 min to 90 °C temperatures in a sauna. Electroencephalograms (EEG) recorded during the night showed deeper sleep in the population exposed to the sauna (72% higher than the control group). The authors suggest that stimulation of noradrenergic mechanisms leads to a temporary depletion of noradrenalin in the central nervous system and to an increased production of serotonin.

Pulmonary System Responses

A humidity level between 15% and 30% at temperatures from 80 °C to 90 °C allows sufficient humidity to be maintained in the mucosa and the upper respiratory tract. In addition, in the same conditions, net heat exchanges between air and water remain low in the respiratory tract (Laitinen and Laitinen 1988). Breathing is more rapid and becomes deeper, while vital capacity, standard volume, peak flow, and FEV1 are increased within the first seconds (Hasan, Karvonen, and Piironen 1966; Kauppinen and Vuori 1986; Laitinen and Laitinen 1988). These physiological variations are minor (about 10%), similar to those caused by rapid walking (Kauppinen 1989). They return to basal levels over the following 2 h (Kauppinen and Vuori 1986). However, a study carried out by Laitinen and Laitinen (1988) on 12 32- to 58-year-old men (average age: 40) showed that sauna use increases lung-function parameters and

induces greater resistance to lung infections. This study of the influence on numbers of pulmonary infections showed significant results; in particular, the number of lung infections in the population regularly using a sauna was reduced by more than half compared to the control group (not exposed to sauna). In addition, a 3-month study of a different population, including 25 subjects exposed to a sauna and 25 control subjects, showed a 50% reduction in colds in those exposed to heat (Laitinen and Laitinen 1988).

Immune System Responses

Traditionally, saunas are used to prevent various infections (Kauppinen and Vuori 1986). Two hypotheses may explain sauna exposure's antimicrobial properties: The first is that the increase in body temperature has an antimicrobial action; the second is that the high temperature in the sauna favors semiconservative viral DNA synthesis, thus resulting in greater resistance to viral infections (Kauppinen and Vuori 1986). Nevertheless, heat therapy must be used carefully, since it can have a negative effect on inflammation (Nadler, Weingand, and Kruse 2004). Indeed, raising body temperature results in an inflammatory response and can lead to swelling that could prolong recovery time after physical exercise (Magness, Garrett, and Erickson 1970; Wallas, Warren, and Kowalski 1979; Coté et al. 1988). During sauna exposure, the number of leucocytes increases, while the number of polynuclear eosinophils decreases (Kauppinen and Vuori 1986).

In addition, Coté and colleagues (1988) observed a significant increase in edema size in 30 subjects suffering from sprained ankles following 20 min immersions in hot water repeated over 3 consecutive days. This phenomenon also appears to apply to exercised muscle (Feibel and Fast 1976). Because of this, sauna use to reduce any type of trauma must be done carefully.

Fatigue Effects

Athletes who are involved in physical activity programs or training increasingly have access to saunas. Consequently, hot-air baths are frequently used shortly after exercise (Paolone et al. 1980). Capacity to perform exercise is reduced in a hot environment compared to colder conditions. Indeed, an increased core-body temperature is a limiting factor for performance, and is accompanied by an increased subjective perception of

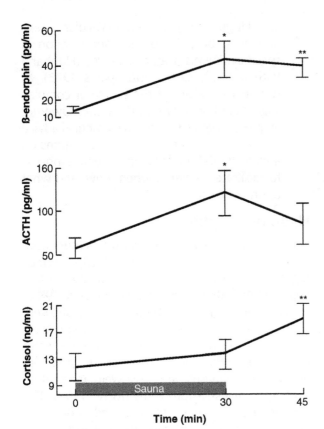

Figure 13.6 Variation in plasma levels of β-endorphin, adrenocorticotropic hormone (ACTH), and cortisol in response to sauna exposure lasting 30 min. Values are indicated as average ± standard deviation.

* Significant difference from basal values, i.e., before sauna (P < 0.05).
** P < 0.01.

Data from Jezová et al. 1989.

fatigue (i.e., RPE, or rating of perceived exertion). Based on this principle, Simmons and associates (2008) compared two techniques during sauna exposure (67.7 ± 0.8 °C at 30% humidity), one consisting of cooling the head and face using vaporizers and ice packs (HC condition), and the second without cooling the head (CON condition). In these two conditions, subjects sat with their torsos covered with plastic in order to maximize the increase in core temperature. Subjects were exposed to heat until their internal body temperature increased by 2 °C, with a maximal exposure of 90 min.

To evaluate the effect of these two practices on fatigue, 9 subjects exercised for 12 min at 70% V̇O₂max before and after each recovery session in a room with a controlled temperature of 34 °C. Following the second exercise session, results indicate a significant reduction in RPE scores, as well as increases in core-body and skin temperatures (P < 0.01) in subjects from the HC group compared to the CON group. The HC group's temperature was 0.5 °C lower than that of the CON group (P < 0.05) at the beginning and end of the second 12 min exercise session. And only 4 of the 8 members of the CON group were able to complete the second exercise session. This recent study shows a significant correlation (r = 0.82, P < 0.01) between increases in RPE score and core-body temperature. In addition, when the temperature reached 39.4 °C, voluntary muscle action and maximal contractile strength were both reduced (Simmons, Mündel, and Jones 2008). This decline in muscular performance is linked to an increase in muscle temperature, which affects contractile properties (Todd et al. 2005).

Contrasting temperatures combining localized hot and cold atmospheres have been suggested to improve athletic recovery through the following:

- Stimulation of blood flow to a specific area
- Increased movement of blood lactate
- Reduced inflammation and edema
- Stimulation of circulation
- Relief of stiffness and muscle pain
- Increased movement amplitude
- Reduced muscle pain and delayed appearance of pain (Wilcock, Cronin, and Hing 2006)

With the introduction of whole-body cryostimulation chambers, we should soon see research protocols based on contrasting ambient temperatures (e.g., combining sauna and WBC).

Athletic Recovery

Many athletes competing in weight categories regularly use a sauna for prolonged periods to rapidly lose body mass through water loss. However, this application of saunas can result in reduced performance. Indeed, prolonged sauna use causes water loss, leading to loss of weight and electrolytes. This loss can hinder sporting performance (Gutiérrez et al. 2003). In a sauna at 80 °C to 90 °C, average water loss is about 0.5 kg (Kukkonen-Harjula and Kauppinen 1988) and can reach 1 kg (Ahonen and Nousiainen 1988). In a study by Gutiérrez and colleagues (2003), 12 subjects (men and women) were exposed to three cycles combining 20 min in a 70 °C sauna and 5 min of rest at room temperature. The results show weight loss of up to 1.6% (± 0.6) in women and 1.8% (± 0.5) in men. For women, despite rehydration after this treatment, squat jump performance was significantly lower than before sauna sessions (23.7 [± 2.2] vs. 25.2 [± 1.4 cm], P < 0.05). The reduced performance in squat jumps for women correlated directly with the amount of weight lost. This correlation was not seen in men.

Small reductions in the body's fluid reserves affect performance. This is particularly the case during prolonged physical exercise, but also when exposures to hot atmospheres (such as a sauna or steam room) are repeated without monitoring. Thus, given the importance of water and the consequences of any deficit, it is necessary to compensate for the losses due to sauna sessions and general daily activities as rapidly and completely as possible. To avoid the physiological perturbations and negative effects on performance linked to dehydration as far as possible, it is necessary to ensure adequate hydration by applying the following advice:

- *Replace lost body fluids as rapidly as possible.* This is done to maintain sweating rate and avoid excessive temperature changes, while maintaining blood volume and heart rate at adequate physiological levels. This is even more important if sauna sessions are repeated and prolonged.

- *Consume a glucose-based energy source.* After exercise or sauna, replace some of the energy spent.

- *Consume mineral elements.* These are required by the body to maintain plasma osmolarity so that sauna use may truly contribute to recovery. Rehydration during multiple sauna-based recovery sessions is essential. The main problem is how fast lost fluids can be replaced; this largely depends on the composition of the rehydration solution consumed. The effects of temperature, moment of ingestion, and quantities consumed will all need to be taken into account for a given drink. The osmolarity of the rehydration solution should be considered (Costill and Saltin 1974; Costill and Sparks 1973), as should its temperature (Costill and Saltin 1974), its sugar concentration, and its osmotic (Costill and Saltin 1974; Hunt and Pathak 1960) and energetic properties (Brener, Hendrix,

and Hugh 1983). Finally, particular attention must be paid to the concentration in specific minerals (Na^+, Cl^-, and K^+) (Hunt and Pathak 1960; Costill and Sparks 1973), as well as the nature of the simple or complex sugars contained within the drink (Owen et al. 1986). We will try to give practical advice in answer to several important questions on athlete hydration following sauna exposure to facilitate recovery: when, how, and what to drink?

Hydration Options

Since sauna use increases both body temperature and heart rate, drinking water will help thermoregulation by lowering the core temperature and adjusting the rate of sweating (Sawka et al. 1984).

As Hubbard and colleagues (1984) showed, lightly flavored cool water at 15 °C is more likely to induce a desire to drink and to increase water intake than other types of drink. In addition, hypertonic solutions are less well assimilated, leading to digestive problems (Riché 1998). The combined presence of glucose and sodium, at appropriate levels, improves water absorption by intestinal cells and its passage into the blood. In addition, exogenous glucose sources also spare endogenous glycogen stores (Flynn et al. 1987; Pallikarakis et al. 1986). The work of Bothorel (1990) has shown that drinking iso-osmotic solutions (osmolarity between 290 and 310 mOsm/kg; total sugars between 50 and 60 g/L) is perfect for maintaining sugar homeostasis. All solutions of this type allow relatively stable glycemia to be maintained (around 1.1 g/L). Owen and associates (1986) also showed that the osmolarity of the dissolved sugars was the main factor in restoring lost water during recovery. The ideal concentration in sugars to optimize the energy provided to the active muscles during the recovery phase is about 45 to 60 g/L. In a hot atmosphere (like a sauna), ingestion of water with low carbohydrate (CHO) concentrations and adequate mineral concentrations should be favored (figure 13.7).

Mineral concentrations allowing optimal restoration of plasma volumes with stable osmolarity are between 0.5 and 0.6 g/L for Na^+, 0.7 and 0.8 g/L for Cl^-, and from 0.1 to 0.2 g/L for K^+. Lamb and Brodowicz (1986) underline the benefit of adding some salt to the water used during rehydration, up to 1.5 g/L, to optimally balance plasma-electrolyte concentrations. The amount

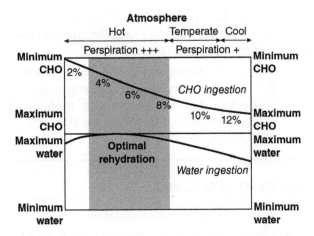

Figure 13.7 Model of choice of drink composition depending on ambient conditions. The upper part of the figure concerns carbohydrate (CHO) ingestion, while the lower part deals with water consumption. In a hot atmosphere (e.g., sauna or steam room), the left part of the figure indicates that water with low carbohydrate concentrations should be favored.

should never exceed 3 g/L, even if sauna exposure is longer than usual, since this concentration (50 mM sodium) is not well tolerated, and induces the opposite effect to that sought.

Hydration Timing

During the different phases of recovery using hot atmospheres, the sequence of fluid intake and the volumes ingested each time are important: Because of its elasticity, the stomach will empty better if it is full. With this in mind, the most efficient process consists of drinking small quantities (about 200 to 250 ml) of an iso- or hypotonic glucose solution at regular intervals (every 15 min) (figure 13.7), at a cool temperature (between 8 °C and 12 °C). Stomach emptying is maximal with water at 5 °C, and it slows significantly with drinks above 25 °C. Cool drinks are recommended because the drink ingested also contributes to body cooling, particularly after a sauna (Snellen and Mitchell 1972). Drinking a cool drink when one is hot will also be perceived as pleasantly refreshing. It improves voluntary fluid intake (Armstrong et al. 1985) and, consequently, the replacement of body fluids, which is essential if the benefits of the sauna are to be felt. On

the other hand, it is not a good idea to drink iced drinks because signals from the deep heat receptors (in the viscera) might have negative effects on the thermoregulatory system, which may then act as if the body did not need to evacuate excess heat (Candas and Bothorel 1989).

Rehydration compensates the effects of hypersomatic hypovolemia. The quality of recovery of physical capacity depends on the quantity of liquid absorbed to compensate fluid losses. With water low in mineral content, rehydration causes osmodilution, which leads to a reduction in glycemia. Better stabilization of physiological parameters is obtained by drinking iso-osmotic drinks containing mainly sugars and sodium. It is of note that in athletes, spontaneous rehydration, guided by thirst, is always insufficient to cover the fluid loss linked to exercise in a warm atmosphere or simple exposure to heat (Bothorel 1990). This is why athletes should drink regularly (200 to 250 ml several times per hour), rather than rehydrate based on their own perception of thirst.

In numerous studies, the sensation of thirst has been clearly shown to be a poor factor to manage optimal rehydration. However, the total amount of fluid an athlete can drink is also determined by the maximum quantity of water absorbed by the intestine (12 to 15 ml · kg^{-1} · h^{-1}, or between 800 ml and 1 L/h for a 70 kg athlete). Several studies suggest that hyperhydration may be beneficial prior to sauna exposure to reduce the delay before sweating, reduce core temperature, and increase sweating rate (Grucza, Szczypaczewska, and Kozlowski 1987). However, these theoretical benefits are not often seen in practice, and pre-exposure hyperhydration has not been proven to improve thermoregulation (Candas et al. 1988; Nadel, Fortney, and Wenger 1980). On the other hand, it is necessary to drink enough before exposure to heat (in a sauna or steam room) so as not to add to the heat stress, and to limit the inevitable increase in heart rate (Davies 1975).

In a more global, daily, hydration context—needed to limit the risk of dehydration faced by the athlete in a warm, or hot and humid atmosphere—fluid intakes should be particularly encouraged during training, recovery, and at the start of the day. A certain level of dehydration is inevitable during physical activity, and recovery must therefore be used to optimize rebalancing of water and mineral levels. This concerns all sporting disciplines, not just those taking place over prolonged periods. To replace water lost as sweat, pure water is not the best drink to consume after exercise, particularly when a sauna is used regularly. It seems clear that postexercise, postsauna rehydration can be achieved only if electrolytes lost through sweating are replaced at the same time as fluids.

Sauna Contraindications

Generally, the risks linked to the use of sauna are relatively low if caution is exercised. The scientific data have allowed us to establish a non-exhaustive list of contraindications and possible risks.

CONTRAINDICATIONS

- Orthostatic hypotension.
- Rheumatoid arthritis in the acute inflammatory phase.
- Fever of any form.
- Skin abrasions and exuding rashes.
- Treatment with medication affecting the nervous system.
- Claustrophobia and epilepsy are contraindications.
- Some skin diseases (cholinergic urticaria) are contraindicated. However; sauna use has been shown to alleviate psoriasis.

POSSIBLE RISKS

- Venous problems (particularly varicose veins), without being a formal contraindication, are an indication that sauna should be avoided.
- Adequately managed hypertension is not a contraindication, but cold showers during and after sauna must be avoided (cold-pressor effect).
- Do not prolong exposure beyond 15 min, as some cases of fainting due to reduced blood pressure have been reported. Sauna use is not recommended for people suffering from heart failure or myocardial infarction, or for those who have a history of heart problems. Some studies were carried out on people suffering from these diseases, because sauna reduces cardiac preload and afterload, and improves endothelial function. But the variations in blood volume are undesirable. In addition, the cold shower may cause coronary spasms.

- No nicotine patches should be worn, since they increase plasma nitroglycerine concentration (headaches).
- Pregnancy: Sauna use is not recommended during pregnancy or when pregnancy is suspected. Finnish women use traditional saunas, which heat the body less than infrared saunas, and only for 6 to 12 min.

NONRISKS AND ADVICE

- Asthma is not a contraindication.
- When using sauna for the first time, it is preferable to limit exposure to a maximum of 5 minutes, interspersed with cool showers.
- Fertility does not seem to be reduced after multiple exposures.

Far-Infrared Therapy

The use of far-infrared (FIR) therapy as a recovery, rehabilitation, or regeneration technique is quite recent. This technique first appeared in the East in the 1950s, before being developed in the West 50 years later. The first users of the technique were the Chinese and Japanese (in particular, Dr. Tadashi Ishikawa), who in 1965 submitted a first patent for ceramic and zirconium infrared heating elements to improve the healing process. Despite the lack of scientific proof of its efficacy, this technique has become widely used, and it should not be ignored. Indeed, more than 700,000 thermal infrared systems for therapeutic applications have been sold in the East.

The scientific studies on the potential benefits of infrared treatment are few and far between (Honda and Inoue 1998; Udagawa and Nagasawa 2000), and none specifically tackles athletic recovery. However, Lehmann and Delateur (1990) gathered together a large collection of data (mainly therapeutic observations or clinical cases) on the therapeutic effects of infrared heat in a wide-ranging review. According to these authors, this heat has several beneficial therapeutic effects, which we will come back to a little later. However, due to the lack of scientific references, the reader must not forget that the effects described here are only observations. They are not the results of rigorous scientific studies. In addition, because no study so far has attempted to dissociate the effects of infrared radiation from the heating effect produced, it is impossible to conclude on the specific effect of a given wavelength. Common sense and objective analysis must be applied when considering the effects of this recovery technique.

Infrared Technology

The electromagnetic spectrum divides all electromagnetic waves, not only according to their wavelength (or frequency), but also according to their physical properties. At one end of the spectrum, gamma rays are the highest-frequency, highest-energy waves. They are used in industry to detect faults in metals. As we move through the spectrum, frequency and energy both decrease. X-rays are still energetic enough to produce the medical images we are all familiar with. Then we have ultraviolet rays, visible light, infrared, and, finally, radio waves. Discovered in 1800 by the English astronomer William Herschel, infrared radiation is an electromagnetic radiation whose wavelength is longer than that of visible light, but shorter than that of microwaves (between 0.7 and 1,000 μm). Infrared (IR) is subdivided into near IR (NIR: from 0.5 to 1.5 μm; i.e., IR-A), mid IR (MIR: from 1.5 to 5.6 μm; i.e., IR-B), and far IR (from 5.6 to 1,000 μm; i.e., IR-C) (Toyokawa et al. 2003).

This classification is not universally applied: The limits vary from one field of expertise to another, and none are more or less valid than others. The split can be based on the wavelength (or frequency) of emitters, receptors (detectors), or even atmospheric transmission bands.

Infrared rays emit heat that can be partially absorbed by molecules: Part of the rays will be reflected, and the remainder is transmitted into the surrounding air (see figure 13.8). The earth's atmosphere ensures that the heat produced as infrared radiation by the solar spectrum is bearable for humans.

Infrared-C radiation (long wavelength) is considerably attenuated in the atmosphere, and it is also absorbed by water and atmospheric water vapor. Infrared-B radiation is, for the main part, also absorbed by water-based media. Thus, infrared-B and -C radiation are naturally filtered (or at least considerably reduced) by air and atmospheric humidity. Infrared-B and -C have very superficial effects and can work either directly through contact with the outer skin layers, or indirectly by heating the surrounding air. Indeed, a significant concentration of water molecules is present in the upper layers of the epidermis, constituting a real barrier for this type of radiation, mimicking the earth's atmosphere. Consequently, with the exception of some very limited parts of the IR-B spectrum, infrared-B and -C rays cannot penetrate the skin, and superficial absorption

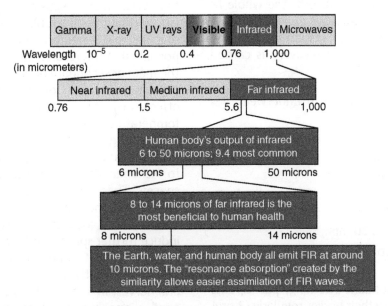

Figure 13.8   This figure shows the promotive effects of far-infrared ray on full-thickness skin wound healing in rats.

Based on Toyokawa et al. 2003.

leads to heating. On the other hand, infrared-A rays are quite compatible with the skin, except for the following absorption bands: 944 nm, 1,180 nm, and 1,380 nm, which are absorbed by aqueous media.

Therapeutic radiation should be safe, but it must still be compatible with the skin's layers so as to penetrate deep into the tissues. Therefore, there is only one alternative: either to use infrared sources emitting a defined wavelength and excluding harmful wavelengths, or to filter the radiation spectrum of a classical infrared source to eliminate the noxious spectral bands and those that are blocked at the skin's surface. The advantage of the latter technology is that it enables the use of intensive radiation over large surface areas. The use of one or another of these technologies depends mainly on the manufacturer. Thus, the technique developed for the HydroSun emitter uses radiation from a halogen lamp filtered through a specific hydrocuvette to block the skin-damaging spectral bands. The resulting radiation is in the infrared-A band, which, because of its compatibility and deep penetration, is appropriate for therapeutic purposes. Other techniques, such as the INOVO, currently used at the INSEP (National Institute of Sport, Expertise and Performance) in France, emit radiation between 4 and 14 μm with a peak at 9 μm corresponding to the spectrum of the human body. When using this technology, the infrared waves are diffused by ceramic powder, which absorbs the whole IR spectrum and emits long IR rays more effectively.

A far-infrared therapy session (using the INOVO system) to optimize recovery is carried out as follows:

- The participant, clothed in underwear, lies down in a chamber.
- The participant's head, laid on a cushion, must not be exposed to the infrared waves or to the heat produced.
- The temperature is set at 45 °C for the torso and lower limbs.
- The session lasts 30 min.
- It is possible to watch a film or to listen to relaxing music.
- In case of abundant sweating, participants are asked to dry their bodies regularly (3 or 4 times per session).

Collagen Tissue Elasticity and Joint Rigidity Effects

Tissues are heated to 45 °C and stretched. They can then be elongated by about 0.5% to 0.9%, and this effect persists after the end of the stretch. It is not possible to stretch these tissues at normal temperatures. Twenty stretching sessions can lead to a 10% to 18% increase in the length of a heated and stretched tissue. Tissue extension, combined with heat, significantly contributes to repair of damaged, thickened, or contracted ligaments, capsulae, tendons, and synovia. When attempting to restore tissue stretching, tension at 45 °C weakens the tissues much less than at normal temperatures. The experiments mentioned here have shown that light tension can induce significant residual elongation when heat is associated with stretching exercises or a range of movement exercises. The effects measured at room temperature were optimal when radiating heat was used.

There was a more significant reduction in joint rigidity at 45 °C (20%) than at 33 °C for rheumatoid fingers. This correlates perfectly with subjective and objective observations of stiffness. According to clinical studies, stiffened joints and thickened conjunctive tissues react in a similar way.

Muscle Spasms and Pain Relief

Many studies have investigated the use of heat to alleviate muscle spasms, in underlying muscles (supporting the skeleton), in muscles that are part of joints, and in neuropathological conditions. It is possible that this result is an effect of heat both on the main or secondary nerves afferent to the spindle cells and on the Golgi tendon organs. Maximal effect was achieved with a range of therapeutic temperatures applied using radiating heat.

Reducing specific or secondary spasms can lead to pain relief. Sometimes pain may be linked to ischemia (lack of blood) due to tension or spasms. These may be improved by heat-induced hyperemia and vasodilation, thus interrupting the feedback loop that causes the ischemia to induce further spasms and more pain. It has been shown that heat can reduce the sensation of pain by acting directly on tissue-specific and peripheral nerves. In a study of dentition, repeated heat applications eliminated the reaction of all the nerves responsible for pain in the dental pulp.

Heat can increase endorphin production and close off what is known as Melzack and Wall's "gate control" (1965), since both heat and endorphins reduce pain. Localized infrared therapy (with lamps adjusted to between 2 and 25 μm) is used to treat and relieve pain in more than 40 reputable Chinese medical institutions. In addition, Masuda and colleagues (Masuda, Kihara, et al. 2005; Masuda, Koga, et al. 2005) showed that repeated heat application using far-infrared therapy significantly reduced pain and fatigue in patients suffering from chronic fatigue syndrome (figure 13.9).

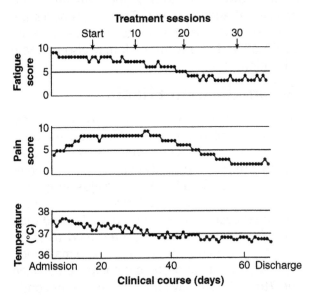

Figure 13.9   Evolution of fatigue and pain in patients suffering from chronic fatigue after repeated treatment with far-infrared therapy.

Based on Masuda et al., 2005.

Blood-Flow Response

Several studies have reported an increase in blood flow when using far-infrared therapy in animals (Yu et al. 2006; Akasaki et al. 2006) or humans (Lin et al. 2007). Indeed, heating a part of the body induces vasodilation in distal areas of the body, even in the absence of a change in body temperature. Increased peripheral blood flow improves edema evacuation, which can help reduce inflammation and contribute to pain relief and more rapid healing. Heating an extremity can also lead to lateral dilation; heating a forearm leads to dilation in the two lower limbs, and heat-

ing the front of the torso induces dilation in the hands. Heating muscles results in increased blood flow similar to that induced by exercise. Heat also stimulates blood flow and dilation of the capillaries, arterioles, and venules, probably through direct action on smooth muscles. Release of bradykinin, as a result of sweat-gland activity, also increases blood flow and vasodilation. Hyperthermia, linked to application of heat, induces vasodilation (through hypothalamic reduction) of the arterio-venous anastomoses. Vasodilation is also produced by axonal reflexes that change the vasomotor balance.

Repair of Light Tissue Damage

In some countries, infrared healing is becoming a common treatment for light tissue damage, to relieve chronic or permanent untreatable cases, and to heal new injuries. Thus, Chinese studies note the positive effects of infrared heat on this type of injury. Researchers mention more than 90% success in a summary of Chinese studies evaluating the therapeutic effects of infrared heat. Similarly, in *La Thérapie Infrarouge*, Japanese teams report that infrared heat acts on the entire human body for the following conditions:

- Rachialgia (spinal pain)
- Rheumatoid diseases
- Muscular fatigue
- Burns (alleviates pain and reduces the healing period with fewer scars)
- Weight loss (due to sweating, energy expenditure to produce sweat, and direct excretion of fat)

Thus, although none of the articles provides scientific proof for these allegations, rigorous measurements do lend some credit to the observations reported. In contrast, some of the aspects mentioned in many documents seem far too speculative. Therefore, these articles must be read and interpreted with extreme caution. In addition, the distinction between the direct effects of infrared waves and those of the heat they produce is never made. To date, no study has, for example, compared recovery using a classical sauna to that using an infrared sauna. The fact that we lack such significant information means that we should be very careful about using this type of technology.

Some manufacturers offer material that allows the head to be outside the treatment zone. This may present an advantage, particularly for those suffering from claustrophobia.

FIR Contraindications

The concerns for restoring water loss through adequate hydration, contraindications, and general precautions linked to the use of FIR would include all of those listed for the sauna technique (see the Practical Application section), as well as the following:

- *Surgical implants.* Metal pins, rods, prostheses, or other surgical implants generally reflect infrared rays and are not heated by them. However, anyone with a surgical implant should consult his or her surgeon before using this type of therapy. In addition, infrared therapy should obviously be interrupted if a person experiences pain in the area of the implant.
- *Silicone.* Silicone absorbs infrared energy. Implanted silicone or silicone prostheses (to replace cartilage in the nose or ears) can be heated by infrared rays. Since silicone only melts above 200 °C, it should not be affected by infrared heating systems.
- *Pain.* No pain should be felt when using infrared heating systems. This type of radiating heat should be discontinued by any person experiencing pain.

Humid-Heat Steam Room

Although heat production due to exercise is a real advantage when athletes are exposed to cold, it is quite the contrary when exercise takes place in a hot, humid atmosphere. In the context, for example, of the Olympic Games in Beijing, where average temperatures were close to 30 °C and humidity was high, metabolic heat production became a considerable load for all the thermoregulatory mechanisms. The physiological changes implemented to combat this heat obviously have a negative effect on sporting performance that is non-negligible.

Exposure to humid heat leads to increased sweating, which depletes the body's fluid reserves and creates a state of relative dehydration. Thus, the efficacy of the body's thermal regulation comes at a high price: loss of intravascular and intracellular liquids. This fluid loss is even more significant in humid atmospheres. Because water evaporation from the skin's surface is minimal due to the significant vapor pressure in the surrounding air, sweating contributes only marginally to cooling of the body.

The temperature in a steam room (around 40 °C) is much lower than in a sauna (between 80 °C and 100 °C), but with a relative humidity close to 100%. The rhythm of exposure to steam rooms is set at once per week. This session, lasting between 10 and 20 min, is generally followed by a cold shower, so as to rinse the body and eliminate sweat and impurities. This step is necessary to cause vasoconstriction after the heat-induced vasodilation, but a warm bath may also promote relaxation.

Acclimatization Effects

To benefit from the effects of exposure to humid heat, it is important to understand how the body adapts to these conditions. Acclimatization, whether through artificial or natural means, seems to set in quite quickly. Indeed, major progress is visible from the first day of acclimatization, and it continues over the next 3 or 4 days. Beyond that, the various parameters measured tend to stabilize. It is generally considered that acclimatization to a hot, humid atmosphere is complete between the 7th and 10th days (Pandolf 1998). In conditions involving humid heat, it is recommended that the intensity of exercise be reduced by about 60% to 70% for the first few days to avoid heat stress and its consequences.

It should also be noted that, according to several authors, a high level of aerobic potential (represented by the athlete's maximal oxygen consumption) improves tolerance to exercises performed in a hot, humid atmosphere and allows for faster acclimatization (Pandolf, Sawka, and Gonzales 1988; Sawka, Wenger, and Pandolf 1996). Finally, it seems important to indicate that the effects of an acclimatization program are gradually lost once the person is no longer exposed and disappear completely after a month, regardless of the humidity level associated with the excess heat. There are also significant differences between individual reactions to heat stress (Pandolf 1998). Some people feel few effects and acclimatize rapidly to the heat, while others encounter insurmountable difficulties. A preliminary appraisal of each athlete's ability to adapt to heat is recommended before going to a hot, humid climate or to the steam room.

All these remarks may be applied to the daily use of a steam room as an effective means of recovery. The notion of acclimatization to humid heat is important so as not to induce heat overload, which would delay recovery. Even if almost no scientific study has evaluated the effects of exposure to humid heat on recovery following exercise, it is necessary to consider the scientific data on hot atmospheres to make recovery plans as coherent as possible. As a general rule, humid heat tends to increase dehydration. Therefore, particular attention must be paid to hydration.

Athletic Recovery

To our knowledge, very few scientific studies have explored the effects of humid heat on the body. Steam rooms (or Turkish baths) have been used since antiquity. They were described first in Greek texts, where the first baths were installed in the year 2000 BC. They were later described in the Maghreb and the Middle East. Since the start of the 21st century, steam baths have been rediscovered by a growing number of French people. But, as early as 1746, Diderot wrote: "If warm steam is to be recommended, it is because it is eminently capable of causing sweat to pour out, opening the skin's blood vessels, relaxing stiffness and tension and even dissolving tenacious and viscous humors" in his *Encyclopédie*. Alain Rousseaux, in his work *Sauna*, published in 1990, reports "The dry or humid sauna leads to elimination from the body through the skin. The latter, stimulated by the heat excretes sweat through the sweat glands, which we might compare to minute kidneys. Sweat effectively contains the same toxic components as urine. Thus, sweating allows the body to directly reject metabolic residues which might hinder the correct functioning of the system."

In reality, we are not simply subjected to heat, but we adjust to the various temperatures by favoring the body's reactions. This is why a steam room session should not be improvised. In a steam room, the humidity level is close to 100% and the

PRACTICAL APPLICATION

Generally, the risks linked to the use of variations in thermal ambience are relatively low if caution is exercised. The scientific data have allowed us to establish a nonexhaustive list of contraindications and possible risks for sauna usage, which applies in general to WBC, FIT, and steam room usage as well.

Contraindications

- Orthostatic hypotension
- Rheumatoid arthritis in the acute inflammatory phase
- Fever of any form
- Skin abrasions and exuding rashes
- Treatment with medication affecting the nervous system
- Claustrophobia and epilepsy
- Some skin diseases (cholinergic urticaria)

Possible Risks

- Venous problems (particularly varicose veins), without being a formal contraindication, are an indication that sauna should be avoided.
- Adequately managed hypertension is not a contraindication, but cold showers during and after sauna must be avoided (cold-pressor effect, coronary spasms).
- Do not prolong exposure beyond 15 min, since some cases of fainting due to reduced blood pressure have been reported.
- Sauna use is not recommended for those who have a history of heart problems.
- No nicotine patches should be worn, since they increase plasma nitroglycerine concentration (headaches).
- Pregnancy: Sauna use is not recommended during pregnancy or when pregnancy is suspected.

temperature does not exceed 45 °C. This contrasts with the sauna, where the air is dry and the temperature very high. The cooling mechanism is therefore very different, since the water-saturated air prevents evaporation.

The steam acts to open the pores and activate the sweat glands, allowing the body to eliminate toxins. The water particles from the steam condense as fine droplets on the cooler body, giving an impression of sweating. At the same time, the humid heat softens the corneous layer and the fatty substances under the skin, to promote the shedding of dead cells.

Steam Room Contraindications

Before the first session, a medical visit can be useful to exclude any contraindications (heart problems, hypertension, venous problems, and so on). Use of steam rooms is not recommended during pregnancy. It is essential to rehydrate adequately after a steam room session.

The concerns for restoring water loss through adequate hydration, contraindications, and general precautions linked to the use of steam room therapy would include all of those listed for the sauna technique (see the Practical Application section).

Summary

Artificial heating or cooling of ambient temperature is an expanding technique, both to prepare athletes for competitions in difficult conditions and to improve the body's recovery capacity.

- The high temperature of the steam room releases muscle tension and stiffness. Thanks to water vapor, the sweat glands secrete and eliminate waste, cleansing the skin deep down. The steam room also provides a sensation of relaxation and well-being.

- During exposure to sauna, body temperature increases in line with the increase in ambient temperature. This rise in body temperature induces physiological changes at the cardiovascular, pulmonary, and neuromuscular levels, as well as to the inflammatory, hormonal, and immune systems.

- When we use FIR therapy at 45 °C, there is a more significant reduction in joint rigidity, and it alleviates muscle spasms in underlying muscles (supporting the skeleton), in muscles that are part of joints, and in neuropathological conditions. Maximal effect was achieved with a range of therapeutic temperatures applied using radiating heat.

- Regarding the WBC data already available in the scientific literature, there are some benefits on some inflammatory parameters, possible improvement of antioxidant status, and improvements in mood and mild depression.

Enhancing Short-Term Recovery in Handball Players

François Bieuzen, PhD, and Christophe Hausswirth, PhD, National Institute of Sport, Expertise and Performance (INSEP)

In many sports (such as soccer, basketball, handball, ice hockey, fencing, judo, American football, and so on) athletes have short periods of rest (from a few minutes to 1 h). During these rest periods, fatigue accumulated during the previous game or round decreases slowly. Athletes who can eliminate it quickly have a performance advantage. However, in most cases, athletes are sedentary, doing nothing and waiting for the next event.

In the best practice, if conditions, space, and money permit, they would use different recovery methods during these times, such as massage, active recovery, or water immersion. One of the main strategies of these recovery modalities is to increase blood flow to eliminate waste products and by-products more quickly. But as previously alluded, these techniques require equipment, space, expert staff, and motivation on the part of the athlete to utilize them.

Recently, a new electrostimulation device (Veinoplus Sport, AdRem Technology, Paris, France) that has been derived from medical applications was shown to cause significant improvement in the total blood flow in the lower limbs. This has been claimed to improve performance. In our view, it was worth testing this device in terms of its capacity to improve short-term recovery. It had the advantages of being small in size, easy for athletes to use by themselves (no staff required), and no effort or discomfort associated with its use.

We had 14 female professional handball players perform two exhaustive tests (yo-yo intermittent test) with a 15 min recovery period in between. During the recovery period, each player used the following techniques:

- Active recovery: 15 min of cycling at 40% of $\dot{V}O_2$max.
- Passive recovery: Players sat quietly in a chair for 15 min with minimal movement.

- 15 min of specific stimulation of calf muscles in a seated position using the aforementioned device.

When players used the device and active recovery modalities, there was no difference between the performances during the two yo-yo tests. However, with passive recovery, distance covered between the first and the second yo-yo test decreased by 16% (figure 1).

Electrostimulation (7.4 ± 1.3) and active (7.5 ± 1.2) recovery methods were associated with higher ratings of recovery perception than passive recovery (4.0 ± 1.1). This was monitored using self-perception of the effectiveness of the recovery method using the Likert scale: 0 = Poor; 10 = High.

During recovery using the device and active recovery, blood lactate returned toward baseline levels more quickly (figure 2). However, with passive recovery, blood lactate concentration is still significantly higher from its baseline level. This

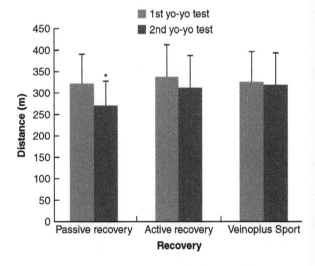

Figure 1 Distance covered during the yo-yo intermittent recovery test.

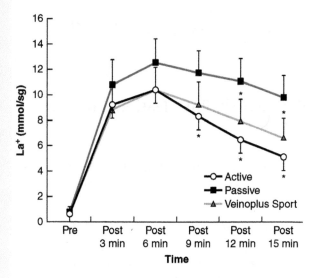

Figure 2 Blood lactate concentration immediately after the first yo-yo test.

suggests that it is less effective than the other methods in assisting clearance of lactate from the blood.

From these results, it appears that stimulation with this device could be an effective recovery method when the period between two events is short or when other recovery methods are impractical. The device is quite passive and easy to use, and it requires little effort or energy demand from the user.

Key Points

No difference between active recovery and recovery using Veinoplus Sport regarding the two following key parameters:

- Restoration of anaerobic performance measured with the yo-yo test
- Blood lactate clearance

CHAPTER 14

Water-Immersion Therapy

François Bieuzen, PhD

The water-immersion recovery technique consists of covering part of the body, or the whole body, in water. The scientific literature describes four main means of immersion, which differ depending on the water temperature used:

- Thermoneutral-water immersion, between 15 °C and 36 °C (TWI)
- Hot-water immersion >36 °C (HWI)
- Cold-water immersion <15 °C (CWI)
- Contrasting water temperature (CWT), which consists of alternating immersion in cold and hot water.

Several aspects of the results of these recovery procedures, including metabolic, neurological, cardiovascular, and muscular aspects, were studied as a function of various indicators of fatigue. Different immersion methods were also studied (Barnett 2006). Athletes generally use either hot- or cold-water immersion rather than a combination of the two (Howatson and van Someren 2003).

Based on the assumption that immersion in varying depths of warm water may be more beneficial than simple immersion, recent work has studied the effects of both water immersion and hydrostatic pressure on early recovery (Wilcock, Cronin, and Hing 2006).

Physiological Responses to Water Immersion

The hypotheses advanced to explain the improved quality of recovery, or the reduced time necessary for recovery, when using immersion are mainly linked to three factors: hydrostatic pressure (Vaile et al. 2008b, c), analgesic phenomena and an anti-inflammatory process (linked to local vasoconstriction in the case of immersion in cold water), and vasomotricity (when contrasting water temperature immersion is used) (Cochrane 2004).

In the first case, hydrostatic pressure under water is higher than in air (at sea level) over the entire surface of the body. This causes gases, substances, and fluids to move toward the thorax; enables improved recovery; and reduces the swelling induced by exercise (Wilcock, Cronin, and Hing 2006). In addition, the nervous influx should be limited by compression of muscles and nerves due to the hydrostatic pressure.

Immersion in thermoneutral water was used to determine how hydrostatic pressure affects recovery. In the literature, this type of immersion consists of submerging all or part of the body in water at a temperature between 16 °C and 35 °C. However, water is considered thermoneutral when body temperature can be maintained when immersed for 1 h (35 °C). Thus, depending on the amount of subcutaneous fat, body temperature can be maintained for 1 h in water between 30 °C and 34 °C. Indeed, no alteration in body temperature is noted for subjects immersed in water at temperatures between 33 °C and 35 °C.

In the second case, the decrease in body temperature, due to immersion in cold water, also has benefits for nerve transmission and inflammatory phenomena. Exposure to cold water may alter nerve transmission by reducing temperature. Nevertheless, other phenomena may be involved, such as local vasoconstriction, which limits both metabolic by-products and the extent of inflammation, thus reducing the pain and discomfort experienced during movement. In addition, exposure to cold reduces the heart rate and increases peripheral resistance as the body seeks to counterbalance the drop in body temperature. To survive, the body favors irrigation of the core rather than the extremities (limbs) (Bonde-Petersen, Schultz-Pedersen, and Dragsted 1992). Historically, cold-water immersion was first used for therapeutic purposes for its analgesic potential, since it plays a major role in the treatment of acute muscle injury. It consists of submerging the part of the body to be treated in water between 4 °C and 16 °C for 5 to 20 min (depending on the study).

In the third case, cold temperatures cause blood vessels in the skin to constrict, while warm temperatures induce vasodilatation (Wilcock, Cronin, and Hing 2006; Bleakley and Davison 2009). The combination of vasodilatation and vasoconstriction stimulates blood flow, and thus reduces both the extent and duration of inflammation. This blood pumping (or *vaso-pumping*) might be one of the mechanisms allowing displacement of metabolic substances, repair of exercised muscles, and reduction of metabolic processes (Cochrane 2004; Hing et al. 2008). The increased blood flow is thought to promote an appropriate response to the gradient, favoring intra- and extracellular exchanges. The increase in blood flow also results in an increase in stroke volume (SV) by increasing cardiac preload. In addition to the increase in SV, a reduction in peripheral resistance during CWT immersion up to the neck has also been observed. By increasing cardiac preload, this can increase blood flow without affecting the heart rate. Conversely, edema can sometimes cause local vaso-compression, which may alter the metabolite transport rate. This means that contrasting water temperature immersion and edema have opposing effects on blood flow and biochemical movements.

While immersion techniques theoretically favor the recovery process, studies in the field have shown very variable results depending on the techniques used (Wilcock, Cronin, and Hing 2006). To offer a clearer, and more practical, view of the situation with regard to the scientific data, we will address the effects of the various immersion techniques based on the type of exercise performed before their application. We will therefore distinguish studies relating to immersion-based recovery following exercises inducing muscle damage from those based on high-intensity exercises (all-out or time trial) and those directly integrated into sporting practice (match, tournament, and so on).

Exercise-Induced Muscle Damage and Immersion

The number of articles covering immersion-based recovery grew considerably between the years 2000 and 2010. However, the percentage of these that specifically investigate recovery following exercise that induces muscle damage remains low. This is in spite of the interest it would attract from trainers and athletes for use in specific muscle-building sessions, for example (Bailey et al. 2007; Burke et al. 2000; French et al. 2008; Eston and Peters 1999; Goodall and Howatson 2008; Howatson, Goodall, and van Someren 2009; Jakeman, Macrae, and Eston 2009; Robey et al. 2009; Sellwood et al. 2007; Skurvydas et al. 2008; Vaille, Gill, and Blazevich 2007; Vaile et al. 2008a, b; Pournot et al. 2010). A common method emerges from these few studies that mirrors the method used in many studies investigating exercise-induced muscle damage (EIMD). It consists of inducing muscle damage based on repeated, mainly eccentric, contractions (jumps, repeated maximal contractions, and so on), followed by immediate application of the recovery method (figure 14.1). Several indicators are then measured and compared with initial values. The most commonly used indicators are pain perception; mechanical markers, such as maximal voluntary contraction (isometric or isokinetic MVC); and biological markers, particularly blood or plasma concentration of creatine kinase (CK) or myoglobin. In contrast, very few in situ performance markers are used.

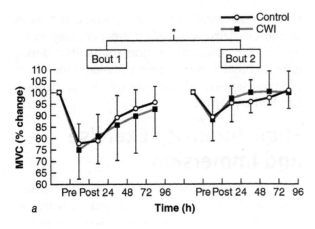

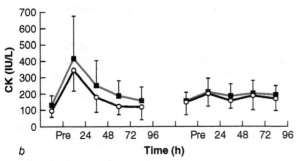

Figure 14.1 *(a)* Percentage change in maximal voluntary isometric contraction before and following bouts 1 and 2. *Denotes significant attenuation of force production in bout 2 compared to bout 1 (P < 0.05). Values are mean ± SD, n = 16. *(b)* Plasma creatine kinase activity before and following bout 1 and bout 2. Values are mean ± SD, n = 16.

Springer and *European Journal of Applied Physiology*, Vol. 105(4), 2009, pgs. 615-621, "The influence of cold water immersions on adaptation following a single bout of damaging exercise," G. Howatson, S. Goodall, K.A. van Someren, figure 1. With kind permission from Springer Science and Business Media.

In addition to the similarity of the methods applied in these studies, the hydrotherapy method is also shared. Indeed, all of the studies use cold-water immersion (or CWT for three of them) as the technique with the potential for the greatest effect on reducing exercise-induced muscle damage. This is not very surprising given the number of studies indicating a positive effect of cold (ice pack, pulsed cold, and so on) on muscle damage. However, the results reported are surprising. Indeed, of the 12 studies cited previously, 8 conclude that recovery based on immersion in cold water is ineffective, while the other 4 observe only a partial effect. Thus, for mechanical indicators, the majority of these studies found that immersion in cold water did not reduce strength loss after tiring exercise. This result has recently

been confirmed by a meta-analysis review dedicated to this topic (Leeder et al. 2012). For example, Howatson and colleagues (2009) showed that 96 h after exercise (100 drop jumps), the control group and the group using immersion were able to reproduce only 96% and 93%, respectively, of their initial isometric strength production for knee extensions. Immediately after exercise, both groups could produce 75% of their initial strength. Similarly, Bailey and associates (2007) observed that recovery of maximal voluntary isometric force for knee extensions was not affected by the type of recovery used after intermittent shuttle exercises for 90 min followed by 10 min of immersion at 10 °C. Like the mechanical indicators, the biological markers of muscle damage are only marginally affected by cold-water immersion. By measuring CK enzymatic activity after eccentric leg exercises, Sellwood and colleagues (2007) showed no difference in effect between recovery based on cold-water immersion (3 × 1 min at 5 °C) and immersion in warm water (24 °C). Similarly, Jakeman and associates (2009) observed no effect on creatine kinase activity following 100 countermovement jumps followed by immersion at 10 °C for 10 min compared to a control group.

The hypotheses explaining this lack of results would be, on one hand, the uncontrollable nature of the inflammatory response and, on the other, the cold-related reduction in nerve conduction. This may, for a time, prevent the athlete from producing maximal power (Rutkove 2001) and voluntary or stimulated maximal strength (Peiffer, Abbiss, Watson, et al. 2009). Indeed, some studies have shown a correlation between reduced muscle temperature and muscle activity (i.e., electromyography signal) (Kinugasa and Kilding 2009). In addition, postimmersion comparison of the two voluntary contractile modes (maximal or stimulated) suggests that the loss of neuromuscular function is likely to be peripheral rather than central (Peiffer, Abbiss, Watson, et al. 2009). Finally, a study on reflexes also concluded that cold reduces muscular performance because of the increased excitability of the motoneuron pool (Oksa et al. 2000).

On reviewing these data, it would be easy to conclude that no form of recovery using immersion-based hydrotherapy has any effect on muscle damage. However, four studies indicate significant positive results (Eston and Peters 1999; Vaile, Gill, and Blazevich 2007; Vaile et al. 2008b; Pournot et

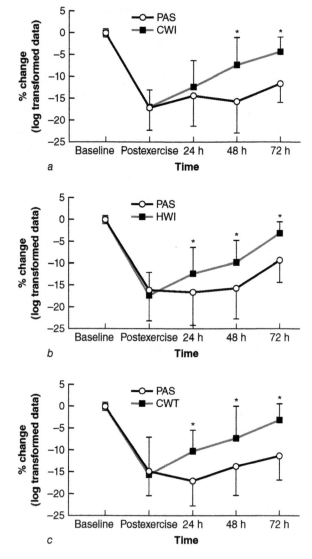

Figure 14.2 Percent change in isometric squat performance (peak force) following CWI *(a)*, HWI *(b)*, and CWT *(c)*. Performance was assessed before and after muscle-damaging exercise as well as at 24, 48, and 72 h postexercise.

*Significant difference between hydrotherapy intervention and passive recovery (PAS).

Springer and *European Journal of Applied Physiology*, Vol. 102(4), 2008, pgs. 447-455, "Effect of hydrotherapy on the signs and symptoms of delayed onset muscle soreness," J. Vaile et al., figure 1. With kind permission from Springer Science and Business Media.

al. 2010). This suggests that it would be an error to conclude too rapidly on the question, especially with regard to the CWT technique. Vaile and colleagues (2008b) showed that, 24 h after muscle-damaging exercise, a 14 min full-body immersion in water at 15 °C limits the increase in CK to 3.6%, compared to more than 300% with passive recovery (figure 14.2). In addition, these studies were generally performed out of the con-

text of real athletic conditions, where the levels of muscle damage involved are generally much milder. The following section shows that during repeated matches or training sessions, the effect of immersion-based recovery on muscle damage can be very different.

High-Intensity Exercise and Immersion

Most of the scientific studies on this theme deal with the effects of immersion-based recovery following high-intensity exercises. Here, we define *high intensity* as all exercises involving an intensity above the anaerobic threshold ($\approx$85–95% of $\dot{V}O_2$max). In the literature, three types of studies can be clearly distinguished based on the type of fatigue-inducing exercise and on the atmospheric conditions: supramaximal, or all-out, exercises such as the 30 s Wingate test; time trials of variable duration and intensity; and exercises performed in a hot climate. These different exercises all induce fatigue, but the principal factor in doing so differs, and the specific consequences of each type of exercise will condition the potential effects of the recovery methods tested. Because of this, we will treat them separately.

Supramaximal Exercise

Metabolic fatigue is, to a great extent, one of the main factors limiting repeat performance after an all-out type exercise. Indeed, the accumulation of metabolic by-products from energy-generating reactions (Pi, H^+, ADP, and HCO_3^-) perturbs the intra- and extracellular media, which limits the muscles' contractile capacity. To accelerate the return to initial performance levels, it might be effective to accelerate clearance of these metabolites by stimulating the blood flow, as with active recovery. Due to the hydrostatic pressure, immersion or alternating vasodilation and vasoconstriction (CWT) could stimulate blood flow. Although the general idea appears simple, the results in the scientific literature do not lead to a straightforward conclusion (Crowe, O'Connor, and Rudd 2007; Al Haddad, Laursen, Ahmaidi, et al. 2010; Al Haddad, Laursen, Chollet, et al. 2010; Buchheit et al. 2008; Buchheit, Al Haddad, et al. 2010; Leal et al. 2010; Morton 2007; Parouty et al. 2010; Schniepp et al. 2002; Stacey et al. 2010). Indeed, several authors present results that, at first reading, appear contradictory (table 14.1). This is particularly the case with Buchheit's results. In a first article (Buchheit,

Table 14.1   All-Out Studies

	Exercise	Recovery methods	Performance	Other indicators
Crowe, O'Connor, and Rudd 2007	2 × 30 s Wingate	10 min ACT + 15 min CWI (13 °C) + 35 min PAS (1 h) vs. 10 min ACT + 50 min PAS (1 h)	CWI < PAS (Peak power, Total work)	↓ $[La]_{Post}$: CWI < PAS
Al Haddad, Laursen, Ahmaidi, et al. 2010	30 s Wingate + 5 min run	5 min face immersion (CWI) vs. 5 min PAS		CWI ↑ parasympathetic reactivation
Al Haddad, Laursen, Chollet, et al. 2010	30 s Wingate + 5 min run	5 min CWI vs. 5 min PAS vs. 5 min TWI		CWI and TWI ↑ parasympathetic reactivation
Buchheit et al. 2008	2 × 1 km time trial	5 min of CWI (14 °C) + 25 min PAS vs. 30 min PAS	PAS ≠ CWI	CWI ↑ parasympathetic reactivation
Buchheit, Al Haddad, et al. 2010	6 × 50 m swimming sprint	Between each sprint: 2 min in water vs. 2 min out of water	IN > OUT	$[La]_{Post}$: PAS ≠ CWI
Leal et al. 2010	30 s Wingate	5 min CWI (5 °C) vs. 5 min LEDT		↓ $[La]_{Post}$: LEDT > CWI ↓ $[CK]_{Post}$: LEDT > CWI
Morton 2007	4 × 30 s Wingate; r = 30 s	30 min PAS vs. CWT (36 °C/12 °C)		↓ $[La]_{Post\ 30\ min}$: CWT > PAS
Parouty et al. 2010	2 × 100 m swimming sprint	5 min of CWI (14 °C) + 25 min PAS vs. 30 min PAS	PAS ≠ CWI	CWI improved subjective perception of recovery
Schniepp et al. 2002	2 × 30 s Wingate	15 min CWI (12 °C) vs. 15 min PAS	CWI < PAS	
Stacey et al. 2010	2 × 50 kJ all-out exercise	20 min CWI (10 °C) vs. 20 min PAS vs. 20 min ACT (cycling 50 Watt)	PAS ≠ CWI ≠ ACT	CWI ↑ immune cell perturbation CWI improved subjective perception of recovery

R = rest between repeat performances; PAS: passive recovery; CWI: cold-water immersion; CWT: contrasting water temperature; TWI: thermoneutral water immersion; LEDT: light emitting diode therapy; ACT: active recovery; [La]: Blood lactate; [CK]: creatine kinase.

Al Haddad, et al. 2010), these authors observed that immersion in warm water for the 2 min recovery between six 50 m swimming sprints leads to a lower reduction in performance compared to passive recovery out of water. In contrast, in a second article simulating the conditions of a swimming competition (Parouty et al. 2010), the same group observed that immersion for 5 min at 14 °C during the 30 min recovery between two 100 m sprints leads to poorer results in terms of performance than passive recovery out of water.

Two important points emerge from these studies: the influence of the duration of recovery between the two all-out exercises and the effect of water temperature. With regard to the duration of recovery between exercises, in one case it was 2 min, while in the other it was 30 min. Based on this literature, when the recovery time between two all-out events is greater than 15 min, immersion does not provide any benefit. In addition, the

majority of studies using cold water for recovery (Crowe, O'Connor, and Rudd 2007; Parouty et al. 2010; Schniepp et al. 2002) observed a negative effect of immersion compared to other recovery techniques, which they attribute mainly to altered nerve transmission due to the reduced temperature. In contrast, when the time between two exercises is very short (<2 min) or the immersion technique used is either warm-water or contrasting temperature immersion (CWT) (Buchheit, Al Haddad, et al. 2010; Morton 2007), positive results for immersion are observed compared to passive recovery. These results must, however, be considered carefully. In Morton's work (2007), the positive effect of contrasting temperature immersion on performance was extrapolated from the kinetics of the plasma-lactate concentration. Some recent work (Cairns 2006) has suggested that a high blood-lactate concentration is not necessarily a factor limiting muscle contraction.

These same teams (Buchheit et al. 2008; Al Haddad, Laursen, Ahmaidi, et al. 2010; Al Haddad, Laursen, Chollet, et al. 2010) also set up another approach to study the effects of immersion on recovery after an all-out type of exercise. This approach is centered on measurement of heart rate variability as a marker of activity of the autonomous nervous system. The autonomous nervous system (or neurovegetative system) controls homeostasis in the body. As the counterparts to neurons voluntarily or consciously control functions such as limb movement, this system controls numerous physiological functions, such as the heart rate, digestion, sweating, and use of energy substrates, without any conscious input. Thus, the contraction rhythm of pacemaker cells in the heart is regulated by two antagonistic elements of the autonomous nervous system: one accelerating the heart (i.e., the sympathetic system), the other inhibiting the heart rate (i.e., the parasympathetic system). This gives the heart rate a rhythm that is not strictly regular.

Many studies have shown chronic fatigue in athletes to be linked to modifications of this nervous system. Some authors have shown that analysis of the heart-rate variability (HRV) can be used to study the effects of training or overtraining (Pichot et al. 2000; Pichot et al. 2002; Borresen and Lambert 2008). Based on this, research has shown that in case of overtraining leading to chronic fatigue in athletes, the HRV generally tends to drop (i.e., there is a reduction in parasympathetic activity). This is followed by a significant rebound in HRV during the tapering/recovery phase (i.e., there is a rise in parasympathetic activity) (Pichot et al. 2000; Pichot et al. 2002). These modifications to HRV also seem to be extensively linked to variations in the athlete's state of fitness (Borresen and Lambert 2008). Recently, work by Buchheit's group indicated that immersion could accelerate reactivation of the parasympathetic nervous system without, however, exerting any positive influence on performance during repeated all-out exercises. Thus, recovery by immersion could be a means to reactivate the parasympathetic system, and through this, a higher HRV. In addition, results also showed that the lower the water temperature, the more effective the treatment.

To summarize, recovery based on immersion after a short all-out bout of exercise appears to present a slight advantage for repeat performances, provided the time between the first and second exercises is short (<2 min) and the immersion either occurs in thermoneutral water or uses contrasting water temperatures. On the other hand, if the time between the two exercises is greater than 15 min or if cold water is used, the effects of immersion on repeat performance are null, or even negative. However, if the aim is not to repeat a performance, but to more rapidly reactivate the parasympathetic nervous system, immersion in cold water should be favored.

Time-Trial Type Exercises

Time-trial (TT) type exercises described in the literature on the theme of recovery using immersion are different from all-out exercises in terms of duration (longer) and intensity (lower). These differences give rise to specific mechanical and physiological adaptations, particularly with regard to fatigue. Thus, while in the case of all-out exercises, the main limiting factors are largely linked to metabolic by-products, in the case of time trials, muscle damage also plays a significant role. Indeed, whether due to mechanical constraints, to energetic and metabolic perturbations, or even to oxidative stress, muscle damage, often revealed by muscle soreness, will appear.

In this context, some of the effects sought when selecting immersion-based recovery techniques are to limit the inflammatory or pain process and to accelerate the blood flow to promote metabolite clearance. To meet these objectives, several studies compared cold-water immersion (Lane and Wenger 2004; Vaile et al. 2011; De Pauw et al. 2010), contrasting temperature (Coffey, Leveritt, and Gill 2004; Versey, Halson, and Dawson 2010), or both (Vaile et al. 2008b; Pournot et al. 2010) to a control situation involving passive or active recovery. In contrast with what is observed after all-out exercise, most studies describing the effects of immersion on recovery after time trials generally reported positive results. Thus, Lane and Wenger (2004) compared the total workload during two 18 min high-intensity cycling sessions with a 24 h interval (resistance = 80 g/kg body weight). One hour after exhausting exercise, the subjects immersed their legs for 15 min in cold water (15 °C), performed active recovery, received massage, or recovered passively. In all groups except the passive recovery group, the next day's performance was identical to the previous day's performance. After passive recovery, the performance was significantly reduced. Similarly, Vaile and colleagues (2008b) (figure 14.3) observed that repeated sprint performance (66 sprints over a total exercise time of

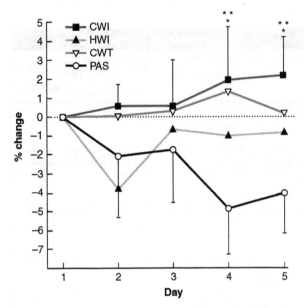

Figure 14.3 Repeated sprint performance following different immersion conditions.

*Significant difference between cold and passive conditions ($P < 0.05$).
**Significant difference between alternating and passive conditions ($P < 0.05$).

Adapted from J. Vaile et al., 2008, "Effect of hydrotherapy on recovery from fatigue," *International Journal of Sports Medicine* 29(7): 539-544. International Journal of Sports Medicine by Georg Thieme Verlag. Reproduced with permission of Georg Thieme Verlag in the format reused in a book/e-book via Copyright Clearance Center.

105 min) could be reproduced daily over 5 days if each session was followed by alternating temperature immersion or immersion in cold water. In contrast, performance was reduced when techniques involving hot water alone or passive recovery were used. Results on soreness and muscle damage must be added to those on performance. Indeed, several studies (Pournot et al. 2010; Vaile et al. 2008b) have shown that immersion in cold water or in water of contrasting temperatures reduced muscle pain and limited the increase in muscle damage markers such as CK. Through this, it contributed to restoring performance levels.

During time trials, recovery based on immersion using cold water or contrasting temperatures seems to provide a real benefit when a performance is to be repeated within the following 24 h. In this context, the effects are more related to the reduction in muscle damage and soreness than to an accelerated clearance of metabolic byproducts that, in any case, will be eliminated in the first hours following the time trial.

Hot Climate Exercise

External temperature has been identified as one of the factors negatively affecting aerobic per-

formance (Cheuvront et al. 2005). Peiffer and colleagues (2008; Peiffer, Abbiss, Watson, et al. 2009; Peiffer, Abbiss, Nosaka, et al. 2009; 2010b) investigated the effects of immersion in cold water on muscle fatigue following time trials in a hot environment. The primary objective was to identify the factors leading to fatigue by manipulating variables such as the external temperature. However, these studies also provide information on how recovery is affected in these temperature conditions. The results of these studies were divided depending on the tiring exercise performed. However, two points emerge, particularly with regard to immersion in cold water (between 5 and 20 min at 14 °C) following exercise in a hot environment (between 32 °C and 40 °C, with 40% humidity). On one hand, the core temperature (rectal temperature) was significantly reduced compared to subjects who recovered without immersion. On the other, immersion did not maintain neuromuscular function (maximal voluntary isometric force). However, in two of these studies, the authors reported that immersion in cold water maintained endurance performance (Peiffer et al. 2008; 2010b) while also reducing perceived fatigue (Halson et al. 2008).

Immersion in cold water in the context of a hot environment may thus facilitate maintenance of performance levels for endurance or time-trial events. However, muscle function will not be restored.

Field-Based Studies

Field-based studies represent a significant proportion of the scientific literature describing the effects of immersion, despite the difficulty of setting up this type of study (Ascensao et al. 2011; Buchheit et al. 2011; Gill, Beaven, and Cook 2006; Parouty et al. 2010; Heyman et al. 2009; Higgins, Heazlewood, and Climstein 2011; Ingram et al. 2009; King and Duffield 2009; Kinugasa and Kilding 2009; Montgomery, Pyne, Cox, et al. 2008; Montgomery, Pyne, Hopkins, et al. 2008; Rowsell et al. 2009, 2010; Tessitore et al. 2007) (table 14.2). All these studies place the constraints generated artificially in the studies described in the previous paragraphs in a wider context. The common objective of these studies was to compare several recovery techniques that are generally used in real athletic situations to quantify their efficacy between two matches or series, or during tournaments. The main difficulty met in

Table 14.2 Field-Based Studies

	Exercise	Recovery method	Performance	Other indicators
Ascensao et al. (2011)	One-off soccer match	10 min CWI (10 °C) vs. 10 min TWI (35 °C)	CWI > PAS (jump, sprint, MVC)	↓ [CRP] and [CK]$_{\text{Post 30 min, 24 h, 48 h}}$: CWI > TWI ↓ [Mb]$_{\text{Post 30 min}}$: CWI > TWI
Buchheit et al. (2011)	2 × soccer matches R = 48 h	PAS vs. spa (sauna + CWI + Jacuzzi)	Spa > PAS (Sprint distance, peak speed)	
Gill, Beaven, and Cook (2006)	Rugby match	9 min CWT (8–10 °C/40–42 °C) vs. 9 min PAS vs. 7 min ACT vs. GAR (12 h)		↓ [CK]: CWT, ACT, GAR > PAS
Ingram et al. (2009)	80 min simulated team sports exercise R = 48 h	15 min CWT (10 °C/40 °C) vs. 15 min CWI (10 °C) vs. PAS	CWI > CWT or PAS (MVC, sprint)	↓ Muscle soreness: CWI > CWT or PAS
Higgins, Heazlewood, and Climstein (2011)	Rugby matches	7 min CWT (10–12 °C/38–40 °C) vs. 5 min CWI (10–12 °C) vs. PAS	CWT > CWI or PAS (300 m test)	
King and Duffield (2009)	2 × simulated netball exercise R = 24 h	15 min CWT (10–10 °C/40 °C) vs. 10 min CWI (10 °C) vs. ACT vs. PAS	CWT or CWI > PAS (RSA, CMJ)	↓ Muscle soreness after recovery session: CWI or CWT > PAS or ACT
Kinugasa and Kilding (2009)	90 min soccer match	15 min CWT (12 °C/38 °C shower) vs. 10 min CWI (12 °C) + ACT vs. PAS	PAS ≠ CWI + ACT ≠ CWT	↑ Subjective perception of recovery: CWI + ACT > CWT or PAS
Montgomery, Pyne, Cox, et al. (2008)	3 × basketball game R ≈ 24 h	5 min CWI (11 °C) vs. GAR (18 h) vs. carbohydrate + stretching		No clear benefits on interleukin-6, FABP, [CK], and [Mb]
Montgomery, Pyne, Hopkins, et al. (2008)	3 × basketball game R ≈ 24 h	5 min CWI (11 °C) vs. GAR (18 h) vs. carbohydrate + stretching	CWI > GAR or carbohydrate + stretching	
Rowsell et al. (2009)	4 × soccer game R ≈ 24 h	5 × 1 min CWI (10 °C) vs. 5 × 1 min TWI (34 °C)	TWI ≠ CWI (CMJ, RSA)	CWI ↑ subjective perception of recovery and ↓ muscle soreness
Rowsell et al. (2010)	4 × soccer game R ≈ 24 h	5 × 1 min CWI (10 °C) vs. 5 × 1 min TWI (34 °C)	CWI > TWI (Total distance run)	CWI ↑ subjective perception of recovery and ↓ muscle soreness
Tessitore et al. (2007)	Soccer training R = 5 h	20 min PAS vs. ACT vs. TWI + ACT vs. EMS	PAS ≠ TWI + ACT ≠ ACT ≠ EMS (anaerobic performance)	↑ subjective perception or ↓ muscle soreness: ACT or EMS > PAS or TWI + ACT
Heyman et al. (2009)	2 × rock climbing R = 20 min	PAS vs. CWI (15 °C) vs. EMS vs. ACT	ACT or CWI > EMS or PAS	↑ [La] removal: ACT, CWI > EMS, PAS
Parouty et al. (2010)	2 × 100 m swimming sprint	5 min of CWI (14 °C) + 25 min PAS vs. 30 min PAS	PAS ≠ CWI	CWI improved subjective perception of recovery

R = time interval between games; PAS: passive recovery; CWI: cold-water immersion; CWT: contrasting water temperature; TWI: thermoneutral-water immersion; ACT: active recovery; GAR: compression garment; EMS: electromyostimulation; [La]: Blood lactate; [CK]: creatine kinase; [CRP]: C-reactive protein; [Mb]: myoglobin; MVC: maximal voluntary contraction; CMJ: countermovement jump; RSA: repeated sprint ability.

this type of study was measurement (in terms of performance) of the potential benefits provided by these techniques. Indeed, comparing the final scores from two football matches played within a 24 h interval does not indicate the players' level of recovery. In contrast, measuring a distance covered in the field, race times, or jump heights during standardized tests and estimating enzymatic concentrations or even perceived recovery or effort will indirectly reflect the effect of recovery.

PRACTICAL APPLICATION

To maintain or limit the decrease of the performance, two major points determine the recovery strategy to use:

- Prior exercise
- Time available between successive exercises

If prior exercise is a time trial, a game, or an exercise that induces muscle damage, contrast water immersion (CWT) is preferred. Athletes must immerse alternately between the two baths—immersion for 60 to 120 s in cold water (8–15 °C), followed immediately by immersion for 120 s in hot water (38–42 °C). Subjects alternated between the two baths for a total of 15 to 20 min. The time between the end of the exercise and the recovery period should be as short as possible. If time between two exercises is less than 20 min and the next exercise requires maximal force rate development, immersion in cold water should be avoided.

In team sports, the studies are almost unanimous in reporting the benefits of immersion-based recovery (CWI or CWT) during tournaments or between matches (Ascensao et al. 2011; Buchheit et al. 2011; Gill, Beaven, and Cook 2006; Ingram et al. 2009; Higgins, Heazlewood, and Climstein 2011; King and Duffield 2009; Kinugasa and Kilding 2009; Montgomery, Pyne, Cox, et al. 2008; Montgomery, Pyne, Hopkins, et al. 2008; Rowsell et al. 2009, 2010; Tessitore et al. 2007). For example, Rowsell and colleagues (2009, 2010) showed that during a football tournament over 4 days, with one match per day, recovery based on immersion in cold water (5 min at 10 °C) compared to immersion in warm water (34 °C) reduced muscle soreness, perceived fatigue, and markers of muscle damage (CK), and increased distance covered. Similarly, Buchheit and associates (2011) observed that a spa treatment (combination of sauna, CWI, and Jacuzzi) compared to passive recovery between two football matches significantly improved distances covered and maximal speeds reached when sprinting during the second match (measured using a GPS system). However, depending on the method used, the effects will be different. Indeed, when CWT and CWI are compared as recovery methods in the same study, it seems that the former is more effective, whether it be for soreness, markers of muscle damage, or recovery of anaerobic performance. Thus, Gill and colleagues (2006) reported a reduced increase in muscle-damage markers (CK) when rugby players used CWT. Some authors, such as Higgins and associates (2011), even suggest that immersion in cold water has a negative effect on anaerobic performance in rugby players compared to contrasting temperature immersion.

In contrast with team sports, few studies have examined the effects of immersion-based recovery on other athletic activities in real conditions. Among these, the study by Heyman and colleagues (2009), which deals with competitive climbing events, presents interesting results. These authors investigated the rest period between two legs of a competition, with an average time of 20 min. They chose to test four commonly used recovery techniques (passive, active, electromyostimulation, and CWI). Their results show that only active recovery and cold-water immersion (5 min at 15 °C) preserved performance levels. In addition, they also observed that lactate was eliminated faster when using these two recovery methods.

Summary

In a practical context, particularly for team sports, recovery using immersion provides a proven benefit. Four main immersion techniques are currently used and studied:

- Thermoneutral-water immersion, between 15 °C and 36 °C (TWI)
- Hot-water immersion >36 °C (HWI)
- Cold-water immersion <15 °C (CWI)
- Contrasting water temperature (CWT), which consists of alternating immersion in cold and hot water.

Doubts still persist as to the method to use, but it seems preferable to favor contrasting temperatures based on the scientific data.

Recovery From Fixture Congestion in Professional Football

David Joyce, MSc, Injury and Performance Specialist

During tournament football or over the Christmas holiday in Europe, it is not uncommon for athletes to play 5 or 6 games in the space of 20 days. Appropriate recovery strategies are probably the most important interventions that sport scientists can provide to reduce performance decrements in this vital make-or-break period.

The concept of recovery must encompass all the various stresses that a player has experienced, be they physical, emotional, or mental. Each athlete is likely to be stressed differently. As such, there cannot be a perfect recovery recipe that equally suits every single player.

Considerations for an individual recovery program:

- *Distances run during the game.* An elite midfielder may cover almost 14 km during a game, whereas a center back is likely to cover around 11.5 km.

- *Number of sprints during the game.* An elite left back would average well over 100 sprints during a game, whereas a right back averages fewer than 50.

- *Number of decelerations.* Body load is much greater when the number of decelerations is high due to the eccentric stress placed on the lower limbs.

- *Emotional stress.* The emotional stress of a striker who has just scored the winning goal is likely to be quite different to that experienced by a fullback who scored a goal that gave a win to the opposition.

- *Injury.* Players who have sustained injuries may not necessarily be able to complete the same recovery program as their teammates.

- *Age.* In general, an older player's body takes longer to recover following a game.

With these factors in mind, an off-the-peg recovery strategy cannot be employed. We do, however, have two non-negotiable elements when it comes to recovery immediately postmatch that must be followed regardless of the individual player: nutrition and sleep.

Our two priorities are rehydration and glycogen resynthesis. In general, we issue a sports drink blend of carbohydrates and electrolytes to rehydrate the players. We aim for a return to pre-match weight within 12 h. Within 2 h after the match, we provide a meal or a meal-replacement drink with about 75 to 90 g of CHO and 20 g of protein. We encourage further feeding at regular intervals up to 5 h afterward. This will, of course, depend on the time the game finishes.

We educate our players on the importance of sleep as a recovery modality and provide advice regarding methods to improve sleep hygiene. Also, we don't schedule an early training the day following a match.

The hydrostatic pressure that pool recovery sessions provide is particularly beneficial the day after a match. Contrast bathing is also helpful. We provide a general program of multidirectional movements and dynamic mobility tasks to perform in the water.

Massage can be helpful to reduce perceptions of muscle soreness, and it can have strong relaxation effects, which are important in reducing psychological stress. However, it is not compulsory. Vigorous massage postmatch is certainly not endorsed because of its capacity to extend any physical trauma sustained in the game.

In terms of scheduling, a 4-day turnaround between games would typically look as follows:

- Day 1: Recovery protocol (hydrotherapy, stretching, massage, active recovery, ice baths, injury treatment) and game analysis

- Day 2: Light small-sided game drills and opposition analysis
- Day 3: Team shape, set piece drills, and specific training for individual roles
- Day 4: Match

Key Points

- When faced with fixture congestion, the main goal is to recover from each game as quickly as possible so that players are able to perform closer to their peak in the coming games.
- It is vital to consider all forms of stress as areas to recover from, not just the physical load from training or playing.
- There can be no one-size-fits-all approach to recovery.
- Proper nutrition and sleep are non-negotiable for recovery.

PART
IV

Unique Considerations
for Recovery

CHAPTER 15

Gender Differences

Christophe Hausswirth, PhD, and Yann Le Meur, PhD

The large majority of exercise physiology research has been performed exclusively on male populations. Until the 1980s, it was widely presumed that the physiological responses to exercise did not truly differ between men and women. From this assumption, the design of training programs and the recommendations for recovery strategies have been generalized for female athletes, without any prior determination of whether such a direct transfer was viable. Since then, numerous studies focusing on gender (Clarkson and Hubal 2001; Tarnopolsky 2000) have uncovered some specificities in women's physiological response to exercise. They have thus determined that gender is an important variable to control for in order to design robust research protocols. The female participants' response to various types of exercise and physical training is now better understood. It appears that the aerobic power and muscular strength of women are naturally lower than those of men, due to differences in body size and composition, hormonal status, sociocultural influences, and dietary habits (Shephard 2000).

Yet despite these factors, well-trained female athletes can deliver performances that are by far superior to those of poorly trained men. Within this context, a growing number of studies have turned their focus toward the effects of gender on recovery in sport, thereby contributing to a better comprehension of the similarities and disparities in the postexercise recovery processes occurring in men and women. Nevertheless, the factors contributing to the lack of a global consensus can be attributed to differences observed in training and nutrition status, to women's hormonal variations over the course of the menstrual cycle, and to the influence of the latter on energy metabolism during exercise (Bonen et al. 1983).

This chapter emphasizes the subtle yet potentially important characteristics observed in female athletes' postexercise physiology and identifies the recovery strategies best suited to meet the demands of their sporting activity. This work will discuss which recovery practices should be prioritized in well-trained female athletes while also critiquing the effectiveness of the principal modalities of recovery. In this chapter, recovery is defined as the return to homeostasis of the various physiological systems presented (Guézennec 1995) following the metabolic, thermoregulatory, and inflammatory challenges and muscle damage incurred by exercise training sessions. Optimal recovery therefore enables the athlete to perform the next training session feeling rested, healthy, and injury free. This chapter also reviews some of the major gender differences in the metabolic, inflammatory, and thermoregulatory response to exercise and its subsequent recovery, focusing on those that may be most pertinent to the design of training programs for female athletes in order to optimize the physiological adaptations sought for improving performance and maintaining health.

Recovery and Maintenance of Energy Stores

Performance in long-duration activities that rely on aerobic metabolism is related to the availability of endogenous energy reserves. In light of this, the fatigue induced by exercise can be linked to the athlete's inability to continue supplying energy to the working muscles, due to the exhaustion of endogenous energy stores (Abbiss and Laursen 2005). This type of fatigue can also occur in athletes training multiple times per day, even those who do not necessarily specialize in endurance activities that cause large quantities of energy to be expended (Snyder 1998). The depletion of energy stores may set in progressively, when daily caloric intake does not compensate for the total energy expenditure linked to both basal metabolism and the practice of a sport. Therefore, recovery strategies must take into account the specificities of female athletes' metabolic response to exercise in order to ensure maintenance of energy stores and to support training workloads.

Metabolic Responses During Prolonged Exercise

Even though female athletes were excluded from participating in the Olympic marathon until 1984, several studies demonstrated that they could actually perform better than men in ultra-endurance events (Bam et al. 1997; Speechly, Taylor, and Rogers 1996). For instance, Speechly and associates (1996) showed that women, despite slower performances in a marathon than a given group of men, performed better than the latter when the race distance exceeded 90 km. Bam and colleagues (1997) showed by way of linear regression analysis that female athletes may potentially hold an advantage over their male counterparts when the distance of a running race reaches or exceeds 66 km. For Tarnopolsky (2000), this phenomenon is linked to gender-based metabolic differences during prolonged exercise, specifically to women's greater capacity for lipid oxidation, which allows them to maintain blood and muscle glucose levels during very long events.

Tarnopolsky explains that the first studies comparing the metabolic responses of men and women during prolonged exercise date from the 1970s and 1980s (Friedmann and Kindermann 1989; Froberg and Pedersen 1984). All reported a gender effect, except that of Costill and associates (1979). However, these results were considered with caution, since the investigations had neither accounted for the occurrence of menstrual cycles, nor precisely evaluated the training status of the subjects involved. Also, exercise intensity was determined relative to maximal oxygen consumption ($\dot{V}O_2max$), without making adjustments for individual body mass. Since the 1990s, many studies have added to these preliminary results by factoring these important methodological details into their protocols (Tarnopolsky et al. 1990; Tarnopolsky et al. 1997).

Estrogen has been repeatedly found to promote lipid oxidation and availability of free fatty acids during exercise, possibly by means of an increased sensitivity to the lipolytic action of catecholamines (Ansonoff and Etgen 2000; Ettinger et al. 1998). Also, even though intramuscular triglyceride content depends on consumption of dietary fat (Vogt et al. 2003), these intramuscular stores are generally much larger in women than in men (Steffensen et al. 2002), and they are utilized to a greater extent in female athletes during prolonged exercise at a moderate intensity (90 min, 58% of $\dot{V}O_2max$) (Roepstorff et al. 2002; Tarnopolsky et al. 2001). During this exercise trial, Roepstorff and colleagues (2002) demonstrated, by measuring the arteriovenous difference in nonesterifed fatty acids, that female athletes consume 47% more nonesterifed fatty acids within the working muscles than their male counterparts. This result was verified by Mittendorfer and associates (2001) during a prolonged exercise protocol (90 min) of moderate intensity (50% $\dot{V}O_2max$). Aside from the greater utilization of plasma nonesterifed fatty acids, Mittendorfer and colleagues (2001) observed greater lipolytic activity in female subjects than in male subjects who were matched by training level and fat mass percentage.

Taken together, these findings convey that women possess larger intramuscular triglyceride stores and rely more heavily on this substrate during exercise than men, as is confirmed by Tarnopolsky (2008) in a meta-analysis of the literature. The gender-based differences in whole-body substrate oxidation during exercise are reflected in the lower respiratory exchange ratio of female participants (compared to male athletes) for a given exercise intensity and duration, as displayed in table 15.1.

Table 15.1 Summary of Studies Where Whole-Body Substrate Metabolism Was Reported in Trained Male and Female Athletes With Exercise Duration > 60 Min

Authors, year	Exercise	SUBJECTS		RESPIRATORY-EXCHANGE RATIO	
		Women	Men	Women	Men
Costill et al. (1976)	60 min run at 70% $\dot{V}O_2$max	12	12	0.83	0.84
Blatchford, Knowlton, and Schneider (1985)	90 min walk at 35% $\dot{V}O_2$max	6	6	0.81	0.85
Tarnopolsky et al. (1990)	15.5 km run at ~65% $\dot{V}O_2$max	6	6	0.88	0.94
Philips et al. (1993)	90 min cycle at 35% $\dot{V}O_2$max	6	6	0.82	0.85
Tarnopolsky et al. (1997)	90 min cycle at 65% $\dot{V}O_2$max	8	8	0.89	0.92
Roepstorff et al. (2002)	90 min cycle at 58% $\dot{V}O_2$max	7	7	0.89	0.91
Melanson et al. (2002)	400 kcal at 40 + 70% $\dot{V}O_2$max	8	8	0.87	0.91
Riddell et al. (2003)	90 min cycle at 60% $\dot{V}O_2$max	7	7	0.93	0.93
Zehnder et al. (2005)	180 min cycle at 50% $\dot{V}O_2$max	9	9	0.86	0.88

Only data from studies using trained subjects are reported in the present table. Significant gender difference was calculated at $P = 0.02$, using a two-tailed independent t-test.

Adapted, by permission, from M.A. Tarnopolsky, 2008, "Sex differences in exercise metabolism and the role of 17-beta estradiol," *Medicine and Science in Sports and Exercise* 40: 648-654.

Tarnopolsky and colleagues (1990) showed that a rise in urinary nitrogen concentration (i.e., an indicator of protein utilization) occurred in male subjects within the 24 h following endurance exercise compared to a control day, while no significant difference was observed in female participants. This reveals that proportionally greater amounts of amino acids are oxidized by men during exercise than by women. These results were then confirmed in a study by Phillips and associates (1993), who demonstrated that male athletes oxidize more protein and leucine than female athletes do. These findings were completed by McKenzie and colleagues (2000), who showed that women utilize leucine as an energy substrate to a lesser extent than men during a 90 min pedaling exercise at 65% $\dot{V}O_2$max, both before and after a 31-day endurance training program. Interestingly, these authors also reported that even though before the training program, leucine oxidation doubled during exercise compared to the resting state, no differences were observed in this variable at the end of the training period.

Metabolic Responses After Prolonged Exercise

Henderson and associates (2007) showed that if greater fatty-acid mobilization occurs in women during prolonged exercise than in men, the inverse is observed during the recovery phase. This could explain why, even though female athletes mobilize lipids to a greater extent than their male counterparts during endurance exercise (Henderson et al. 2007) they lose less fat mass during a physical training program, as several studies demonstrated (Donnelly and Smith 2005). Some experimental data have indeed revealed that some metabolic disturbances caused by an exercise session could still be observed several hours after its completion (Henderson et al. 2007; Horton et al. 1998; Kuo et al. 2005; Phelain et al. 1997). Henderson and colleagues (2007) showed that the rate of lypolysis was still elevated 21 h after a 90 min pedaling exercise at 45% of $\dot{V}O_2$max, or 60 min at 65% of $\dot{V}O_2$max in male subjects, whereas significant differences in lipolysis are no longer observed in female participants by that point (figure 15.1).

To further elucidate the effect of gender on the evolution of postexercise metabolism, Henderson and associates (2008) compared the evolution of glycemia in sedentary men and women over the 3 h before a 90 min bout of pedaling at 45% of $\dot{V}O_2$max and after pedaling for 60 min at 65% of $\dot{V}O_2$max, using labeled glucose. Because euglycemia is influenced by the time of day, a control situation during which the sedentary subjects continued to go about their normal routine was included. Results revealed an increase in the rates of blood glucose appearance and disappearance, as well as a greater metabolic clearance during the

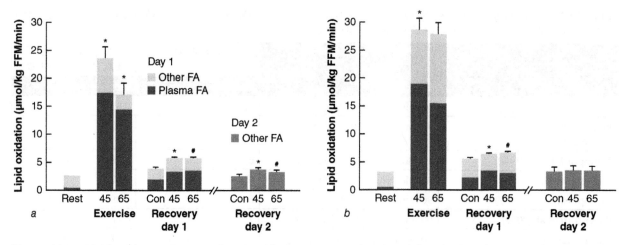

Figure 15.1 Lipid oxidation. Fatty acid (FA) oxidation rate in male *(a)* and female *(b)* subjects. FA oxidation by a combination of tracer-derived measurement and indirect calorimetry on day 1, and solely by indirect calorimetry on day 2. Values are means ± SEM. In male subjects, n = 10 for days 1 and 2. In female subjects, n = 8 for day 1 and n = 6 for day 2. CON: control trial; 45: 45% $\dot{V}O_2$peak trial; 65: 65% $\dot{V}O_2$peak trial. Recovery, day 1: average from 30 min postexercise. Recovery, day 2: the next day following exercise bouts. *Total FA oxidation in 45% $\dot{V}O_2$peak trial significantly different from corresponding time points in control trial, $P < 0.05$. FA oxidation was significantly elevated above control during exercise at either intensity in both sexes ($P < 0.05$). Plasma FA oxidation was elevated above control during exercise and recovery for both exercise intensities, while other (nonplasmatic) FA oxidation was significantly elevated above ($P < 0.05$), but not during postexercise recovery.

Reprinted, by permission, from G.C. Henderson et al., 2007, "Lipolysis and fatty acid metabolism in men and women during the postexercise recovery period," *Journal of Physiology* 584: 963-981.

two exercise situations compared to the control setting in both sexes. Differences between male and female subjects appeared, however, when comparing the metabolic responses postexercise compared to the control situation. Three hours after stopping exercise, male subjects were found to have higher rates of blood glucose appearance and disappearance, higher metabolic clearance, and lower blood-glucose levels, while no differences were found in female participants for any of these parameters between the postexercise and the control situations. These results suggest that compared to men, women have a greater ability to maintain glycemia during the recovery period following prolonged exercise, which could explain why lipolysis occurs to a lesser degree in female subjects during this phase. They also confirm their previous results (Henderson et al. 2007), which revealed a weaker reliance on postexercise lipolysis and a more precise glucose regulation in women than in men.

In all, given that female athletes mobilize lipids to a greater extent during exercise, that their lipid stores are greater, and that they show a better propensity to spare glycogen, women have a greater ability than men to maintain constant energy substrate stores, during exercise as well as during the recovery period (Henderson et al. 2008). These metabolic specificities imply the necessity for gender-specific nutrition recovery strategies (Boisseau 2004; Tarnopolsky 2000).

Gender-Specific Carbohydrate Intake Strategies

Kuipers and associates (1989) studied glycogen resynthesis following a bike ergometer exercise to exhaustion in endurance-trained athletes (i.e., 7 men and 9 women). During the 2.5 h following the end of exercise, subjects consumed a 25% maltodextrin-fructose solution (carbohydrate: 471 ± 5 g and 407 ± 57 g for male and female subjects, respectively). Glycogen repletion occurred in similar proportions in male and female participants.

Different studies demonstrated improved glycogen repletion when carbohydrate (or carbohydrate and protein) are consumed immediately after exercise instead of a few hours later (Ivy et al. 2002; Ivy, Lee, et al. 1988). Tarnopolsky and colleagues (1997) compared the rate of glycogen resynthesis in male and female subjects following a 90 min exercise bout at 65% of $\dot{V}O_2$max and the ingestion of three solutions, immediately after and 1 h following exercise: one a placebo; one containing 1 g/kg of carbohydrate; and one containing a combination of 0.7 g/kg of carbohydrate, 0.1 g/kg of protein, and 0.02 g/kg of lipids. Glycogen resynthesis occurred faster for both sexes with both test solutions than with the placebo, with no differences between male and female subjects (figure 15.2). Finally, a study by Roy and associates (2002) showed, in 10 young women, that the postexercise intake of 1.2 g/kg of carbohydrate, 0.1 g/kg of protein, and 0.02 g/kg of lipids during a period of training (4 training sessions per week) resulted in an increased time to exhaustion at 75% of $\dot{V}O_2$max and tended to diminish protein oxidation over the week following the protocol.

These data therefore show that, when carbohydrate intake is proportional to body mass, no significant differences appear between male and female subjects in their capacity to replenish their glycogen stores. In light of these findings, when the time allotted between two training sessions is less than 8 h, both female and male athletes must consume carbohydrate as soon as possible after exercise in order to maximize the replenishment of glycogen stores. Athletes should break up the total intake of carbohydrate into smaller portions in the early phase of recovery, especially when it

is not possible or practical for them to eat a full meal shortly after a training session (Ivy, Lee, et al. 1988). Carbohydrate-rich meals are recommended during recovery. Adding 0.2 to 0.5 g of protein per day and per kg to carbohydrate in a 3:1 (CHO:PRO) ratio is recommended. This holds particular importance when the athlete's training sessions occur twice daily or are very prolonged (Berardi et al. 2006; Ivy et al. 2002). Let us recall that no difference is observed in terms of glycogen repletion when carbohydrate is consumed in the form of solids and liquids (Burke, Kiens, and Ivy 2004). Carbohydrates with a moderate to high glycemic index supply energy quickly for glycogen resynthesis during recovery and must therefore be prioritized in energy-replenishing snacks following exercise (Burke, Collier, and Hargreaves 1993).

In all, the daily intake of female athletes must reach 5 g of carbohydrate per kg of body weight per day (Manore 1999). If the training volume is high and the training sessions occur at least once daily, this can increase to 6 to 8 g/kg of body weight per day. Given the tendency of female athletes to limit their daily caloric intake, it appears necessary to ensure that these needs be met daily in order to maintain adequate energy balance (Manore 2002).

Gender-Specific Lipid-Intake Strategies

Despite the importance of sufficient daily intake in essential fatty acids, many female athletes considerably limit their consumption of lipids (down to lipids providing only 10% to 15% of daily caloric intake), believing that the latter are susceptible to compromise performance and increase body fat mass (Larson-Meyer, Newcomer, and Hunter 2002). However, in addition to compromising their health, these very-low-fat diets reduce intramuscular triglyceride stores, which are crucial to supply free fatty acids to the muscles during recovery, especially for athletes in long-duration disciplines or those training multiple times per day. Larson-Meyer and colleagues (2002) showed that the lipid intake for female athletes specializing in long events must reach 30% of daily caloric intake to ensure a rapid resynthesis of intramuscular triglyceride stores. If postexercise lipid intake is insufficient, the depletion of intramuscular triglyceride stores is still observed 2 days following exercise (Larson-Meyer et al. 2008; Larson-Meyer,

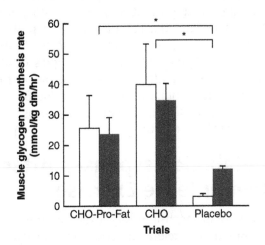

Figure 15.2 Rate of glycogen resynthesis for each trial over the first 4 h postexercise (Tarnopolsky et al. 1997). Open bars: male subjects; solid bars: female subjects.

*P < 0.05 compared with placebo.

Reprinted, by permission, from M.A. Tarnopolsky et al., 1997, "Postexercise protein-carbohydrate and carbohydrate supplements increase muscle glycogen in men and women," *Journal of Applied Physiology* 83: 1877-1883.

Newcomer, and Hunter 2002). Since some studies have demonstrated the very low lipid intake in this population (Beals and Manore 1998), it has also been hypothesized that a very low lipid intake (i.e., 10–15% of caloric intake) coupled with a high-endurance training volume could be detrimental to performance by compromising intramuscular triglyceride stores (Muoio et al. 1994).

Also, even though this study by Larson-Meyer and associates (2008) did not report any decrease in performance during a 2 h running trial following a short-term low-lipid diet (3 days, 10% of caloric intake), this diet did yield a modification of the athletes' lipid profiles, which had already been proven in another study by Leddy and colleagues (1997). They showed that a low-lipid diet increases triglyceride concentration and decreases the concentration of HDL cholesterol, leading to a lowered ratio of total cholesterol (HDL cholesterol is an adverse consequence in terms of cardiovascular health). Even though no study to date has yet demonstrated this, this observation suggests, according to Larson-Meyer and associates (2008), that endurance athletes must make sure to consume a sufficient amount of lipids daily, particularly if they, or members of their family, have any problems linked with their lipid profiles. Future studies are needed to observe the effect of a chronic hypolipidic diet on prolonged exercise performance.

In conclusion, it appears that while carbohydrate and protein intake should be favored immediately postexercise, in the scope of the athlete's general diet, lipid intake should be maintained at a sufficient level. If fat consumption supplies less than 15% of daily caloric intake, they are at risk of developing a deficiency in essential fatty acids and vitamin E (Manore 2002). Female athletes should therefore be advised to consume vegetable oils, nuts, and fatty fish, such as tuna and salmon, on a regular basis (Manore 2002).

Gender-Specific Protein-Intake Strategies

Chronic aerobic exercise is well known to diminish amino-acid utilization. However, the metabolic oxidation capacity of amino acids augments as a result of an improved activity of the limiting enzyme, glutamate dehydrogenase (Lamont, Lemon, and Bruot 1987). For this reason, male and female athletes must increase their daily protein intake.

Concerning proteins, it is recommended that female athletes maintain a daily energy intake superior to those recommended for the standard population in order to ensure that the mechanisms of recovery, on a structural standpoint, are able to occur (1.2 to 1.4 g/kg of body weight per day, compared to 0.8 g/kg of body weight) (Manore 1999). Phillips and colleagues (1993) demonstrated that a daily protein intake of 0.8 g/kg leads to a negative nitrogen balance in female athletes. Many studies performed on female athletes put in evidence the fact that this intake is often at the lower end of these recommendations, if not below them, particularly in amenorrheic athletes (Howat et al. 1989; Pettersson et al. 1999). For female athletes specializing in strength sports, these recommendations reach 1.4 to 1.8 g/kg of body weight, given the extent of muscular damage caused by this type of activity. Foods containing proteins with a high biologic availability (a high retention and utilization rate by the body) should be emphasized. Meats, fish, eggs, and dairy products offer complete sources of protein (providing all essential amino acids). It is particularly important for vegetarian or vegan athletes to pay attention to their protein sources to ensure that all essential amino acid requirements are met, usually by combining different sources such as nuts, whole grains, and legumes (Hoffman and Falvo 2004). Also, protein-rich animal foods are an important source of dietary iron, which helps limit the risk of anemia, a problem of particular concern for female athletes (Akabas and Dolins 2005). Volek and colleagues (2006) assert that female athletes should be aware that animal protein sources provide vitamins B_{12} and D, thiamin, riboflavin, calcium, phosphorus, iron, and zinc (Manore 2005).

Metabolic Responses After Brief, Intense Exercise

During brief, high-intensity exercise, the work performed by skeletal muscle relies heavily on the breakdown of phosphocreatine (PCr) stores and the recruitment of glycolysis to ensure the resynthesis of energy. Under these conditions, the resulting decrease in intramuscular pH and the accumulation of inorganic phosphate (P_i) play an important role in the development of muscular fatigue (Allen 2009). For this reason, it has been suggested that a rapid clearance of lactate and hydrogen (H⁺) ions after intense exercise is essential for recovery in order to restore work capacity quickly, particularly in the context of sport activities involving repetitive, short, and intense exercise bouts (Ahmaidi et al. 1996). Several studies performed on the effect of gender on the metabolic response during and after exercise contribute interesting clues regarding the methods of recovery that must be prioritized in female athletes after such brief, high-intensity exercise.

The pre-exercise phosphagen (ATP, ADP, and PCr) concentrations do not differ between men and women (Esbjornsson-Liljedahl, Bodin, and Jansson 2002). Still, Ruby and Robergs (1994) reported a greater glycolytic activity in men than in women during exercise. In this regard, Jacobs and associates (1983) had already shown that male athletes reach higher concentrations of intramuscular lactate than female ones following anaerobic exercise tests. These results were then confirmed by the in vivo study of Russ and colleagues (2005), which showed male subjects relying more heavily on glycolysis than female participants, even though their reliance on the flux of PCr and aerobic pathways for force production did not show parallel gender effects. This difference could be explained by a greater cross-sectional area of Type II muscle fibers or greater glycolytic enzyme activity (Coggan et al. 1992). Recently, Wüst and associates (2008) reinforced this hypothesis by demonstrating that men's lower resistance to fatigue would be more strongly linked to sex differences in distribution of muscle-fiber type than to differences in motivation, muscular size, oxidative capacity, or local blood flow.

In this context, Esbjornsson-Liljedahl and associates (2002) showed that over three all-out 30 s sprints separated by 20 min of passive recovery, female subjects were able to maintain a steadier level of performance than male ones. Despite similar fitness levels (physically active, noncompetitive) female participants showed a lower accumulation of metabolic products than their male counterparts, particularly in regard to Type II fibers (Esbjornsson-Liljedahl, Bodin, and Jansson 2002). During the recovery phase, women have a greater capacity for recycling their ATP supply than men, making them better equipped to take on the next intense, brief exercise bout.

Chronic Fatigue and Daily Energy Balance

In female athletes, the maintenance of energy balance is a topic in need of particular attention, due to the frequently observed concerns with physical appearance and the social pressures pushing them to maintain a low percentage of fat mass (Burke 2001; Manore 1999, 2005; Volek, Forsythe, and Kraemer 2006). A positive energy balance is indeed necessary to ensure optimal metabolic recovery and to promote the muscular regeneration processes in order to withstand the training loads (Snyder 1998). However, the literature reports the daily caloric intake as often insufficient in high-level female athletes, particularly those involved in sports in which a thin silhouette confers an advantage (Deutz et al. 2000; Manore 2002). Figure 15.3, issued from an article by Deutz and colleagues (2000) shows that the energy balance of high-level gymnasts remains almost always negative throughout the day. Maintaining the equilibrium between daily energy intake and expenditure by way of well-adapted nutrition recovery methods concerns all athletes

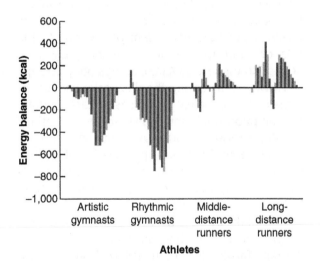

Figure 15.3 Comparison of within-day energy balance in four groups of elite female athletes (from Deutz et al. 2000). Each group has 24 bars, beginning with waking and ending 24 h later. The bars represent 4 h. Energy surpluses and deficits are represented, respectively, by variations above and below the zero (0) energy-balance line.

in any sport who must train daily or multiple times per day.

A negative energy balance is associated with a chronic state of fatigue, decreased alertness, sleep disturbances, and a rise in protein catabolism, which may compromise the mechanisms for muscle resynthesis (Snyder 1998). It therefore appears necessary to inform female athletes of the importance of maintaining an adequate daily energy intake, especially since two-thirds of them express the desire to lose weight and to limit their daily caloric intake (Hinton et al. 2004). These dietary behaviors are not always, however, associated with weight loss (due to a decrease in resting metabolism) (Deutz et al. 2000), but rather to disturbances in the menstrual cycle (Loucks 2003).

The majority of female athletes require a minimum of 2,300 to 2,500 kcal per day (9,263–10,460 kJ) in order to maintain their body mass (45–50 kcal · kg^{-1} · day^{-1}) (Manore 1999, 2002). Those involved in endurance sports, such as marathon running or triathlons, have energy needs susceptible to reach as much as 4,000 kcal per day (16,736 kJ) (Manore 2002). The American Dietetic Association, the Dietitians of Canada, and the American College of Sports Medicine revealed that a relatively high carbohydrate intake is usually recommended for athletes during the recovery period, independent of sex, due to the role of carbohydrates in maintaining glycogen stores (Rodriguez, DiMarco, and Langley 2009).

However, increasing the daily caloric intake of female athletes solely by increasing carbohydrate consumption does not appear to be an optimal solution (Boisseau 2004). A female athlete consuming about 2,000 kcal per day is not able to significantly increase her glycogen stores solely by increasing the percentage of carbohydrate in her diet. In fact, she would have to boost her caloric intake by 30% over the course of 4 days in order to reach a carbohydrate intake superior to 8 g/kg (Tarnopolsky et al. 2001). From a practical standpoint, this is neither habitual nor acceptable among athletes watching their weight permanently. Moreover, such a large shift of caloric sources toward carbohydrate is likely to lead to deficiencies in some essential proteins and lipids. It may also compromise nitrogen balance, which is necessary for maintaining normal levels of sexual steroid hormones (Volek and Sharman 2005).

In this context, several authors (Larson-Meyer, Newcomer, and Hunter 2002; Volek, Forsythe, and Kraemer 2006) recommend a lipid intake of 30% of daily energy needs for female athletes.

Recovery and the Musculoskeletal Regeneration Process

Certain exercises that induce important local mechanical constraints can be accompanied, in the hours and days following their execution, by muscle aches. This delayed-onset muscle soreness (DOMS) is symptomatic of a number of physiological disturbances at the level of the muscle cell and interstitial fluid: Inflammatory processes, edema, and muscle-fiber lysis are a few examples. The recovery methods following any exercise able to generate this type of response in female athletes must take into account the extent of the muscle damage caused and the inflammatory response observed.

Inflammatory Responses to Long-Duration Exercise

Several studies revealed a smaller elevation in plasma creatine kinase (CK, an enzyme indicating the severity of muscular damage) in women than in men following aerobic exercise (Apple et al. 1987; Shumate et al. 1979). Shumate and colleagues (1979) examined the evolution of plasma CK after graded exercise on a bike ergometer. The mean increase in CK levels was 664 U/L in male subjects and 153 U/L in female subjects. Similarly, Janssen and associates (1989) showed that male marathon runners displayed a higher plasma-CK elevation than female runners. Apple and colleagues (1987) also demonstrated that the rise in plasma-CK levels in female runners was 5.5 times inferior to their male counterparts after a marathon. By contrast, Nieman and associates (2001) did not observe any gender differences in the postexercise inflammatory response following a marathon. However, these results should be considered with caution, since this type of field study limits the ability to quantify the actual workload that such an event imposes on each participant. Some of these discrepancies could be due to the lower workload performed by the female athletes.

Strength Exercise Responses

Training-induced specificity of neuromuscular performance is characterized by changes in maximal strength and isometric force–time or force–velocity curves. These changes are dependent on the type or duration of training. The magnitudes and time courses of the adaptations in the neuromuscular system give valuable information as regards the attempts to develop more effective and more specific training regimens, as well as for men or women. Recent scientific interest has been focused on examination of the mechanisms that lead to understanding the differences between male and female athletes in terms of strength exercise responses and recovery.

Muscle Function

Studies comparing the evolution of strength following a bout of strength exercise in male and female athletes report conflicting results. Some show no gender-based difference, and others show a better recovery among female participants (Clarkson and Hubal 2001).

Borsa and Sauers (2000) examined the effect of gender on recovery following a traumatizing strength exercise consisting of 50 eccentric and concentric maximal contractions of the elbow flexors. After the exercise bout, participants from both sexes revealed important muscle aches, a reduced amplitude of joint movement, and a significant decrease in strength. After adjusting for the pre-exercise value, the degree of strength loss was similar for both groups (–15.3% vs. –18.2% for male and female subjects, respectively). In the same manner, Rinard and colleagues (2000) did not report any gender differences in strength loss and recovery kinetics after 70 maximal eccentric contractions. Hakkinen (1993) had earlier observed that after a strenuous leg-extensor exercise (20 maximal contractions), even though the percentage of strength lost over the repetitions did not differ between groups, the recovery of maximal force 1 h postexercise was greater in female participants. Later, Linnamo and associates (1998) demonstrated the same trend toward a greater central fatigue in male subjects after an explosive strength-loading exercise. Sayers and Clarkson (2001) also did not observe any gender

differences in relative strength loss after 50 maximal voluntary eccentric contractions, but they did find a greater incidence of extreme strength losses (>70%) in women than in men. These authors also added that the female athletes showing these largest losses actually recovered faster than the male ones, who had shown the same degree of strength loss postexercise. This finding suggests that women can recover to a functional level more rapidly than men. This hypothesis has also been strengthened by Sewright and colleagues (2008), who revealed a greater strength loss in female than in male subjects immediately following 50 maximal eccentric contractions. This significant difference, however, disappeared 6 h postexercise (figure 15.4).

Taken together, the heterogeneity of these results does not allow us to state that women differ significantly from men in terms of the strength aspects of recovery (Clarkson and Hubal 2001). Still, these findings show that female athletes are subjected to muscle damage from the practice of strength exercises. Therefore, this must be taken into account in the planning of recovery methods following this type of exercise.

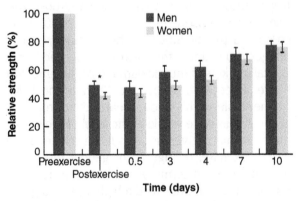

Figure 15.4 Relative strength loss and recovery over time (at baseline, immediately after exercise, and 0.5, 3, 4, 7, and 10 days after exercise) expressed as mean ± SEM (from Sewright et al. 2008). Female subjects exhibit significantly greater strength loss immediately after exercise, but there is no significant effect of gender at any other point in time.

* $P \leq 0.05$.

Reprinted, by permission, from K.A. Sewright et al., 2008, "Sex differences in response to maximal eccentric exercise," *Medicine and Science in Sports and Exercise* 40: 242-251.

Muscle Damage and Inflammatory Responses

The protective effect of estrogen on skeletal muscle inflammation and repair has been well documented in animal models (Enns and Tiidus 2010; Tiidus et al. 2001). Studies have reported that 17-beta estradiol exerted a protective effect on the extent of total muscle damage (Stupka and Tiidus 2001; Tiidus et al. 2001), in part by reducing infiltration of leucocytes into damaged muscle cells, preventing additional or excess release of oxidizing agents. The few studies performed on human subjects report differing results. MacIntyre and associates (2000) examined muscle aches, strength loss, and the intramuscular accumulation of neutrophils (which play a role in the immune system) in premenopausal female and male subjects after 300 maximal voluntary knee extensions. After this exercise, the evolution of muscle pain and strength loss differed between sexes: Female participants generally showed a greater degree of damage after 20 to 24 h compared to male subjects. They showed a greater neutrophil accumulation than male participants at + 2 h (but not at + 4 h), despite the fact that the male participants had accomplished a larger workload. These results suggest that women have a stronger neutrophil response than men in the early postexercise period. Stupka and associates (2000) have compared the evolution of the granulocyte (i.e., phagocyte) count in the plasma, of CK, and histological markers of muscle damage 24, 48, and 144 h after an eccentric exercise at 120% of the 1RM.

The results of this study reveal that if disruptions of the Z lines were visible at the level of the sarcomeres within the vastus lateralis muscle involved in that exercise, no significant gender difference was observed in terms of postexercise damage. But this study also revealed a higher plasma granulocyte count in male subjects 48 h after the exercise, suggesting that the inflammatory response was more attenuated in female subjects by that point in time. While these findings could appear to conflict with those of MacIntyre and colleagues (2000), the absence of any difference 4 h postexercise in the latter's study could imply a different evolution of postexercise infiltration of inflammatory cells in female and male subjects.

It is important to note however, that the female participants in Stupka and associates' study (2000) all used oral contraceptives (OC), and tested during the equivalent of the late follicular phase, whereas MacIntyre's protocol (MacIntyre et al. 2000) did not appear to account for menstrual cycle phase. Estrogen levels in women taking oral contraceptives are known to be lower than non-OC users, and this could have accentuated the discrepancy between the findings of the two studies. Further, Tiidus and colleagues (2005) demonstrated that in rats, the protective effect of estrogen on neutrophil infiltration did not occur with synthetic estrogen supplementation.

In light of these results, Clarkson and Hubal (2001) stated that women demonstrate an earlier postexercise inflammatory response, but that this response remains weaker than men's over the long term. For Peake and associates (2005), a potential explanatory hypothesis would involve the leucocytes' infiltration of muscle cells postexercise, which could differ in male and female subjects because of the differences they show in membrane permeability following a muscle-damaging exercise.

From this perspective, recovery strategies making use of cold exposure to reduce the postexercise inflammatory response appears particularly beneficial to athletic women. It has been proven that vasoconstriction and the reduction of metabolic activity had the effect of decreasing tissue swelling, inflammation, the immediate sensation of pain, and the severity of an injury (Enwemeka et al. 2002). Experimental evidence shows that the local application of cold, started promptly after a muscle injury and maintained for a prolonged time period, limits the process of cell destruction by leucocytes and improves nutrient perfusion through the tissues (Schaser et al. 2006). In the acute treatment of musculoskeletal trauma, cold application is an adequate method by which to improve cellular survival against local hypoxia generated by the inflammatory process and the formation of edema (Merrick et al. 1999).

Future studies will need to determine whether recovery by cold-water immersion or whole-body cryostimulation would be susceptible to yield greater benefits to female athletes, given the specificity of their inflammatory response compared to that of male athletes.

Bone Turnover and Stress Fracture

The majority of physical activities exert strong pressures on the skeletal system, thereby increasing bone-cell turnover. The mechanisms for osteogenesis and bone recovery are largely dependent on calcium intake and overall caloric intake (Manore 1999; Nattiv 2000) as well as hormonal status (Kameda et al. 1997; Weitzmann and Pacifici 2006).

Howat and colleagues (1989) report that the diets of female athletes are frequently poor in calcium, especially when dairy-product consumption is low or nonexistent. This contributes to compromising their bone health and increasing their risk for stress fractures (Manore 1999; Nattiv 2000). It therefore seems essential, during recovery, to promote calcium consumption to aid the bone remodeling processes (Nattiv 2000). In spite of this, female athletes often fail to maintain a sufficient calcium intake, especially when dairy products are not included in their diet. Several studies have thus shown that daily calcium intake can vary between 500 and 1,623 mg in female athletes, with most of them not even reaching 1,000 mg/day (Kopp-Woodroffe et al. 1999). By contrast, the recommended daily intake of calcium is 1,300 mg for girls aged 9 to 18 years, and 1,000 mg for women 19 to 50 years old.

The results of a recent study by Josse and colleagues (2010) illustrate the beneficial effect of milk consumption on bone turnover in young female subjects, over the course of a 12-week strength training program. Five times per week, subjects consumed either two 500 ml doses of fat-free milk (MILK group, received 1,200 mg per day of calcium and 360 IU/day of vitamin D), or an isoenergetic carbohydrate drink (CON group) immediately following each strength training session and 1 h after (figure 15.5a and b). The MILK group showed a larger rise in serum 25-hydroxyvitamin D (vitamin D acts to promote calcium absorption and bone turnover) levels than the CON group. Parathyroid hormone (responsible for bone resorption by enhancing the release of calcium into the bloodstream) levels decreased in the MILK group only. The greater lean mass gains observed in the MILK group also suggest larger gains in bone mass. Finally, fat mass decreased in the MILK group only. These results emphasize the important role of calcium as part of nutritional strategies for optimal bone health and recovery.

Inadequate caloric intake also undermines bone health by suppressing activity of the hypothalamic-pituitary-gonadal axis, resulting in low estrogen production and menstrual-cycle disturbances (De Souza and Williams 2005; Laughlin and Yen 1996). Estrogen plays a multifactorial role in maintaining bone health in premenopausal women, both by slowing bone resorption (Kameda et al. 1997) and stimulating its formation (Weitzmann and Pacifici 2006). The elevated occurrence of osteopenia and increased rate of bone fractures is well documented in premenopausal female athletes with hypothalamic amenorrhea and in anorexic female subjects (Grinspoon et al. 1999; Marcus et al. 1985). Athletes who restrict caloric intake

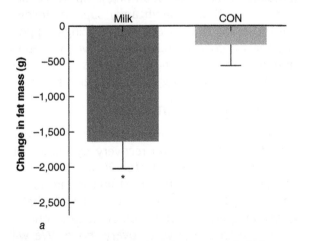

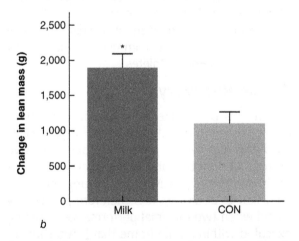

Figure 15.5 Evolution of *(a)* fat mass and *(b)* lean mass before and after 12 weeks of strength training in female athletes having consumed milk following training sessions (MILK, n = 10) or an isoenergetic carbohydrate drink (CON, n = 10) (from Josse et al. 2010).

*Significantly different from CON in the pre- to posttraining measures ($P < 0.05$). Values are mean ± SE.

despite important training loads compromise the hormonal processes involved in bone remodeling, placing themselves at a risk for bone loss comparable to postmenopausal women. In this regard, maintaining adequate energy availability holds particular importance through puberty and until late adolescence, when young women reach their peak bone mass (Eastell 2005).

Vitamin D status is another important factor for preserving bone health. Athletes living in Nordic countries and training mostly indoors often show a poor vitamin D status. A level ≥ 80 nM is necessary for maintaining good bone health (Grant and Holick 2005). Studies show that the daily supply in vitamin D averages 5 μg per day (200 IU/day) in women between 19 and 50 years of age. It is therefore advisable to compensate for this deficit with a well-adapted diet or with luminotherapy. Finally, adequate protein intake should be included in the nutrition recovery strategies for preserving bone health in order to limit, over the short term, the risks of stress fractures, and over the long term, the onset of osteoporosis.

Targeted Recovery Strategies

The time course and severity of muscle soreness, muscular dysfunction, and the appearance of markers of muscle damage in the systemic circulation can vary considerably depending on the duration, intensity, and type of exercise performed. Many investigations have attempted to alleviate or prevent exercise-induced muscle damage using active recovery or water immersion, specifically for male and female athletes.

Active Recovery

An analysis of the literature on the benefits of active recovery reveals the latter as especially advantageous when performed between two exercise sessions that are closely scheduled in time (<1 h) and that rely heavily on anaerobic processes (Heyman et al. 2009). It is therefore particularly useful when two maximal performances must be executed within a time frame that is too short to allow a return to homeostasis by way of passive recovery.

Many studies have shown that an active recovery yields a more rapid return to resting blood-lactate concentrations, compared to a passive recovery (Watts et al. 2000). Similarly, Yoshida

and associates (1996) discerned that active recovery limits the accumulation of P_i after an intense, 2 min exercise bout compared to a passive mode of recovery. Other results showed that maintaining a submaximal intensity of exercise enables a quicker return of intramuscular pH to its resting value after a high-intensity effort (Fairchild et al. 2003; Yoshida, Watari, and Tagawa 1996). It appears appropriate to recommend that women maintain a moderate intensity of exercise between two high-intensity exercise bouts that are separated by a relatively short time span for recovery (i.e., <1 h).

Carter and colleagues (2001) showed a greater decrease in arterial blood pressure after exercise in women than in men, such that it may decrease below its pre-exercise value during passive recovery. Given that the return to homeostasis after an effort relying heavily on glycolytic processes appears linked to maintaining good venous return (Takahashi and Miyamoto 1998), it is conceivable that female athletes would find a greater advantage to active recovery modes than their male counterparts, at the circulatory level (Carter et al. 2001). Further, a recent study by Jakeman and associates (2010) has proven the effectiveness of wearing compression socks during the recovery period in female subjects. These authors demonstrated that muscle aches were significantly diminished by their use, and that the loss in knee extensor strength was reduced. In addition, jumping performance (squatting jump with countermovement) was significantly superior following this mode of recovery. Such findings support strongly the concept that improving venous return during recovery in female athletes is an essential parameter contributing to muscle performance.

Recovery by Immersion

To our knowledge, only one study has examined the effect of gender on recovery by immersion after aerobic exercise (Morton 2007). The results of this study show that a 30 min immersion in alternating hot (36 °C) and cold (12 °C) baths generates a greater drop in blood-lactate concentrations than a passive recovery mode. We will mention, however, that even though this method appears to accelerate postexercise lactate clearance significantly, these differences remain small. New studies are necessary to determine whether this type of recovery can yield positive effects on the level of performance of female athletes, when used between repeated anaerobic exercise sessions. Still, it appears that basic indications can be provided for female athletes regarding this recovery mode, not only from traumatizing exercises, but for a variety of activities.

Hydration Strategies for Recovery

In their review of the literature on the effects of gender on thermoregulation during exercise, Kaciuba-Uscilko and Grucza (2001) report that women differ from men in their exogenous heat storage, their thermolytic capacities, and their endogenous heat production during exercise. These differences are mainly attributed to a lower ratio of body mass to surface area, to greater subcutaneous fat stores, and to a lower exercise capacity in women. A higher fat-mass percentage augments thermal stress by increasing the metabolic cost of performing a given exercise task (Bar-Or, Lundegren, and Buskirk 1969; Nunneley 1978). In addition, the larger subcutaneous fat layer in women increases the distance between the active, heat-producing tissues (i.e., skeletal muscle) and the skin, reducing the rate of nonevaporative heat loss and decreasing the ratio of body surface area to body weight (Saat et al. 2005).

Nevertheless, when these differences are accounted for in research studies, differences between the sexes are still observed. Women sweat less than men to attain an identical subjective thermal rating, while maintaining the same body temperature as male subjects due to a greater efficiency of the sweat evaporation process. Grucza (1990) also showed that female subjects sweated less than male ones, without demonstrating a greater concomitant rise in rectal and body temperature. This author showed that this difference can be explained by a greater amount of evaporated sweat (58.7% vs. 52.0% of total sweat lost, respectively) and a lower amount of dripping sweat in women than in men (21.2% vs. 34.1%) of total sweat loss in female and male subjects, respectively. Even though these experimental data suggest that the extent of water loss associated with exercise is proportionally less in women, they should, nonetheless, compensate for these losses with well-adapted rehydration strategies.

Recommendations for postexercise rehydration are similar for men and women, and both aim to compensate for the water and electrolyte losses that occur during exercise. Few studies have, how-

ever, compared the hydration strategies adopted by both sexes. On this topic, Broad and colleagues (1996) have quantified the sweat losses and fluid intake of female soccer players. These authors reported that female players drank, on average, 0.4 L/h during training and competition periods, which is lower than that reported for male players. This difference may be mainly linked to morphological differences between sexes, but also to lower metabolic heat production rates in female athletes. However, electrolyte losses measured in the sweat of soccer players younger than 21 years old and senior on a British team over the course of one practice and one game appeared identical to those measured in their male counterparts (Shirreffs, Sawka, and Stone 2006).

In light of these results, it appears that the same recommendations for rehydration during recovery given to male athletes may be well adapted for women. Burke and associates (2001) do not bring forth any postexercise hydration strategy specific to female athletes. We therefore retain that in the postexercise period, it is necessary to replace, as rapidly as possible, the fluids lost during exercise. With subjects who lost 3% of body mass due to dehydration, Shirreffs and colleagues (2006) showed that ingesting 150% to 200% of the water-volume deficit, in small fractions of a 23mM sodium solution, was the most effective method to compensate for the fluid losses caused by exercise. To speed the rehydration process in the context of multiple practices or competitions performed in one day, it appears advantageous to consume cool water (i.e., 12–15 °C) that is slightly flavored and that contains 2% of carbohydrate and 1.15 g/L of sodium (Maughan and Shirreffs 1997). When the amount of time separating the performances is greater, female athletes can recover fluids with a combination of water and solid food intake. In this context, urine clarity may serve as the most practical method for athletes to evaluate their hydration status.

Recovery and Thermoregulation

Reaching a high core temperature as a result of an imbalance between the mechanisms of thermogenesis and thermolysis during prolonged exercise has been well identified in scientific literature as a physiological factor that limits performance (Armstrong et al. 2007). The rise in core temperature during exercise is a normal consequence to any form of exercise over long durations. It is controlled by the interaction of vascular, muscular, and metabolic responses (Maughan and Shirreffs 2004). In response to these, the body adapts by augmenting both peripheral blood flow and sweat loss. The research described in this section indicates that postcooling methods during recovery can offer specific benefits to the female athlete.

Kenny and Jay (2007) showed that compared to men, women reveal a diminished ability to lower their temperature, at the level of the esophagus as well as that of the muscles previously activated by a given effort (figure 15.6). This difference appears linked to women's higher threshold for postexercise cutaneous vasodilation and to a greater drop in postexercise arterial pressure compared to men (Kenny et al. 2006).

In addition, the hormonal variations of the menstrual cycle modify the thermoregulatory response to exercise in female subjects, such that the core temperature during rest and exercise may differ according to the phases of the menstrual cycle. In this way, Kaciuba-Uscilko and Grucza (2001) explain that the internal temperature fluctuates by more than 0.6 °C depending on the menstrual cycle phase, at rest (Hessemer and Brück 1985b) as well as during exercise (Hessemer and Brück 1985a), and even under warm conditions for exercise (Hessemer and Brück 1985b). The rise in progesterone levels during the luteal phase (i.e., the postovulation phase) triggers a rise in body and skin temperature, which delays the activation of perspiration mechanisms, in both ambient and warm environmental conditions. If Marsh and Jenkins (2002) report, on the other hand, that no tangible proof exists to claim that the female athlete is confronted with greater risks of heatstroke during exercise than the male athlete, the findings gathered here still suggest that strategies of postcooling are likely to be at least as beneficial for women as they may be for men.

Given the lower thermolytic capacities of women in the postexercise period, the cooling-off methods performed after an exercise session (postexercise cooling), particularly those involving cold-water immersion or wearing a cooling jacket, appear as a particularly interesting recovery method for women. Indeed, cold-water immersion and the use of a cooling jacket both seem to reduce body temperature and heart rate efficiently (Vaile et al. 2008a). Moreover, cold-water immersion may cause an adjustment of

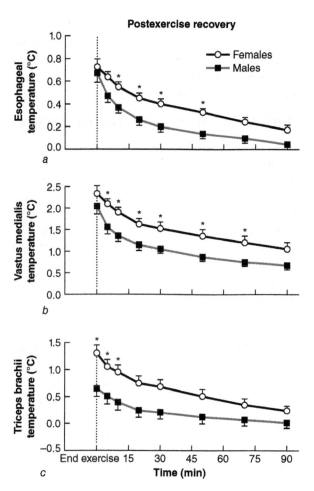

Figure 15.6 Changes from pre-exercise rest in esophageal temperature (a), vastus medialis muscle temperature measured 10 mm from the deep femoral artery and femur (b), and triceps brachii muscle temperature measured 10 mm from the superior ulnar collateral artery and humerus (c) (from Kenny and Jay 2007). Data separated according to male (□) and female (O) subjects. Values are means ± SEM.

*Significant difference between sexes at an alpha level of 0.05.

Reprinted, by permission, from G.P. Kenny and O. Jay, 2007, "Sex differences in postexercise esophageal and muscle tissue temperature response," *American Journal of Physiology: Regulatory and Integrative, Comparative Physiology* 292: R1632-1640.

blood-flow distribution, which could benefit the quality of performance delivered during the next exercise session (Vaile et al. 2008a). The benefits of postexercise cooling are most apparent following exercise performed in the heat (Duffield 2008; Halson et al. 2008). A high internal thermal load can lead to an immediate drop in the level of performance and can slow the recovery to an optimal state of functioning (Wendt, van Loon,

PRACTICAL APPLICATION

- Consume carbohydrate as soon as possible after exercise in order to maximize glycogen store repletion.
- Maintain an adequate level of dietary lipid intake (30%). This point is particularly important for female athletes specializing in long-duration events.
- Ensure that protein balance remains positive during periods of caloric restriction, especially when working with female athletes who show a tendency to limit their caloric intake on a daily basis.
- Use cooling recovery methods following exercise, such as cold-water immersion or a cooling vest. These appear particularly beneficial for female athletes who display lower thermolytic capacities.
- When performances must be repeated in a short period (<30 min), active recovery should be planned because it accelerates the return to homeostasis. This is particularly interesting because a greater decrease in arterial blood pressure is observed after exercise in women than in men. In contrast, no clear benefit appears for maintaining submaximal exercise intensity when maximal exercises are interspersed with longer recovery periods.

and Lichtenbelt 2007). In this way, the degree of thermal stress induced by exercise, or the inability to tolerate an imposed workload, may result in a prolonged drop in performance.

Previous studies highlighted the reduction in voluntary muscle activation following an exercise protocol to exhaustion due to whole-body hyperthermia, implying a selective reduction by the central nervous system of muscle activity following exercise (Martin et al. 2005). According to these results, a decrease in maximal force could be the consequence of decreased command of the central nervous system on the active muscles, acting as a protection mechanism that limits metabolic heat production and further rise in body temperature (Martin et al. 2005). Therefore, maintaining a high internal thermal load can impair performance during a subsequent exercise bout. While contradiction exists in the literature, postexercise cooling is certainly beneficial as far as reducing the internal thermal load after exercise (Duffield 2008). Even though future studies will need to confirm this hypothesis, it is conceivable that the reduction of the thermal stress tied to exercise through postcooling methods will demonstrate benefits of greater magnitude in female subjects, given their lesser capacity to lower their internal temperature after exercise.

Summary

Performing elevated training workloads can yield deleterious effects if the body cannot recuperate completely between training sessions. In this context, the recovery phase must be considered an inherent component of the training process. Therefore, it must be granted the same degree of attention in its programming and management as the exercise sessions themselves. Given the effects of gender on the physiological responses to exercise and during the postexercise recovery period, recovery methods must be customized for men and women to optimize the processes of physiological recovery and supercompensation while minimizing the risks of injury for female athletes.

CHAPTER 16

Temperature and Climate

Frank E. Marino, PhD

The human ability to exercise in a wide range of ambient temperatures is the hallmark of a system developed in concert with the environment and the need for adaptation at all levels of functioning. Much of our capacity for thermal tolerance can be traced to the evolutionary processes that provided the stimulus for such adaptations. Beginning with the primordial soup, where heat combined with methane, nitrogen, ammonia, and hydrogen to give rise to the molecules that eventually formed amino acids and the building blocks of life, high temperatures were indeed a necessary element for starting life (Miller, Schopf, and Lazcano 1997). It follows that in order for life to survive, the ability to handle temperature fluctuations was also essential. In fact, as far back as 1885, it was shown that paramecia reacted to varying degrees of temperature according to what seemed to be a preferred environment of 24 °C to 28 °C (Mendelssohn 1895). These observations led to the conclusion that there must have been an environmental temperature that suited optimal function, which manifested as behavioral changes. The advent of upright posture and bipedalism increased the need of early hominids to tolerate high heat loads during locomotion and persistence hunting (Carrier et al. 1984). It was this challenge, rather than tool making (Fialkowski 1978, 1986; Lynch and Granger 2008), that provided the stimulus for large brain development. Another factor was the early hominid environment, which changed from forest to grassland, resulting in increased distance between food patches and exposure to a savannah habitat (Isbell and Young 1996).

Interestingly, among the animal kingdom (mammals and birds), there is a set core temperature of about 37 °C. Few understand at present why an organism would choose 37 °C as its set temperature, although some speculation exists (Schmidt-Nielsen 1995). The mathematical model proposed by Gisolfi and Mora (2000) and based on the Law of Arrhenius suggests that humans have a preference for ambient temperatures around 25 °C. The calculation of this mathematical model is beyond the scope of this chapter, so the reader is referred to the appropriate text (Gisolfi and Mora 2000, 114–116). In brief, working backward from the assumption that 25 °C is the preferable ambient temperature, the mathematical model arrives at a core temperature of 37 °C. Although somewhat simplistic, this model indicates that 37 °C might be an ideal temperature where mechanisms for heat gain and loss can achieve equilibrium. Furthermore, a set temperature of this magnitude permits heat loss rather than gain in a range of environments. Conversely, if the set temperature were lower than 37 °C, it would necessitate a lower sweating threshold and increase the reliance on water. In essence, core temperature and the interaction with the environment are exquisitely balanced. Whatever the mechanisms responsible for the development of human thermoregulation, they have allowed for the survival of the human species.

Thermoregulation During Exercise

A vast amount of literature describes the human thermoregulatory response under varying conditions. The cornerstone of our understanding in this area is based on the classical work of Marius Nielsen (1938). This work showed how thermal balance was maintained during exercise of various intensities over 1 h. Essentially, as body temperature increases during physical work, heat production and energy output are regulated by heat loss. This study clearly showed that heat exchange by evaporation increased, whereas heat exchange by convection and radiation decreased during exercise of constant intensity when room temperature increased from 5 °C to 35 °C (see figure 16.1). Neilsen's observations underpin the general understanding of human thermoregulation during exercise and, more generally, form the basis on which virtually all thermoregulatory studies interpret their findings. The key aspect of this thermoregulatory model is the regulation of heat balance relative to a set-point core temperature of ~37.0 °C to a maximum that is close to 40 °C during high exercise work rates.

Since this early work, a multitude of studies have established that exercise performance is compromised when ambient temperature rises (Galloway and Maughan 1997). Generally, time to exhaustion increases as ambient temperature moves from 4 °C to 11 °C, decreasing as ambient temperatures reach 31 °C (see figure 16.2). This observation is confirmed when one considers the number of finishers in the Olympic marathon, which is reduced to about 54% of the starting field when ambient conditions rise above 25 °C (Marino 2004).

The reason for the reduction in performance due to rising ambient temperature is still not fully understood, although studies from the 1990s through the early 2000s indicated that a critical limiting temperature was associated with a reduction in motor drive by the central nervous system (CNS), likely as a result of increasing brain temperature (González-Alonso et al. 1999; Nielsen et al. 1993; Nybo and Nielsen 2001a, b; Nybo, Secher, and Nielsen 2002). It is now apparent that critical temperature may not be the limiting factor, since field studies show that the fastest athletes can achieve higher than normal critical temperatures without succumbing to heat illness (Byrne et al. 2006). In addition to thermal strain as a factor limiting exercise performance, there has been a

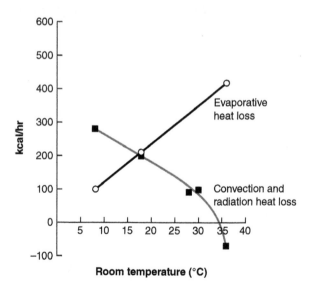

Figure 16.1 Heat exchange during physical work at different room temperatures ranging from 5–40 °C. Physical work was set at 900 kpm/min. Note the increase in required evaporative heat loss (*open circles*) and the reduction in heat loss by convection and radiation (*filled squares*) over the course of rising room temperature.

Data are redrawn from Nielsen 1938.

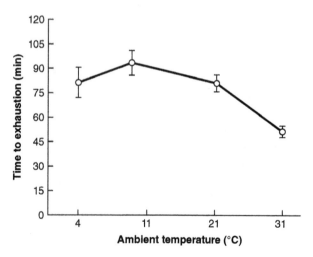

Figure 16.2 Time to exhaustion as ambient temperature increases from 4 °C to 31 °C. Note that time to exhaustion improves when moving from 4 °C to 11°C and then decreases as temperature rises from 11 °C to 31°C. This suggests that the optimum temperature for endurance events is likely to be in the low end of ambient temperatures.

Data are redrawn from Galloway and Maughan 1997.

recent surge in the popularity of recovery strategies to either assist in the return to optimal performance or training earlier than anticipated. This chapter provides a contemporary understanding of the thermoregulatory responses and adaptations to exercise heat stress. Specifically, it considers the interventions used to alleviate thermal strain and examines the limitations of the various recovery strategies following exercise performance in the heat.

Exercise-Induced Hyperthermia and Performance

The classic study by Nielsen and colleagues (1993) showed that when heat acclimated subjects exercising at 60% of maximum oxygen consumption ($\dot{V}O_2$max) were compared to their nonacclimated counterparts, their time to exhaustion was increased by ~32 min, although terminal rectal temperatures were identical for both conditions. These authors concluded that it was the high core temperature rather than the circulatory strain that was the critical factor for the shorter exercise duration in the nonacclimated subjects. Since the publication of this study, others have also shown that exercise was always terminated at ~40 °C, coinciding with reduced motivation for efferent drive (Martin et al. 2005; Nybo and Nielsen 2001a). However, some evidence suggests that efferent drive is reduced relative to whether or not skeletal muscle was used in the preceding exercise bout (Saboisky et al. 2003). Nevertheless, the consistent observation that increasing core temperature precipitates the reduction in neural drive suggests that the CNS is either sensitive to the rising body temperature, and thus downregulates the ability to sustain motor unit recruitment, or that the brain itself is heat sensitive. The sensitivity of the CNS continues to be an unanswered problem due to the methodological complexities associated with the measurement of brain temperature in humans. No available data conclusively show that the brain is more susceptible to thermal injury than any other organ or tissue, at least in humans.

Almost a century ago, it was shown that the irritability of frog nervous tissue decreased as temperature rose from 36.5 °C to 44 °C (Halliburton 1915). In fact, it has been shown that other body tissues can withstand heating up to 45 °C

(Jung 1986). Thus, it seems somewhat counterintuitive for an organ such as the brain to be more heat sensitive than any other tissue. The question also exists whether humans can invoke selective brain cooling, which is an apparent mechanism in some, but not all, animals (Jessen 1998, 2001; Jessen et al. 1994; Mitchell et al. 2002). The purpose of such a mechanism would be to attenuate the rise in brain temperature by countering current exchange between the warm and cooler blood entering and leaving the head (Cabanac 1986). The existence of this mechanism has been debated ever since the initial description in the goat (Caputa, Feistkorn, and Jessen 1986), but consensus about its universal existence is still under debate (Nybo and White 2008). Regardless of its existence, it remains unclear how selective brain cooling could complement a physiological phenomenon such as the critical limiting temperature. That is, by invoking selective brain cooling, it follows that a critical limiting temperature might never be attained, and downregulation of skeletal muscle recruitment might never occur (Marino 2011).

Although there is wide consensus that exercise-induced hyperthermia invokes centrally mediated reductions in skeletal-muscle activation, the purpose of this reduction remains unclear. The most obvious reason could be the need to abate the possibility of thermal injury or irreversible cellular damage (Marino, Lambert, and Noakes 2004), which would intuitively manifest itself as a reduction in skeletal-muscle recruitment just prior to or at the moment of reaching a critical or threshold core temperature or heat load. Indeed, studies that assess the voluntary activation (VA) at the termination of exercise coinciding with high core temperatures confirm this hypothesis (Nybo and Nielsen 2001a). However, when subjects were passively heated from a core temperature of about 37.5 °C to 39.5 °C, with maximal voluntary contraction (MVC) and VA measured at each 5 °C interval, rising core temperature reduced both MVC and VA from the start to the end of passive heating noticeably well before the attainment of a critical core temperature (Morrison, Sleivert, and Cheung 2004). In fact, MVC and VA returned to normal values once core temperature was restored by cooling the subjects. These findings show that by elevating core temperature alone, there is a direct effect on the ability to recruit skeletal muscle. However, this is not a consistent finding.

Others have shown that core temperature eleva-tion by up to 1 °C did not significantly alter the VA of the wrist flexors (Bender et al. 2011). Thus, the influence of hyperthermia on skeletal muscle acti-vation is confounded by a number of variables: whether hyperthermia is induced by exercise or if it results from passive heating or the musculature involved, and a combination of these factors.

A further confounding issue in the relationship between exercise performance and heat strain is the development of hyperthermia relative to exercise modality. The most popular method for determining the effect of heat strain on exercise performance in laboratory settings has tradition-ally been to compare the time that an exercise bout at a fixed intensity or constant load can be sustained in different ambient temperatures. The typical finding from these studies is that higher thermal strain or increased rate of heat accumula-tion results in shorter exercise duration and, by extension, an earlier onset of fatigue compared to more favorable conditions or other types of inter-ventions (Galloway and Maughan 1997; Nielsen et al. 1993; Nybo and Nielsen 2001a). The typi-cal conclusion drawn from these findings is that the earlier onset of fatigue is primarily due to the systemic physiological limitations (cardiovascular, respiratory, thermal, and metabolic). Although attractive, this classical understanding assumes that physiological limitations achieve their maxi-mum under given conditions. When the termina-tion of exercise is viewed in this way, the body's capacity is thought to be governed by systemic and cellular limitations that, when achieved, lead to catastrophic cellular events, thereby limit-ing exercise performance. This is known as the catastrophe model of exercise limitation (Edwards 1983; Noakes and St Clair Gibson 2004).

An alternative view has been established whereby exercise is regulated by anticipating the rate at which the thermal load develops, rather than by reaching a critical limit (Marino, Lambert, and Noakes 2004). For example, when subjects were required to sprint for 60 s at regular intervals between bouts of low-intensity self-paced exer-cise over 60 min, the power output and related integrated electromyography (iEMG) of the vastus lateralis muscle were significantly reduced within the initial sprints, only to return to near initial values at the end of the exercise (Kay et al. 2001) (see figure 16.3). In addition, heart rate remained unchanged during the sprints, leading

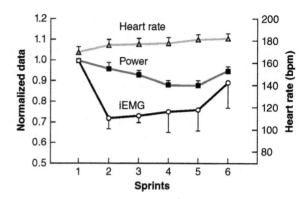

Figure 16.3 Early reduction in power output accompanied by reductions in iEMG (left ordinate). Heart rate (right ordinate) during the high-intensity efforts remained unchanged. Note the return of power and iEMG for the final high-intensity effort, indicating an end spurt and anticipatory regulation of motor unit recruitment.

Spinger and *European Journal of Applied Physiology*, Vol. 84. 2001, pgs. 115-121, "Evidence for neuromuscular fatigue during high-intensity cycling in warm, humid conditions," D. Kay et al., figure 3. With kind permission from Springer Science and Business Media.

the authors to conclude that the subjects gave a conscious effort while muscle recruitment was subconsciously attenuated.

Similarly, in a subsequent study, it was found that running speed was reduced before a criti-cal core temperature was achieved, indicating an anticipatory regulatory response, the purpose of which was to avoid a lethal hyperpyrexia (Marino, Lambert, and Noakes 2004). Extending these findings, others have shown that during a 20 km cycling time trial, power output and iEMG were reduced in hot (35 °C) compared with cool (15 °C) conditions, despite there being no difference in the subjects' rectal temperature, heart rate, or rating of perceived exertion (RPE). Again, the reduction in efferent command was shown to occur well in advance of reaching a critical core temperature of 40 °C. Astonishingly, however, both the power output and iEMG increased significantly in the final kilometer in both hot and cool conditions and when core temperatures were at their highest. Therefore, the hypothesis that centrally mediated reductions in skeletal-muscle recruitment occur at the attainment of a critical core temperature is flawed, since the change in iEMG and power output preceded the theoretical limit.

More recent attempts to disprove the antici-patory regulation paradigm have been unable to conclusively show that other factors, such as cardiovascular strain, could be the predominant cause of impaired performance in the heat. To evaluate the influence of thermal strain on cardiovascular function, subjects cycled 40 km in self-paced mode in hot (35 °C) and thermoneutral (20 °C) conditions (Périard, Cramer, et al. 2011). The time-trial duration was about 4.5 min longer in the heat, and it was accompanied by significantly depressed stroke volume, cardiac output, and mean arterial pressure, compared with thermoneutral conditions.

The authors concluded that the thermoregulatory-mediated rise in cardiovascular strain was associated with reductions in power output during self-paced exercise in the heat. In a follow-up study, this group also showed that cycling at a constant load in the heat led to severe cardiovascular strain, as evidenced by significantly higher heart rate and reduced stroke volume and cardiac output (Périard, Caillaud, and Thompson 2011). The effect of thermal strain on the cardiovascular response cannot be underestimated as a likely correlate of reduced exercise performance. However, when data from these studies are redrawn so that rises in rectal temperature, power output, and cardiac output are simultaneously compared, the question remains as to how it is possible for power output to increase and return to initial values or higher when both rectal temperature and cardiac output are at their limit (see figure 16.4). If cardiac output and rectal temperature were truly limiting the ability to increase power and, by extension recruit a greater amount of skeletal muscle, this should not be possible.

In summary, the development of hyperthermia has been shown to reduce exercise performance. The factors that contribute to that reduction are not completely understood, although the current evidence strongly suggests that a reduction in CNS drive, either as body temperature rises or approaches its limit, might initiate a cascade of events that contribute to the cessation of exercise.

Reducing Thermal Strain and Improving Performance

In order to alleviate the consequences of heat strain and thereby confirm the existence of deleterious consequences of rising core temperature on exercise performance, studies have conclusively shown that precooling by either cold-water immersion or by cold-air exposure will improve the physical work capacity (Booth, Marino, and Ward

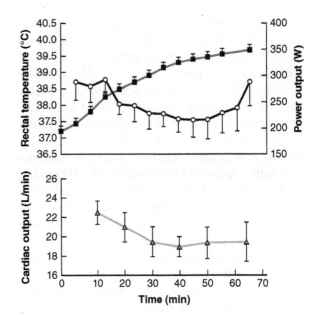

Figure 16.4 Top panel: simultaneous comparison of rectal temperature response (*left ordinate*), and power output (*right ordinate*) during a simulated (self-paced) 40 km time trial in the heat (35 °C). Bottom panel: cardiac output at the same time points during the trial. Although time-trial performance was reduced and physiological strain was higher in the heat compared to thermoneutral conditions (20 °C, data not shown), the simultaneous comparison of these responses indicates that cardiac output and rectal temperature are not limiting, since power output increased when these responses were apparently at their limit.

Data are redrawn from Périard, Cramer, and colleagues 2011.

1997; Brück and Olschewski 1986; Hessemer et al. 1984; Lee and Haymes 1995; Olschewski and Brück 1988). Although the benefits of the precooling method for alleviating thermal strain were shown more than 60 years ago (Bazett et al. 1937), studies within the last 15 years have made precooling a popular intervention strategy for improving athletic performance. This was necessitated by the fact that many sporting competitions were either conducted or planned to take place in less-than-ideal climates (Nielsen 1996; Sparling 1995, 1997). However, the practical limitations in employing precooling maneuvers have constrained their application in the field (Marino 2002).

To address the practical limitations of the precooling maneuver, researchers have studied a variety of applications, either in combination or separately. The use of cooling jackets has been shown to have good practical application in reducing thermal strain and improving exercise performance (Cotter et al. 2001; Duffield and Marino 2007), albeit less than the effect of direct cold-water immersion. More recently, a study evaluating the effect of precooling volume equivalent of either 10%, 15%, or 35% of body surface area during 85 min of free-paced intermittent sprint exercise found that larger coverage of surface area improved exercise capacity in conjunction with a reduced physiological load (Minett et al. 2011). These findings confirm at least two of the limitations of precooling: The improvement in exercise performance depends on the surface area exposed to the cooling. The greater the surface area required for cooling, the greater the practical limitation of the precooling maneuver.

Nevertheless, researchers have endeavored to use the obvious theoretical advantage of the application to develop more useful and practical precooling methods. The use of endogenous cooling in the form of a cold solution or ice slurry has been found to improve exercise performance (Siegel et al. 2010). The reasons for the improvement in performance with ice-slurry ingestion are not eminently clear, but they could include the creation of a larger heat sink because of the phase change from ice to liquid. A sensory effect associated with the ingestion of a cold solution may also provide the perception of a more comfortable internal environment (Siegel et al. 2010).

These authors have posited a very attractive physiological mechanism by suggesting that the ingestion of ice could have a direct effect on internal thermoreceptors. This might allow for the adjustment of afferent feedback relative to the thermal state of the body, causing athletes to perceive exercise as easier for a given thermal load and allowing them to work out longer. Thus, they would end with a higher core temperature than those in the control condition. These authors further hypothesized "that when the body is in a hyperthermic state, cold stimulation of internal thermoreceptors in the mouth, throat, and stomach regions via ice ingestion may improve neuromuscular function by way of lowering central inhibition of higher brain regions" (Siegel et al. 2010, 2518). Although evidence exists for the presence of thermoreceptors in various body cavities, it is not clear if stimulation of these thermoreceptors leads to a lowering of central inhibition in higher brain regions. However, this remains a plausible hypothesis.

In summary, the use of precooling maneuvers in both laboratory and field settings, albeit limited in their applicability, indirectly show that reducing thermal strain or creating an artificial heat sink improves the likelihood of performance, particularly when ambient conditions are less than favorable. Several questions remain to be answered with respect to any precooling maneuver. First, the length of exposure to achieve the desired effect still remains unclear. In addition to this, there is no clear understanding of how long the effect might last and whether it depends on the intensity of exercise. Second, the volume of the surface area exposed to the cooling maneuver might play a significant role in the performance outcome: The higher the exposure, the greater the effect of the cooling. Finally, the mechanism by which precooling improves performance is still not understood, but the development of techniques such as ice-slurry ingestion suggests that neurological structures may exist that directly affect either perception or CNS drive for skeletal-muscle activation.

Morphology and Heat Strain

Performance in the heat also depends on the morphology, or the size, of the person. This has been known for some time, and it is apparent in the results and secular trends of long-distance events, which show that long-distance (marathon) runners have maintained short stature for more than a century, in contrast to the increasing height of the general population (Norton and Olds 2001). The advantage of small stature in reducing heat strain during endurance events is related mainly to the ratio of body surface area to mass. That is, the more surface area per unit of mass available for evaporation of sweat, the less the thermal strain (Dennis and Noakes 1999; Havenith 2001; Marino et al. 2000). However, the caveat is that potential evaporation is reduced when ambient temperature is warm and relative humidity (RH) is greater than 60%. Therefore, heavier runners cannot compensate for their increasing heat storage by sweating more, since the required evaporation is significantly reduced, leading to greater heat accumulation and high core temperatures being reached sooner than desirable (Marino et al. 2000). Essentially, when ambient temperature remains between 15 °C and 25 °C with an RH of 60% or less, there is very little difference in the observed running speed for highly trained runners matched for peak speed of varying mass (55–90 kg). However, as ambient temperature rises toward 35 °C, there is a significant decline in the running speed as body mass increases over 60 kg. This has important implications for recovery, since a greater heat strain for athletes above this body-mass threshold might require further intervention in the immediate postexercise recovery stage, simply due to the higher sweat rates and body fluid deficits that they may incur. However, no evidence at present indicates that higher thermal strain would induce greater central or peripheral fatigue within the 24 h period postexercise. The best available evidence indicates that immediately and up to 4 h following a marathon, there is a significant reduction in maximum voluntary contraction and activation of the skeletal muscle used (tibialis anterior) in the preceding exercise, with moderate restoration by 24 h postmarathon (Ross et al. 2007). Although there was likely some heat strain developed over the course of the marathon, the authors do not report this. Therefore, it remains unclear whether or not recovery from different levels of heat strain requires different or specific intervention and recovery strategies.

Heat Stress and Exercise Recovery

The recovery from strenuous exercise has always been a consideration for athletes, coaches, and sport scientists, since the ability to recover quickly from any exercise bout allows athletes either to resume their normal training routine earlier than would otherwise be the case or to perform at the best possible level in subsequent competitions. Performing in the heat implies that higher physiological strain has occurred. This requires more time for recovery than for a given performance in a more favorable environment. This view is strengthened by studies that show passive heating to promote a reduction in peak voluntary muscle force, due to increases in core temperature rather than increases in specific muscle temperature (Morrison, Sleivert, and Cheung 2004; Thomas et al. 2006). Since optimal muscle performance was hastened when cooling procedures reversed the core temperature toward normal values, this finding suggests that there may indeed be a heat-strain effect on ensuing exercise bouts if thermal load is maintained rather than reduced. The general observation that reversal of core temperature

restores skeletal-muscle performance (Morrison, Sleivert, and Cheung 2004) has likely advanced the concept of body cooling to hasten recovery.

The development of recovery strategies in recent years has entailed the use of such methods as cold-water immersion, contrast water therapy (cold and warm water), ice baths, and a combination of these. Although in clinical settings, the use of cold-water immersion is standard practice for the treatment of heat illness (Casa et al. 2007; Clements et al. 2002), it is not particularly clear why the use of these strategies would be of benefit in conditions where heat strain was not a factor. However, the notion of recovery from heat strain has previously been studied in order to prevent the assumed acute cerebral hyperthermia associated with high core temperature (39.77 °C) (Germain, Jobin, and Cabanac 1987). By employing face fanning during exercise and continuing this at the end of the trial for a 15 min recovery period, these authors showed that the rise in tympanic and esophageal temperatures was significantly blunted. Thus, it is reasonable to conclude that by blunting the rise in postexercise core temperature, further physical work could be possible. It was not known or perhaps was not even considered at the time of these studies whether the attenuated postexercise response in core temperature would be a useful recovery strategy to enhance the prospects of subsequent performance.

Normal Course of Recovery From Heat Stress

Before considering the efficacy of a recovery intervention, it is useful to highlight the normal course of recovery from exercise heat stress with little or no intervention. Stimulated by the incidences of heat illness in American football players, the normal course of recovery from exercise-induced hyperthermia was described more than 40 years ago (Fox et al. 1966; Murphy and Ashe 1965). When comparing exercise of 30 min duration in 25 °C and 33% RH done by players wearing either shorts, a football uniform, or shorts and a backpack equivalent to the mass of the uniform, it was shown that heat load as measured by increases in rectal temperature was greatest with the uniform, followed by the backpack and shorts, and then by the shorts alone (see figure 16.5). However, recovery of rectal temperatures over the following 30 min rest period, compared to pre-exercise values, were within 0.42 °C, 0.65 °C, and 1.20

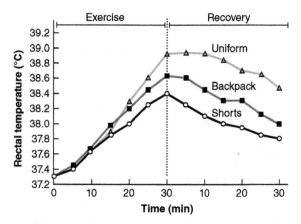

Figure 16.5   Increase in rectal temperature in American football players wearing either shorts, a uniform, or shorts plus a backpack equivalent to the mass (6.2 kg) of the uniform while exercising in the heat (25 °C, 33% RH) and subsequent reduction in rectal temperature during a 30 min recovery period.

Data from Matthews et al. 1969.

°C for the shorts, uniform, and backpack, respectively (see figure 16.5). These data indicate that recovery is graded according to the level of rectal temperature achieved during the exercise without specific intervention other than rest. Notably, in this study, heart rates recovered to approximately 100 to 105 beats per minute (bpm) after 30 min compared to a starting value of 70 bpm. Thus, the question remains as to whether or not an additional intervention beyond passive rest would be of significant benefit for subsequent exercise. A further confounding factor is the level of training and acclimation of the individual athlete. Again, this relationship between heat tolerance and acclimation was described more than 40 years ago, and it is clear that physical training alone over a 6-week period can improve heat tolerance by significantly attenuating the heart rate, skin, and rectal temperature responses (Gisolfi and Robinson 1969; Gisolfi 1973).

Proactive Recovery From Heat Stress

Returning to the issue for the potential to add value to the recovery from exercise heat stress, early studies in this area showed very little benefit under specific conditions. When comparing the effect of 60 min of passive-recovery protocols in either neutral (22 °C) or hot (35 °C) conditions following high-intensity exercise (5 × 15

s sprints), there was no significant change in the ensuing postrecovery high-intensity efforts relative to ambient temperatures (Falk et al. 1998). These authors concluded that heat exposure did not appear to affect the physiological responses during 60 min of recovery. More recently, however, 5 min of cold-water immersion between bouts of endurance exercise (25 min at 65% $\dot{V}O_2$max + 4 km time trial) in the heat attenuated the rise in rectal temperature, resulting in maintenance of the performance in subsequent time trials (Peiffer et al. 2010a). Similarly, when comparing cold-water immersion, hot-water immersion, contrast water therapy, and passive recovery, their individual effects on time-trial performance showed that cold-water immersion and contrast water therapy improve the overall recovery from high-intensity cycling compared to the other two treatments by maintaining the subsequent performance throughout 5 consecutive days of testing (Vaile et al. 2008b). These authors suggested that a similar recovery strategy could provide improvements for athletes in situations where there is a short time between competitive events. In a follow-up study, this group also showed that cold-water immersion is an effective intervention when repeated cycling performance is required in the heat. Importantly, their data show that core temperature and limb blood flow could serve as significant physiological changes associated with the improved performance (Vaile et al. 2011). That is, the authors suggest that cold-water immersion resulted in a reduction in limb blood flow during the immersion period in the nonactive limbs (arms) compared with the active limbs (legs), which indicates redistributed blood flow. This redistribution of blood flow from the periphery to the core could be one mechanism whereby venous return is improved, leading to better cardiac efficiency and exercise performance.

An issue that needs to be addressed by researchers studying the effects of recovery strategies for subsequent exercise performance is the timing of the recovery strategy immediately after a performance and before the next performance. For example, when comparing 15 min of cold-water immersion in different temperatures (10 °C, 15 °C, 20 °C, 20 °C+) with active recovery, combined with 40 min of passive recovery in the heat before a subsequent performance, the range of cold-water immersion protocols reduced thermal strain and proved more effective in maintaining the subsequent high-intensity cycling performance compared to the active recovery method (Vaile et al. 2008a). However, researchers must further clarify to what extent an experimental design of this type, where cold-water immersion is employed relatively soon after one performance and shortly before the subsequent performance, can be thought of as a recovery strategy rather than simply a precooling maneuver.

A practical issue with respect to the use of cold-water immersion as a recovery strategy is the potential cold adaptive response that occurs with repeated exposure. To date, this has not been studied with respect to recovery interventions; however, the available data on cold adap-

tation indicate that physiological changes can occur within a relatively short period of time, and that those changes are possibly detrimental to exercise performance. For example, sympatho-adrenal response due to cold adaptation could also invoke metabolic changes that could limit the available fuels for exercise (Marino, Sockler, and Fry 1998).

In addition to this issue is the potential effect of exposure time. In comparing cold-water immersion of either 5, 10, or 20 min with passive recovery following cycling time to exhaustion in the heat (40 °C, 40% RH), muscle temperature decreased more significantly in the 10 and 20 min exposures than in the 5 min one. However, muscle contractile

PRACTICAL APPLICATION

Unequivocal evidence exists that environmental temperatures above 25 °C significantly reduce the capacity to perform exercise at an optimal level. This is confirmed in both laboratory and field-based studies and by the results of competitive performances in events such as the marathon. The reasons for the deterioration in performance are not self-evident, although evidence indicates that activation of the central nervous system may play a critical role. Nevertheless, because many competitions are either held in geographical locations where environmental temperatures are less than optimal or during the day when temperatures are relatively high, it is up to officials, sport scientists, and athletes to implement strategies that help maintain performance and protect individual athletes or teams from heat-related illness.

The practical approaches to reduce thermal strain and avoid heat-related illness could include the following:

- Precooling maneuvers aimed at reducing the starting core temperature can increase either the time or intensity of exercise for ambient temperatures over 21 °C and approaching 31 °C (see figure 16.2).

- When precooling maneuvers by water immersion or cold-air exposure are limited by the immediate surroundings or the availability of cooling resources, then ice-slurry

ingestion could provide a useful intervention strategy by increasing the available heat sink.

- Coaches and sport scientists involved in the preparation of athletes who will compete in less than favorable conditions should take note of morphology and its possible limitations with respect to thermal strain. Larger athletes generate heat faster, and therefore may need specific cooling interventions before and following exercise to hasten recovery.

- It is not clear whether higher than normal thermal strain resulting from exercise requires a recovery intervention that would bring the core temperature close to its pre-exercise values; however, if subsequent performance is required within a short period, then body cooling between exercise bouts (e.g., during breaks in games) may be a useful and practical strategy.

- When employing successive cooling interventions for recovery either over several days or even weeks, caution should be exercised, since successive cooling maneuvers could result in cold adaptation that ultimately interferes with the thermal or metabolic demands of the competition for which recovery was sought.

function as measured by changes in isokinetic and peak isometric torque were similar across all cold-water interventions (Peiffer, Abbiss, Watson, et al. 2009). In contrast, this group also found that recovery by cold-water immersion for 20 min, following 90 min of cycling at a constant power output with a subsequent 16.1 km time trial in the heat (32 °C), resulted in reduced maximal-voluntary isometric torque compared with a control condition (Peiffer, Abbiss, Nosaka, et al. 2009).

Although cold-water immersion recovery significantly reduced rectal temperature, there was a clear detrimental effect on neuromuscular function. The authors suggested that because there was no significant increase in force output with electrical stimulation for either the cold-water immersion or control condition, the loss of strength was not likely due to central inhibition, but rather to decreases in peripheral contractility.

In addition to this paradox is the finding that intermittent sprint exercise in the heat attenu-ates the recovery of MVC for at least 60 min following exercise (Duffield, King, and Skein 2009). The mechanism for the apparent slow recovery of MVC is not well understood. However, when muscle fiber–conduction velocity (MFCV) and motor-unit recruitment (measured as the root mean square of EMG) of the quadriceps muscle following 50 min of exercise in either 40 °C or 19 °C conditions was compared to passive heating in 40 °C, it was found that the MFCV decreased less than motor-unit recruitment when core temperature was highest (40.3 °C) as a result of exercise alone (Hunter, Albertus-Kajee, and St Clair Gibson 2011). This finding provides evidence that indicates that reduced motor-unit recruitment rather than reduced MFCV is responsible for significant declines in torque output in exercise-induced hyperthermia. Therefore, cold-water immersion might be a potential recovery strategy that could maintain motor-unit recruitment by blunting the rise in core temperature.

Summary

The various recovery strategies that employ methods to reduce body temperature following exercise in the heat have thus far been unable to unequivocally establish the efficacy of such strategies. However, this reflects the uncertainty surrounding the establishment of evidence-based practice in an area that has yet to fully develop its applicability and practicality. Although the findings are equivocal with respect to the efficacy of recovery interventions using cold-water immersion or similar strategies, further research is warranted to understand its limitations and its usefulness. In particular, researchers need to establish guidelines that entail the duration of exposure, the temperature of the exposure agent, the surface area to be exposed, and the proximity of the exposure at the end of exercise and subsequent performance.

CHAPTER 17

Recovery at Altitude

Charles-Yannick Guézennec, PhD

With: Nicolas Bourrel

Since the first great Olympic summer sport event at altitude, the Mexico Olympic Games, took place 40 years ago, altitude training has attracted consistent interest. Numerous studies have been conducted to evaluate how this type of training benefits performance. It is clear that achieving optimal physical performance at altitude requires adequate acclimatization. Altitude training has been used for decades by endurance athletes to improve sea-level performances. However, the potentiating effects on performance after the return to sea level remain controversial (De Paula and Niebauer 2010).

Recovery management is critical for assessing the global effect of altitude training on performance (at altitude or sea level). As altitude rises, several physical characteristics of the environment are modified. The main change is a progressive decrease in barometric pressure, which leads to a decreased partial oxygen pressure in the ambient air. Atmospheric pressure decreases as a function of altitude according to an approximately exponential law. Partial oxygen pressure and barometric pressure both decrease according to the same law. Exposure to midrange altitude can be considered to start from 1,500 m, while *high altitude* refers to locations above 2,600 m. These limits are determined by the physiological consequences of progressive hypoxia. Because intense physical exercise increases the need for oxygen supply, the physiological effects of high altitude may be observed in highly trained athletes at altitudes below 2,600 m. Maximal oxygen consumption is significantly reduced at an altitude as low as 600 m above sea level in elite-endurance athletes. These elite athletes develop more severe levels of arterial hypoxemia during maximal and submaximal exercise than sedentary controls both under normoxic and hypoxic conditions (Gore et al. 1996; Anselme et al. 1992). Since the majority of altitude training centers for Olympic sports throughout the world are located at heights between 1,600 m and 2,400 m, it is assumed that some of the athletes trained at these altitudes will present the physiological reactions observed for higher altitudes.

Physiological Responses to Altitude Exposure

The physiological response to altitude is centrally governed in response to oxygen availability, which is determined at several checkpoints in the body. Oxygen pressure drops progressively between the air inspired and the tissues. We can thus distinguish a lung stage, between the mouth and alveolar cells; then a blood stage, when oxygen is transported in the blood; and, finally, a tissue stage, when oxygen is extracted and used by the cells. Since all the cellular systems in the body require oxygen, it is logical that the physiological response to hypoxia involves mechanisms that tend to decrease the tissue's oxygen contribution. These mechanisms take place in three steps:

respiratory, circulatory, and tissue based. We will distinguish two types of responses: short-term responses, which are immediate reactions to hypoxia and which do not require adaptation, and long-term adaptations, which lead to modifications of the structures involved in the transport and peripheral use of oxygen.

- *The early stage is ventilatory.* The body reacts quickly to hypoxia by increasing the ventilatory drive, as reviewed by Schoene (1997). Hyperventilation accounts for approximately two-thirds of the increase in tidal volume and for approximately one-third of the increase in breathing frequency. Hyperventilation increases the output of carbon dioxide without increasing its metabolic production. It results in a decrease in alveolar carbon dioxide pressure and a corresponding increase in alveolar oxygen pressure. This change in composition of alveolar gas is one of the most important reactions of the body in the early stages of acclimatization to altitude.

- *The second stage is circulatory.* The body then reacts by increasing cardiac output and oxygen transport capacity per volume of blood (Vogel et al. 1974). Convection of oxygen by the circulatory system depends on the cardiac output and the blood's capacity to transport oxygen. As soon as a person arrives at altitude, the resting cardiac output is increased in line with the level of hypoxia. Therefore, at altitudes above 5,000 m, cardiac output can be doubled. The resting heart rate is also increased by some tenths of a percent. For an equivalent submaximal workload, the cardiac output is slightly higher at altitude than at sea level, and is almost identical for maximal workloads performed at up to 4,000 m. During acclimatization to altitude, cardiac output and resting heart rate decrease gradually and return to values close to those observed at sea level. Blood pressure generally stays within the normal limits. Subjects who are born at altitude or live there for a long time have high pulmonary blood pressure as a result of the effects of hypoxia on the pulmonary blood vessels.

- *The third stage involves adaptations of the blood.* The first adaptation in the blood is an increase in the erythrocyte and hemoglobin concentrations. This increase is rapid the first few days at altitude, but then it slows down. The progressive hemoconcentration can extend over several months or even years (Berglund 1992). Adjustments of the number of red blood cells and hemoglobin are controlled by the altitude level and the duration of exposure, but interindividual variability is also observed. The increased concentration of red blood cells observed at altitude is related to erythropoietin hyperactivity. Hypoxia induces the kidney to release erythropoietin (EPO), which stimulates proliferation of erythrocytes. These blood modifications increase hemoglobin-based oxygen transport. Another mechanism to increase oxygen transport by erythrocytes is based on an increase in 2,3-diphosphoglycerate, a metabolic by-product that influences the oxygen binding curves of hemoglobin and improves oxygen extraction at the tissue level.

Muscular Structure Adaptations

Skeletal muscles undergo a number of changes at altitude, depending on the duration and intensity of altitude exposure. These changes compensate for the lack of oxygen in the atmosphere. It is well established that hypoxia stimulates capillary growth around the muscle cells due to the effects of several growth factors. These structural modifications lead to an increased capillary network around the muscle that favors extraction of oxygen from the blood. After prolonged exposure to high altitudes, a decrease in muscular-fiber surface has been observed. This reduces the distance between the capillary blood and the center of the muscle fiber to further improve oxygen diffusion toward the muscles' metabolic machinery. This adaptation, however, it is not entirely positive. The decrease in muscle surface area reduces the muscle mass and, consequently, the force-production capacity. This type of effect is mainly observed during exposure to altitudes above 4,000 m, but it is possible that this phenomenon may begin during very intense training at altitudes between 2,000 and 3,000 m, contributing to the reduction in maximal force capacity.

The effect of altitude on the subcellular structure responsible for aerobic energy production, the mitochondrion, is more controversial. It was previously suggested that, in addition to stimulating the capillary network, hypoxia leads to an increased mitochondrial density. However, more recent data, reviewed by Hoppeler and colleagues (2008), indicate that exposure to a very high altitude could result in mitochondrial degradation. Physical training at midrange altitudes leads to tighter metabolic control through improved delivery and conversion of metabolic substrates within the muscle fibers, combined with more efficient mitochondrial metabolism and coupling. These effects could be exploited to optimize performance at altitude, and they may also be beneficial after the return to sea level.

The genomic mechanisms involved in muscle adaptation were recently identified. Thus, capillary growth appears to be governed by vascular endothelial growth factor (VEGF), while the mechanism involved in the muscle mitochondrial response to hypoxia is linked to the formation of reactive oxygen species (ROS) in mitochondria during oxidative phosphorylation (Flueck 2009). Given that recovery is the time allowed for protein synthesis in response to physical training, it is possible that the adaptive process at muscle level occurs during the early recovery phase after exercise.

Metabolic Adaptations

Exposure to altitude leads to numerous metabolic changes during exercise. The main effect is an enhanced contribution of carbohydrate (CHO) oxidation to total energy expenditure. Hypoxic exercise at the same absolute intensity as normoxic exercise has been shown to increase the oxidation of glucose during exercise (Brooks et al. 1991). It was suggested that this greater dependence on glucose could be explained by an adaptation to optimize aerobic metabolism in hypoxic conditions, since it provides the highest ATP yield per mole of O_2 (McCleland, Hochachka, and Weber 1998). However, the oxygen advantage provided by consumption of CHO must be balanced against the potential depletion of CHO stores. This metabolic adaptation suggests that the energetic constraints related to limited CHO reserves outweigh this critical substrate's advantage of saving O_2.

The increased reliance on CHO at altitude is combined with a reduction in fat oxidation (Roberts et al. 1996). This shift from fat to glucose use was also observed in previous studies investigating cardiac metabolism in hypoxic conditions. It was suggested that acclimatization to altitude could enhance fat oxidation and, consequently, reduce the rate of muscle glycogen use. However, even with prolonged acclimatization, the rate of glucose metabolism remains above sea-level values (Brooks et al. 1991). Several mechanisms have been suggested to explain this

phenomenon, reviewed by Braun (2008). The main mechanism involves stimulation of the sympathoadrenal system, as evidenced by an immediate rise in circulating epinephrine at the beginning of altitude exposure. However, short-term acclimatization was also shown to enhance insulin sensitivity and to improve muscle glucose transport. More recently, it was also suggested that hypoxia upregulates glucose transport by AMP-activated protein kinase (AMPK) signaling pathways. AMPK is a protein kinase that responds to a decrease in cellular energy status by stimulating ATP production and glucose transport.

Another metabolic change induced by exposure to high altitudes has been extensively studied, although it appears to contradict the enhanced glucose metabolism. Chronic hypoxia has been shown to reduce blood-lactate concentrations at both submaximal and maximal exercise intensities. This phenomenon was termed the "lactate paradox." It is transient, and after prolonged acclimatization, the blood lactate response is similar to that measured at sea level (Van Hall et al. 2001). From a practical point of view, the metabolic adaptations to physical training at altitude indicate the crucial role of supplying carbohydrates during the recovery phase.

Effects of Altitude on Performance

The main interest for altitude training is its potential role on performance. On one hand, altitude training is absolutely necessary to complete performance in competitions at altitude. On the other hand, the effect of altitude training on performance after returning to sea level remains an area of discussion. Meanwhile, the interaction between altitude and maximal work capacities are sustained by the basic knowledge on the physiological effects of altitude.

Aerobic Performance

The limited adaptation of the gaseous exchange system has repercussions on maximal oxygen consumption ($\dot{V}O_2max$). The reduction of $\dot{V}O_2max$ is about 1.5% to 3.5% per 300 m increase in altitudes above 1,200 m. As an example, the maximal oxygen consumption decreases by one-third at 5,000 m, and by three-quarters at the top of Mount Everest (8,848 m). Exposure to a modest altitude may also limit aerobic performance, although the variations affect only a small percentage of athletes. A significant decline in $\dot{V}O_2max$ is also observed from 600 m in highly trained subjects (Gore et al. 1996), while the threshold rises to 1,200 m for sedentary subjects (Terrados 1992). This premature limitation of maximal aerobic power is linked to arterial oxygen desaturation during maximal exercise. This desaturation already exists at sea level, and it is exacerbated at low altitudes. Consequently, when athletes are exposed to hypoxia equivalent to an altitude greater than 1,200 m, their aerobic potential becomes limited. Thus, it is very important to take this constraint into account in the timing of recovery, since the oxygen consumed during recovery allows the resynthesis of the energy substrates consumed during exercise. Therefore, reduced oxygen availability could delay glycogen resynthesis.

Anaerobic Performance

The resting muscle phosphate concentration and the reserve of oxygen bound to myoglobin do not appear to be modified by hypoxia. The blood-oxygen reserve is decreased, but this represents only approximately 10% of the alactic oxygen debt. Thus, the alactic anaerobic capacity is practically unchanged by altitude (Ceretelli and Di Prampero 1985). Acute altitude exposure does not modify lactate production at the end of a supramaximal exercise. On the other hand, for equivalent submaximal power, blood lactate is higher at altitude than at sea level. However, if the exercise is performed at the same relative intensity with regard to $\dot{V}O_2max$, blood lactate is identical at all altitudes. As previously mentioned, under the effect of chronic hypoxia, the lactate concentration is lower than observed for an equivalent power output at sea level. This change should be taken into account when using blood lactate kinetics to manage maximal exercise and recovery in the field. In contrast to aerobic performance, anaerobic power is only slightly modified at altitude (Coudert 1992; Richalet et al. 1992). Thus, an athlete's capacity to perform a brief, intense exercise is protected. In the case of repeated sprint exercises, lack of O_2 does not significantly affect performance. Short-term performances are not altered, and they may even be improved, for example, in speed events where the performance is favored by the reduced air resistance.

Effects of Altitude on Recovery

Metabolic recovery after exercise is generally assumed to involve restoring glycogen stocks and eliminating metabolic by-products. The metabolic changes observed at altitude could be thought to interfere with recovery, mainly due to the increase in carbohydrate use for the same level of energy expenditure. This results in a greater consumption of glycogen (Guézennec et al. 1986). Consequently, the time required for full glycogen resynthesis is longer. Recent data obtained during recovery after maximal exercise at a simulated altitude of 2,000 m confirm enhanced reliance on carbohydrate during exercise. They clearly show that this substrate is used during recovery and that lipolysis occurs at a lower rate than during sea-level recovery (Katayama et al. 2010) (figure 17.1). For recovery from maximal exercise performed with a predominance of anaerobic metabolism, restoring muscle energy stores requires an equivalent O_2 volume to the anaerobic energy expended, defined as the excess postexercise O_2 consumption (EPOC). The rapid EPOC component during early recovery mainly contributes to the replenishment of the high-energy phosphate stores, as well as to the cost of increased circulation and ventilation and glycogen resynthesis. Thus, if a second supramaximal effort is performed shortly after an initial maximal effort, the rate of energy release and, consequently, power output will decrease (Bogdanis et al. 1995; Balsom et al. 1994). These authors also demonstrated a greater reduction in performance when 10 successive sprints were performed in acute hypoxic conditions. Nevertheless, in both cases, postexercise O_2 uptake was lower in hypoxic conditions, suggesting a reduced rate of oxidative processes during early recovery (Robach et al. 1997).

Effects of Altitude on Body Composition, Nutrition Patterns, and Sleep

It is well established that appropriate nutrition and adequate sleep periods determine good recovery for athletes training at sea level. Both these factors can be disturbed by exposure to high altitudes.

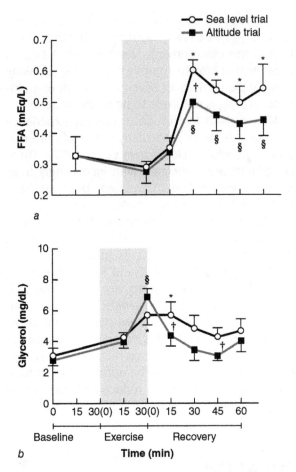

Figure 17.1 Serum free fatty acids (a) and glycerol (b) concentrations during exercise (vertical rectangle) and recovery at 2,000 m altitude and sea level. Values are means ± SE.

* P < 0.05 versus baseline at sea level. § P < 0.05 versus baseline at altitude. † P < 0.05 between moderate altitude and sea level.

Reprinted from *Metabolism: Clinical and Experimental*, Vol. 59, K. Katayama et al., "Substrate utilization during exercise and recovery at moderate altitude," pgs. 959-966. Copyright 2010, with permission from Elsevier.

For a number of years, it has been known that prolonged exposure to high altitudes results in significant loss of body mass and profound changes in physical composition. This was well illustrated by the data from the American medical expedition to Everest in 1981. During this study, all climbers who were born at sea level reported an average 2 kg of weight loss during the approach march from 1,000 m up to the base camp at 5,400 m. The loss of mass increased with the duration of exposure, and subjects lost an average of 6 kg (Boyer and Blume 1984). The reasons for this altitude-related loss of body mass are still under debate. Among

the suggested mechanisms, we can list an imbalance between energy intake and expenditure, dehydration, poor absorption of certain macronutrients, and a significant and very reproducible reduction in food intake. Prolonged exposure to altitude leads to significant effects on nutrition behavior that affect both performance and tolerance to physical training. One critical consequence is a decrease in muscle-protein synthesis, which results both from a negative energy balance and from a specific effect of hypoxia on protein metabolism (Bigard et al. 1996; Bigard et al. 1993). The objectives of the nutrition follow-up at altitude can have two complementary orientations: on the one hand, to limit the extent of the loss of body mass and the loss of lean mass, and, on the other hand, to ensure preservation of performance levels by optimizing the body's stores of energy substrates. Since the relative contribution of carbohydrate as a fuel source may increase at altitude, it is crucial to maintain an adequate carbohydrate intake.

In the short term (3 weeks or less), the caloric deficit may not impair performance, provided that glycogen stores are maintained. Because of this, it is recommended that 60% or more of a mountaineer's caloric intake be in the form of carbohydrate. This helps ensure adequate glycogen storage (Kechijian 2011). The spontaneous nutrition changes observed at altitude indicate that diet should be strictly monitored to promote recovery. Athletes should focus on carbohydrate-rich foods, since they require less oxygen to metabolize than fat- or protein-rich foods. They help replace glycogen stores and have a protein-sparing effect. In addition, mountaineers show a preference for carbohydrate and an aversion to fat, since high-altitude environments blunt the sense of taste. This may contribute to an inadequate energy intake.

However, at altitude, other things cause mass to be lost. Fluid loss at altitude is attributed to increased sweating from exertion and ventilatory changes occurring in cold, dry air. Cold weather also increases diuresis. This loss of water is not directly perceived by the subject because the cold environment at altitude decreases the perception of thirst. Given the consequences of dehydration on performance in dynamic or muscular exercises, special attention should be paid to the level of hydration, particularly during the recovery period.

Based on the available research, it is recommended that athletes working at altitude consume 3 to 5 L of fluid per day to replace losses due to respiration, urination, sweating, and thermal regulation (Richardson, Watt, and Maxwell 2009).

Athletes performing in these environments should make a conscious effort to eat at frequent intervals. An inadequate energy intake leads to body mass loss and negatively affects strength and endurance. The goal should therefore be to consume an adequate volume of food to provide sufficient calories. Carbohydrate should account for more than 50% of the total energy supplied by the diet. This is more important at altitude than at sea level. According to recent data, consuming carbohydrate and protein in the same meal during the early recovery phase can help promote protein synthesis. This type of diet can also be used during recovery at altitude (Koopman et al. 2005).

Before exposure to high altitudes, it is generally recommended that athletes increase their intake of vitamins and minerals. However, no direct evidence exists that vitamin supplementation enhances altitude tolerance. However, improved iron status may be useful before altitude training, since oxygen-carrying capacity is stretched to its limit in this environment. Nevertheless, it takes a great deal of time to improve iron status, so taking iron supplements over several weeks during altitude training is not likely to be of much benefit.

New arrivals to altitude commonly experience poor-quality sleep, particularly those submitted to a high workload (Hoshikawa et al. 2007). These complaints are associated with increasingly fragmented sleep linked to frequent brief awakenings, which are in turn connected with breathing difficulties. Changes in sleep architecture include a shift toward lighter sleep stages, with marked decreases in slow-wave sleep. Increased hypoxic ventilatory responsiveness and loss of breathing regularization during sleep are also observed. This results in sleep fragmentation and increased periods of wakefulness. These sleep disturbances occur in trained athletes for midrange altitude exposure and lead to poor recovery. Thus, the quality of lodging should be optimized in altitude training centers so as to reduce disturbances to sleep as much as possible.

PRACTICAL APPLICATION

When arriving at altitude, it is necessary to evaluate the different athletic capacities: aerobic, anaerobic, speed, and strength.

To this end, tests to be administered include the following:

- Progressive maximal running test for aerobic capacity: It is interesting to use a test without interruption between each step. This is associated with heart rate monitoring.

- Wingate test to evaluate anaerobic capacity.

- Squat jump (SJ) height test with no arm swing, countermovement jump (CMJ) height test with no arm swing to measure maximal leg strength.

- One-repetition maximum (1RM) strength test of upper body movement: Find the maximal weight the subjects can raise, with different specific movements used according the type of sport.

The progressive maximal running test can be used to evaluate aerobic capacity and short-term recovery until maximal running speed is regained. It is useful to calculate the same relative workload as at sea level, since it gives an indication of the recovery time required between bouts of exercise. On arrival, the principle is to double the recovery time that would be provided for an equivalent relative workload at sea level. Repeat this test periodically at various intervals during altitude acclimatization to check the effect of altitude training on both maximal aerobic running speed and the change in recovery for each running speed. The data are used to define the recovery time necessary during interval training. Normal acclimatization is associated with a progressive increase in maximal speed and a decrease in heart rate at the end of recovery for each equivalent submaximal effort. This method allows recovery time to be progressively reduced. From a practical point of view, it seems that active recovery at a very slow pace is more efficient at altitude than at sea level.

In contrast to aerobic capacity, anaerobic capacity, maximal strength, and speed are not directly influenced by altitude until a high level of hypoxia is reached. According to the field experience of many coaches, it seems that recovery time is also unaffected at altitude. It could be slightly decreased in many conditions. The basic principles are to measure maximal force production on arrival at altitude and to repeat this test with the same periodicity as the tests of aerobic capacity. This can help determine to what extent the potential negative influence of altitude on muscle mass occurs. It may also facilitate early detection, as indicated by a decrease in strength performance. For the protocol oriented toward strength endurance with numerous repeats at a low workload, a longer recovery time than at sea level is required between each series (table 17.1).

During strength-promoting sessions, based on the data obtained for maximal repetitions, we apply a recovery time according to the various constituents of the strength sought (maximal, power, stamina, explosivity), as shown in table 17.2. Practical observations show that at altitude, for a workload that is already 80% of 1RM, the recovery time is lower than the usual data used at sea level for a well-acclimatized subject. For example, for strength work at 80% of 1RM on 5 series of 6 repetitions, it is necessary to set a basic recovery time of 2.5 min. Practice subjects who are acclimatized to altitude recover faster (1.5–2 min).

Table 17.1 Recovery for Strength Endurance Protocols

	Long-duration strength endurance	Medium-duration strength endurance	Short-duration strength endurance
Intensity	20–40%	40–55%	55–70%
Duration of effort	>3 min	90–180 s	20–45 s
Recovery	Duration of effort	60–120 s	60–120 s
Series	30–10, depending on the sport		
Repetitions	40 and above	20–40	10–20

Table 17.2 Recovery Time Used During Strength Training

	Work intensity	Number of repetitions	Total number of repetitions	Recovery / series	Recovery / session
Maximal efforts	90–100% of 1RM	1–3	Between 10 and 18 (3 × 6; 5 × 2)	2–4 min	24–48 h
Repeated efforts	60–75%	6–12	30–60	2.5–3.5 min	24–48 h
Dynamic efforts	50–70%	7 10 s reps	80–90	2.5–4 min	24–48 h
Submaximal efforts	80–90%	4–5	20–30	2.5–3.5 min	24–48 h

Summary

Achieving optimal physical performance at altitude requires adequate acclimatization and recovery management. Research presented in this chapter provided the following guidelines:

- Increase the recovery time provided in the training session between each bout of intense exercise to allow sufficient recovery of energy substrates.

- Increase the period of reduced training intensity between two maximal-intensity training sessions.

- Increase the total energy supplied by the diet and the proportion of carbohydrate, as well as the quality of food, to reduce the negative effect of altitude on appetite.

- Propose the use of a mixed carbohydrate-protein source during the early recovery period after physical training.

- Increase total daily fluid intake.

- Take all the environmental factors that could improve the quality of sleep into account.

References

Abbiss, C.R., and P.B. Laursen. 2005. Models to explain fatigue during prolonged endurance cycling. *Sports Med* 35:865–898.

Abernethy, P.J., R. Thayer, and A.W. Taylor. 1990. Acute and chronic responses of skeletal muscle to endurance and sprint exercise. A review. *Sports Med* 10:365–389.

Acevedo, E., K. Rinehardt, and R. Kraemer. 1994. Perceived exertion and affect at varying intensities of running. *Res Q Exercise Sport* 65(4):372–376.

Achten, J., and A.E. Jeukendrup. 2003. Heart rate monitoring: Applications and limitations. *Sports Med* 33:517–538.

Adam, C.L., and J.G. Mercer. 2004. Appetite regulation and seasonality: Implications for obesity. *Proc Nutr Soc* 63:413–419.

Adie, W.J., J.L. Duda, and N. Ntoumanis. 2008. Autonomy support, basic need satisfaction and the optimal functioning of adult male and female sport participants: A test of basic needs theory. *Motiv Emotion* 32:189–199.

Afaghi, A., H. O'Connor, and C.M. Chow. 2007. High-glycemic-index carbohydrate meals shorten sleep onset. *Am J Clin Nutr* 85:426–430.

Ahmaidi, S., P. Granier, Z. Taoutaou, J. Mercier, H. Dubouchaud, and C. Prefaut. 1996. Effects of active recovery on plasma lactate and anaerobic power following repeated intensive exercise. *Med Sci Sports Exerc* 28:450–456.

Ahonen, E., and U. Nousiainen. 1988. The sauna and body fluid balance. *Ann Clin Res* 20(4):257–261.

Akabas, S.R., and K.R. Dolins. 2005. Micronutrient requirements of physically active women: What can we learn from iron? *Am J Clin Nutr* 81:S1246–S1251.

Akasaki, Y., M. Miyata, H. Eto, T. Shirasawa, N. Hamada, Y. Ikeda, S. Biro, Y. Otsuji, and C. Tei. 2006. Repeated thermal therapy up-regulates endothelial nitric oxide synthase and augments angiogenesis in a mouse model of hindlimb ischemia. *Circ J* 70(4):463–470.

Aldemir, H., G. Atkinson, T. Cable, B. Edwards, J. Waterhouse, and T. Reilly. 2000. A comparison of the immediate effects of moderate exercise in the late morning and late afternoon on core temperature and cutaneous thermoregulatory mechanisms. *Chronobiol Int* 17:197–207.

Al Haddad, H., P.B. Laursen, S. Ahmaidi, and M. Buchheit. 2010. Influence of cold water face immersion on post-exercise parasympathetic reactivation. *Eur J Appl Physiol* 108(3):599–606.

Al Haddad, H., P.B. Laursen, D. Chollet, F. Lemaitre, S. Ahmaidi, and M. Buchheit. 2010. Effect of cold or thermoneutral water immersion on post-exercise heart rate recovery and heart rate variability indices. *Auton Neurosci* 156(1-2):111–116.

Ali, A., M.P. Caine, and B.G. Snow. 2007. Graduated compression stockings: Physiological and perceptual responses during and after exercise. *J Sports Sci* 25:413–419.

Allemeier, C.A., A.C. Fry, P. Johnson, R.S. Hikida, F.C. Hagerman, and R.S. Staron. 1994. Effects of sprint cycle training on human skeletal muscle. *J Appl Physiol* 77:2385–2390.

Allen, D.G. 2009. Fatigue in working muscles. *J Appl Physiol* 106:358–359.

Allen, D.G., G.D. Lamb, and H. Westerblad. 2008. Skeletal muscle fatigue: Cellular mechanisms. *Physiol Rev* 88:287–332.

American College of Sports Medicine (ACSM). 1998. ACSM position stand. The recommended quantity and quality of exercise for developing and maintaining cardiorespiratory and muscular fitness, and flexibility in healthy adults. *Med Sci Sports Exerc* 30(6):975–991.

Ames, B.N., R. Cathcart, E. Schwiers, and P. Hochstein. 1982. Uric acid provides an antioxidant defense in humans against oxidant and radical caused aging and cancer: A hypothesis. *Proc Natl Acad Sci* 78:6858–6862.

Ames, C. 1992. Achievement goals, motivational climate, and motivational processes. In *Motivation in sport and exercise,* ed. G.C. Roberts, 161–176. Champaign, IL: Human Kinetics.

Andersson, H., T. Raastad, J. Nilsson, G. Paulsen, I. Garthe, and F. Kadi. 2008. Neuromuscular fatigue and recovery in elite female soccer: Effects of active recovery. *Med Sci Sports Exerc* 40:372–380.

Angus, R.G., R.J. Heslegrave, and W.S. Myles. 1985. Effects of prolonged sleep deprivation, with and without chronic physical exercise, on mood and performance. *Psychophysiology* 22:276–282.

Anselme, F., C. Caillaud, I. Courret, and C. Prefaut. 1992. Exercise induced hypoxemia and histamine excretion in extreme athletes. *Int J Sports Med* 13:80–81.

Ansonoff, M.A., and A.M. Etgen. 2000. Evidence that oestradiol attenuates beta-adrenoceptor function in the hypothalamus of female rats by altering receptor phosphorylation and sequestration. *J Neuroendocrinol* 12:1060–1066.

Anthony, J.C., T.G. Anthony, and D.K. Layman. 1999. Leucine supplementation enhances skeletal muscle protein metabolism in human. *Am J Physiol* 263:E928–E934.

Apple, F.S., M.A. Rogers, D.C. Casal, L. Lewis, J.L. Ivy, and J.W. Lampe. 1987. Skeletal muscle creatine kinase

MB alterations in women marathon runners. *Eur J Appl Physiol Occup Physiol* 56:49–52.

Armstrong, L.E., D.J. Casa, M. Millard-Stafford, D.S. Moran, S.W. Pyne, and W.O. Roberts. 2007. American College of Sports Medicine position stand. Exertional heat illness during training and competition. *Med Sci Sports Exerc* 39:556–572.

Armstrong, L.E., R.W. Hubbard, J.P. DeLuca, E.L. Christensen, and W.J. Kraemer. 1987. Evaluation of a temperate environment test to predict heat tolerance. *Eur J Appl Physiol* 56(4):384–389.

Armstrong, L.E., R.W. Hubbard, P. Szlyk, W. Matthew, and I. Sils. 1985. Voluntary dehydration and electrolyte losses during exercise in the heat. *Aviat Space Environ Med* 56:765–770.

Armstrong, L., and J. Van Heest. 2002. The unknown mechanisms of the overtraining syndrome. Clues from depression and psychoneuroimmunology. *Sports Med* 32:185–209.

Arngrimsson, S.A., D.S. Petitt, M.G. Stueck, D.K. Jorgensen, and K.J. Cureton. 2004. Cooling vest worn during active warm-up improves 5-km run performance in the heat. *J Appl Physiol* 96:1867–1874.

Arnulf, I., P. Quintin, J.C. Alvarez, L. Vigil, Y. Touitou, A.S. Lebre, A. Bellenger, O. Varoquaux, J.P. Derenne, J.F. Allilaire, C. Benkelfat, and M. Leboyer. 2002. Midmorning tryptophan depletion delays REM sleep onset in healthy subjects. *Neuropsychopharmacol* 27:843–851.

Ascensao, A., M. Leite, A.N. Rebelo, S. Magalhaes, and J. Magalhaes. 2011. Effects of cold water immersion on the recovery of physical performance and muscle damage following a one-off soccer match. *J Sports Sci* 29(3):217–225.

Attar-Levy, D. 1998. [Seasonal depression]. *Therapie* 53:489–498.

Aubert, A.E., B. Seps, and F. Beckers. 2003. Heart rate variability in athletes. *Sports Med* 33:889–919.

Baar, K. 2009. The signaling underlying FITness. *Appl Physiol Nutr Metab* 34(3):411–419.

Backx, K., L. McNaughton, L. Crickmore, G. Palmer, and A. Carisle. 2000. Effects of differing heat and humidity on the performance and recovery from multiple high intensity, intermittent exercise bouts. *Int J Sports Med* 21(6):400–405.

Bailey, D.M., S.J. Erith, P.J. Griffin, A. Dowson, D.S. Brewer, N. Gant, and C. Williams. 2007. Influence of cold-water immersion on indices of muscle damage following prolonged intermittent shuttle running. *J Sports Sci* 25(11):1163–1170.

Baldari, C., M. Videira, F. Madeira, J. Sergio, and L. Guidetti. 2004. Lactate removal during active recovery related to the individual anaerobic and ventilatory thresholds in soccer players. *Eur J Appl Physiol* 93:224–230.

———. 2005. Blood lactate removal during recovery at various intensities below the individual anaerobic threshold in triathletes. *J Sports Med Phys Fitness* 45:460–466.

Ballantyne, C. 2000. An off-season preparatory program for women lacrosse athletes. *Strength Cond J* 22:42–47.

Balsom, P.D., J.Y. Seger, B. Sjodin, and B. Ekblom. 1992a. Maximal-intensity intermittent exercise: Effect of recovery duration. *Int J Sports Med* 13:528–533.

———. 1992b. Physiological responses to maximal intensity intermittent exercise. *Eur J Appl Physiol Occup Physiol* 65:144–149.

Balsom, P.G., C. Gaitanos, B. Ekblom, and B. Sjödin B. 1994. Reduced oxygen availability during high intensity intermittent exercise impairs performance. *Acta Physiol Scand* 152:279–285.

Bam, J., T.D. Noakes, J. Juritz, and S.C. Dennis. 1997. Could women outrun men in ultramarathon races? *Med Sci Sports Exerc* 29:244–247.

Bandopadhyay, P., and W. Selvamarthy. 2003. Respiratory changes due to extreme cold in the Arctic environment. *Int J Biometeorol* 18:178–181.

Banfi, G., M. Krajewska, G. Melegati, and M. Patacchini. 2008. Effects of whole-body cryotherapy on haematological values in athletes. *Br J Sports Med* 42:558.

Banfi, G., G. Melegati, A. Barassi, G. Dogliotti, G. Melzi d'Eril, B. Dugué, and M.M. Corsi. 2008. Effects of whole body cryotherapy on serum mediators of inflammation and serum muscle enzymes in athletes. *J Thermal Biology* 34:55–59.

Bangsbo, J., T. Graham, L. Johansen, and B. Saltin. 1994. Muscle lactate metabolism in recovery from intense exhaustive exercise: Impact of light exercise. *J Appl Physiol* 77:1890–1895.

Bangsbo, J., L. Norregaard, and F. Thorso. 1991. Activity profile of competition soccer. *Can J Sport Sci* 16:110–116.

Banister, E. 1991. Modeling elite athletic performance. In *Physiological testing of elite athletes*, ed. H. Green, J. McDougall, and H. Wenger, 403–424. Champaign, IL: Human Kinetics.

Barbiche, E. 2006. *Intérêt de la cryothérapie du corps entier dans la rééducation du sportif de haut niveau, en phase de renforcement, à distance d'une ligamentoplastie du genou.* Thesis. Capbreton, France: CERS.

Barnett, A. 2006. Using recovery modalities between training sessions in elite athletes: Does it help? *Sports Med* 36(9):781–796.

Baron, R.A., and M.J. Kalsher. 1998. Effects of a pleasant ambient fragrance on simulated driving performance: The sweet smell of . . . safety? *Environ Behav* 30:535–552.

Bar-Or, O., H.M. Lundegren, and E.R. Buskirk. 1969. Heat tolerance of exercising obese and lean women. *J Appl Physiol* 26:403–409.

Barron, G., T. Noakes, W. Levy, C. Smith, and R. Millar. 1985. Hypothalamic dysfunction in overtrained athletes. *J Clin Endocr Metab* 60:803–806.

Bartosz, C. 2003. *Another face of oxygen: Free radicals in nature*. Warsaw, Poland: PWN.

Bazett, H., J. Scott, M. Maxfield, and M. Blithe. 1937. Effect of baths at different temperatures on oxygen exchange and on the circulation. *Am J Physiol* 119:93–110.

Bazett-Jones, D.M., J.B. Winchester, and J.M. McBride. 2005. Effect of potentiation and stretching on maximal force, rate of force development, and range of motion. *J Strength Cond Res* 19:421–426.

Beals, K.A., and M.M. Manore. 1998. Nutritional status of female athletes with subclinical eating disorders. *J Am Diet Assoc* 98:419–425.

Beckers, E.J., N.J. Rehrer, F. Brouns, and W.H.M. Saris. 1992. Influence de la composition des boissons et de la fonction gastro-instestinale sur la biodisponibilité des liquides et des substances nutritives pendant l'exercice physique. *Science & Sports* 7:107–119.

Beckett, J.R., K.T. Schneiker, K.E. Wallman, B.T. Dawson, and K.J. Guelfi. 2009. Effects of static stretching on repeated sprint and change of direction performance. *Med Sci Sports Exerc* 41:444–450.

Beersma, D.G. 1998. Models of human sleep regulation. *Sleep Med Rev* 2:31–43.

Belcastro, A.N., and A. Bonen. 1975. Lactic acid removal rates during controlled and uncontrolled recovery exercise. *J Appl Physiol* 39:932–936.

Belitsky, R.B., S.J. Odam, and C. Hubley-Kozey. 1987. Evaluation of the effectiveness of wet ice, dry ice, and cryogenic packs in reducing skin temperature. *Phys Ther* 67:1080–1084.

Bender, A.L., E.E. Kramer, J.B. Brucker, T.J. Demchak, M.L. Cordova, and M.B. Stone. 2005. Local ice-bag application and triceps surae muscle temperature during treadmill walking. *J Athl Train* 40:271–275.

Bender, R.W., T.E. Wilson, R.L. Hoffman, and B.C. Clark. 2011. Passive-heat stress does not induce muscle fatigue, central activation failure or changes in intracortical properties of wrist flexors. *Ergonomics* 54:565–575.

Bender, T., G. Nagy, I. Barna, I. Tefner, E. Kadas, and P. Geher. 2007. The effect of physical therapy on beta-endorphin levels. *Eur J Appl Physiol* 100:371–382.

Berardi, J.M., T.B. Price, E.E. Noreen, and P.W.R. Lemon. 2006. Post-exercise muscle glycogen recovery enhanced with a carbohydrate–protein supplement. *Med Sci Sports Exerc* 38:1106–1113.

Berger, R.J., and N.H. Phillips. 1995. Energy conservation and sleep. *Behav Brain Res* 69:65–73.

Berglund, B. 1992. High-altitude training. Aspects of haematological adaptation. *Sports Med* 14:289–303.

Berglund, B., and H. Safstrom. 1994. Psychological monitoring and modulation of training load of world-class canoeists. *Med Sci Sport Exer* 26:1036–1040.

Berry, E.M., J.H. Growdon, J.J. Wurtman, B. Caballero, and R.J. Wurtman. 1991. A balanced carbohydrate: Protein diet in the management of Parkinson's disease. *Neurology* 41:1295–1297.

Berry, M.J., S.P. Bailey, L.S. Simpkins, and J.A. TeWinkle. 1990. The effects of elastic tights on the post-exercise response. *Can J Sport Sci* 15:244–248.

Berry, M.J., and R.G. McMurray. 1987. Effects of graduated compression stockings on blood lactate following an exhaustive bout of exercise. *Am J Phys Med* 66:121–132.

Bigard, A.X., P. Douce, D. Merino, F. Lienhard, and C.Y. Guézennec. 1996. Changes in dietary protein intake fail to prevent decrease in muscle growth induced by severe hypoxia in rats. *J Appl Physiol* 80(1):208–215.

Bigard, A.X., and C.Y. Guézennec. 2003. *Nutrition du sportif*. Paris: Masson.

Bigard, A.X., H. Sanchez, and G. Claveyrolas. 2001. Effects of dehydration and rehydration on EMG changes during fatiguing contractions. *Med Sci Sport Exerc* 33:1694–1700.

Bigard, A.X., P. Satabin, P. Lavier, F. Canon, D. Taillandier, and C.Y. Guézennec. 1993. Effects of protein supplementation during prolonged exercise at moderate altitude on performance and plasma amino acid pattern. *Eur J Appl Physiol Occup Physiol* 66(1):5–10.

Billat, L.V. 2001a. Interval training for performance: A scientific and empirical practice. Special recommendations for middle- and long-distance running. Part I: Aerobic interval training. *Sports Med* 31:13–31.

———. 2001b. Interval training for performance: A scientific and empirical practice. Special recommendations for middle- and long-distance running. Part II: Anaerobic interval training. *Sports Med* 31:75–90.

Biolo, G., K.D. Tipton, S. Klein, and R.R. Wolfe. 1997. An abundant supply of amino acids enhances the metabolic effects of exercise on muscle protein. *Am J Physiol Endocrinol Metab* 273:E122–E129.

Bishop, D. 2004. The effects of travel on team performance in the Australian national netball competition. *J Sci Med Sport* 7:118–122.

Bishop, D., and M. Spencer. 2004. Determinants of repeated-sprint ability in well-trained team-sport athletes and endurance-trained athletes. *J Sports Med Phys Fitness* 44:1–7.

Blagrove, M., C. Alexander, and J.A. Horne. 1995. The effects of chronic sleep reduction on the performance of cognitive tasks sensitive to sleep deprivation. *Appl Cogn Psychol* 9:21–40.

Blatchford, F.K., R.G. Knowlton, and D.A. Schneider. 1985. Plasma FFA responses to prolonged walking in untrained men and women. *Eur J Appl Physiol Occup Physiol* 53:343–347.

Bleakley, C.M., and G.W. Davison. 2009. What is the biochemical and physiological rationale for using cold water immersion in sports recovery? A systematic review. *Br J Sports Med* 44(3):179–187.

Blom, P.C., A.T. Hostmark, and O. Vaage. 1987. Effect of different post-exercise sugar diets on the rate of muscle glycogen synthesis. *Med Sci Sports Exerc* 19:491–496.

Blomstrand, E., P Hassmén, and E.A. Newsholme. 1991. Effect of branched-chain amino acid supplementation on mental performance. *Acta Physiol Scand* 143:225–226.

Bloomer, R.J., M.J. Falvo, A.C. Fry, B.K. Schilling, W.A. Smith, and C.A. Moore. 2006. Oxidative stress response in trained men following squats or sprints. *Med Sci Sports Exerc* 38:1436–1442.

Bobbert, M.F., A.P. Hollander, and P.A. Huijing. 1986. Factors in delayed onset muscular soreness of man. *Med Sci Sports Exerc* 18:75–81.

Bogdanis, G.C., M.E. Nevill, L.H. Boobis, H.K.A. Lakomy, and A.M. Nevill. 1995. Recovery of power output and muscle metabolites following 30 s of maximal sprint cycling in man. *J Physiol* 482:467–480.

Bohe, J., J.F. Low, R.R. Wolfe, and M.J. Rennie. 2001. Latency and duration of stimulation of human muscle protein synthesis during continuous infusion of amino acids. *J Physiol* 532:575–579.

Boileau, R.A., J.E. Misner, G.L. Dykstra, and T.A. Spitzer. 1983. Blood lactic acid removal during treadmill and bicycle exercise at various intensities. *J Sports Med Phys Fitness* 23:159–167.

Boisseau, N. 2004. Gender differences in metabolism during exercise and recovery. *Science & Sports* 19:220–227.

Bompa, T.O. 1999. *Periodization: Theory and methodology of training.* Champaign, IL: Human Kinetics.

Bonde-Petersen, F., L. Schultz-Pedersen, and N. Dragsted. 1992. Peripheral and central blood flow in man during cold, thermoneutral, and hot water immersion. *Aviat Space Environ Med* 63(5):346–350.

Bonen, A., and A.N. Belcastro. 1976. Comparison of self-selected recovery methods on lactic acid removal rates. *Med Sci Sports* 8:176–178.

Bonen, A., F.J. Haynes, W. Watson-Wright, M.M. Sopper, G.N. Pierce, M.P. Low, and T.E. Graham. 1983. Effects of menstrual cycle on metabolic responses to exercise. *J Appl Physiol* 55:1506–1513.

Bonen, A., G.W. Ness, A.N. Belcastro, and R.L. Kirby. 1985. Mild exercise impedes glycogen repletion in muscle. *J Appl Physiol* 58:1622–1629.

Bonnet, M.H. 1980. Sleep, performance and mood after the energy-expenditure equivalent of 40 hours of sleep deprivation. *Psychophysiology* 17:56–63.

Bonnet, M.H., and D.L. Arand. 1992. Caffeine use as a model of acute and chronic insomnia. *Sleep* 15:526–536.

Booth, F.W., W.F. Nicholson, and P.A. Watson. 1982. Influence of muscle use on protein synthesis and degradation. *Ex Sport Sci Rev* 10:27–48.

Booth, J., F. Marino, and J.J. Ward. 1997. Improved running performance in hot humid conditions following whole body precooling. *Med Sci Sports Exerc* 29:943–949.

Borresen, J., and M.I. Lambert. 2008. Autonomic control of heart rate during and after exercise: Measurements and implications for monitoring training status. *Sports Med* 38(8):633–646.

Borsa, P.A., and E.L. Sauers. 2000. The importance of gender on myokinetic deficits before and after micro-injury. *Med Sci Sports Exerc* 32:891–896.

Bostic, T.J., D.M. Rubio, and M. Hood. 2000. A validation of the subjective vitality scale using structural equation modeling. *Soc Indic Res* 52:313–324.

Bothorel, B. 1990. *Thermorégulation chez l'homme exposé à la chaleur: Influence de l'état d'hydratation, de la charge thermique endogène, et de stimulations thermiques locales.* Thesis. Strasbourg, France: Strasbourg University of Science.

Boyer, S.J., and F.D. Blume. 1984. Weight loss and changes in body composition at high altitude. *J Appl Physiol* 57:1580–1585.

Brandenberger, G., V. Candas, M. Follenius, and J.N. Kahn. 1989. The influence of the initial state of hydration on endocrine responses to exercise in the heat. *Eur J Appl Physiol* 58:674–679.

Braun, B. 2008. Effects of high altitude on substrate use and metabolic economy: Cause and effect? *Med Sci Sports Exerc* 40:1495–1500.

Brawn, T.P., K.M. Fenn, H.C. Nusbaum, and D. Margoliash. 2008. Consolidation of sensorimotor learning during sleep. *Learn Mem* 15:815–819.

Brener, W., T. Hendrix, and P. Hugh. 1983. Regulation on the gastric emptying of glucose. *Gastoenterology* 85:76–82.

Brewer, B. 2003. Developmental differences in psychological aspects of sport-injury rehabilitation. *J Athl Training* 38:152–153.

Bringard, A., R. Denis, N. Belluye, and S. Perrey. 2007. Compression élastique externe et fonction musculaire chez l'homme. *Science & Sports* 22:3–13.

Bringard, A., S. Perrey, and N. Belluye. 2006. Aerobic energy cost and sensation responses during submaximal running exercise: Positive effects of wearing compression tights. *Int J Sports Med* 27:373–378.

Broad, E.M., L.M. Burke, G.R. Cox, P. Heeley, and M. Riley. 1996. Body weight changes and voluntary fluid intakes during training and competition sessions in team sports. *Int J Sport Nutr* 6:307–320.

Brooks, G.A., G.E. Butterfield, R.R. Wolfe, B.M. Groves, R.S. Mazzeo, J.R. Sutton, E.E. Wolfel, and J.T. Reeves. 1991. Increased dependence on blood glucose after acclimatization to 4,300 m. *J Appl Physiol* 70:919–927.

Broughton, R.J. 1989. Chronobiological aspects and models of sleep and napping. In *Sleep and alertness: Chronobiological, behavioral, and medical aspects of napping,* ed. R.J.E. Broughton, 71–98. New York: Raven Press.

Brouns, F., E.M.R. Kovacs, and J.M.G. Senden. 1998. The effect of different rehydration drinks on post-exercise electrolyte excretion in trained athletes. *Int J Sports Med* 19:56–60.

Brown, L.E. 2001. Nonlinear versus linear periodization models. *Strength Cond J* 23(1):42–44.

Brück, K., and H. Olschewski. 1986. Body temperature related factors diminishing the drive to exercise. *J Physiol Pharmacol* 65:1274–1280.

Brukner, P., and K. Khan. 2001. *Clinical sports medicine* (2nd ed.). Sydney: McGraw-Hill.

Buchheit, M., H. Al Haddad, A. Chivot, P.M. Lepretre, S. Ahmaidi, and P.B. Laursen. 2010. Effect of in- versus out-of-water recovery on repeated swimming sprint performance. *Eur J Appl Physiol* 108(2):321–327.

Buchheit, M., C. Horobeanu, A. Mendez-Villanueva, B.M. Simpson, and P.C. Bourdon. 2011. Effects of age and spa treatment on match running performance over two consecutive games in highly trained young soccer players. *J Sports Sci* 29(6):591–598.

Buchheit, M., A. Mendez-Villanueva, G. Delhomel, M. Brughelli, and S. Ahmaidi. 2010. Improving repeated sprint ability in young elite soccer players: Repeated shuttle sprints vs. explosive strength training. *J Strength Cond Res* 24:2715–2722.

Buchheit, M., J.J. Peiffer, C.R. Abbiss, and P.B. Laursen. 2008. Effect of cold water immersion on post-exercise parasympathetic reactivation. *Am J Physiol Heart Circ Physiol* 296:H421–427.

Buchheit, M., and P. Ufland. 2011. Effect of endurance training on performance and muscle reoxygenation rate during repeated-sprint running. *Eur J Appl Physiol* 111:293–301.

Budgett, R., E. Newsholme, M. Lehmann, C. Sharp, D. Jones, T. Peto, D. Collins, R. Nerurkar, and P. White. 2000. Redefining the overtraining syndrome as the unexplained underperformance syndrome. *Br J Sports Med* 34:67–68.

Bulbulian, R., J.H. Heaney, C.N. Leake, A.A. Sucec, and N.T. Sjoholm. 1996. The effect of sleep deprivation and exercise load on isokinetic leg strength and endurance. *Eur J Appl Physiol Occup Physiol* 73:273–277.

Burke, D.G., L.E. Holt, and R.L. Rasmussen. 2001. Effects of hot and cold water immersion and modified proprioceptive neuromuscular facilitation flexibility exercise on hamstring length. *J Athl Train* 36:16–19.

Burke, D.G., S.A. MacNeil, L.E. Holt, N.C. Mackinnon, and R.L. Rasmussen. 2000. The effect of hot or cold water immersion on isometric strength training. *J Strength Cond Res* 14(1):21–25.

Burke, L.M. 2001. Nutritional practices of male and female endurance cyclists. *Sports Med* 31:521–532.

Burke, L.M., G.R. Collier, P.G. Davis, P.A. Fricker, A.J. Sanigorski, and M. Hargreaves. 1996. Muscle glycogen storage after prolonged exercise: Effect of the frequency of carbohydrate feedings. *Am J Clin Nut* 64:115–119.

Burke, L.M., G.R. Collier, and M. Hargreaves. 1993. Muscle glycogen storage after prolonged exercise: Effect of the glycemic index of carbohydrate feedings. *J Appl Physiol* 75:1019–1023.

Burke, L.M., B. Kiens, and J.L. Ivy. 2004. Carbohydrates and fat for training and recovery. *J Sports Sci* 22:15–30.

Buroker, K.C., and J.A. Schwane. 1989. Does post-exercise static stretching alleviate delayed muscle soreness? *Physician Sportsmed* 17:65–83.

Butterfield, D.L., D.O. Draper, M.D. Ricard, J.W. Myrer, S.S. Schulthies, and E. Durrant. 1997. The effects of high-volt pulsed current electrical stimulation on delayed-onset muscle soreness. *J Athl Train* 32:15–20.

Byrne, C., J.K.W. Lee, S.A.N. Chew, C.L. Lim, and E.Y.M. Tan. 2006. Continuous thermoregulatory responses to mass-participation distance running in heat. *Med Sci Sports Exerc* 38:803–810.

Cabanac, M. 1986. Keeping a cool head. *News Physiol Sci* 1:41–44.

Cadefau, J., J. Casademont, J.M. Grau, J. Fernandez, A. Balaguer, M. Vernet, R. Cusso, and A. Urbano-Marquez. 1990. Biochemical and histochemical adaptation to sprint training in young athletes. *Acta Physiol Scand* 140:341–351.

Cafarelli, E., and F. Flint. 1992. The role of massage in preparation for and recovery from exercise. An overview. *Sports Med* 14:1–9.

Cairns, S.P. 2006. Lactic acid and exercise performance: Culprit or friend? *Sports Med* 36(4):279–291.

Callaghan, M.J. 1993. The role of massage in the management of the athlete: A review. *Br J Sports Med* 27:28–33.

Callister, R., R. Callister, S. Fleck, and G. Dudley. 1990. Physiological and performance responses to overtraining in elite judo athletes. *Med Sci Sport Exer* 22(6):816–824.

Campbell, K.S. 2009. Interactions between connected half-sarcomeres produce emergent mechanical behavior in a mathematical model of muscle. *PLoS Comput Biol* (doi:10.1371/journal.pcbi.1000560).

Candas, V., and B. Bothorel. 1989. *Hydratation, travail et chaleur.* Vandœuvre-lès-Nancy, France: Institut National de Recherche et de Sécurité.

Candas, V., J.P. Libert, G. Brandenberger, J.C. Sagot, and J.M. Kahn. 1988. Thermal and circulatory responses during prolonged exercise at different levels of hydration. *J Physiol* 83:11–18.

Candas, V., J.P. Libert, and J. Vogt. 1983. Sweating and sweat decline of resting men in hot humid environments. *Eur J Appl Physiol* 50:223–234.

Caputa, M., G. Feistkorn, and C. Jessen. 1986. Effects of brain and trunk temperatures on exercise performance in goats. *Pflügers Arch* 406:184–189.

Carling, C., J. Bloomfield, L. Nelsen, and T. Reilly. 2008. The role of motion analysis in elite soccer: Contemporary performance measurement techniques and work rate data. *Sports Med* 38:839–862.

Carraro, F., W.H. Stuart, W.H. Hartl, J. Roenblatt, and R.R. Wolfe. 1990. Effects of exercise and recovery on muscle protein synthesis in human subjects. *Am J Physiol* 259:E470–E476.

Carrier, D., A. Kapoor, T. Kimura, M. Nickels, E. Scott, J. So, and E. Trinkaus. 1984. The energetic paradox of human running and hominid evolution. *Curr Anthropol* 25:483–495.

Carrithers, J.A., D.L. Williams, P.M. Gallagher, M.P. Godard, K.E. Schulze, and S.W. Trappe. 2000. Effects of postexercise carbohydrate-protein feedings on muscle glycogen restoration. *J Appl Physiol* 88:1976–1982.

Carter, S., S. McKenzie, M. Mourtzakis, D.J. Mahoney, and M.A. Tarnopolsky. 2001. Short-term 17 beta-estradiol decreases glucose R(a) but not whole body metabolism during endurance exercise. *J Appl Physiol* 90:139–146.

Casa, D.J., B.P. McDermott, E.C. Lee, S.W. Yeargin, L.E. Armstrong, and C.M. Maresh. 2007. Cold water immersion: The gold standard for exertional heatstroke treatment. *Exerc Sport Sci Rev* 35:141–149.

Castagna, O., X. Nesi, J. Briswalter, and C. Hausswirth. 2010. Comparaison entre deux gilets refroidissants: Effets sur le rendement énergétique lors d'un exercice de pédalage en condition chaude et humide, et lors de ka période de récupération. In *Récupération et performance en sport*, ed. C. Hausswirth, 379–389. Paris: INSEP.

Castle, P.C., A.L. Macdonald, A. Philp, A. Webborn, P.W. Watt, and N.S. Maxwell. 2006. Precooling leg muscle improves intermittent sprint exercise performance in hot, humid conditions. *J Appl Physiol* 100:1377–1384.

Ceretelli, P., and P. Di Prampero. 1985. Aerobic and anaerobic metabolism during exercise at altitude. *Med Sport Sci* 19:1–19.

Cermakian, N., and D.B. Boivin. 2009. The regulation of central and peripheral circadian clocks in humans. *Obes Rev 10 Suppl* 2:25–36.

Cheing, G.L., and C.W. Hui-Chan. 2003. Analgesic effects of transcutaneous electrical nerve stimulation and interferential currents on heat pain in healthy subjects. *J Rehabil Med* 35:15–19.

Chesley, A., J.D. MacDougall, M.A. Tarnopolsky, S.A. Atkinson, and K. Smith. 1992. Changes in human muscle protein synthesis after resistance exercise. *J Appl Physiol* 73:1383–1388.

Cheung, K., P. Hume, and L. Maxwell. 2003. Delayed onset muscle soreness: Treatment strategies and performance factors. *Sports Med* 33:145–164.

Cheung, S., and A. Robinson. 2004. The influence of upper body pre-cooling on repeated sprint performance in moderate ambient temperatures. *J Sports Sci* 22:605–612.

Cheuvront, S.N., R. Carter, K.C. Deruisseau, and R.J. Moffatt. 2005. Running performance differences between men and women: An update. *Sports Med* 35:1017–1024.

Chiu, L., and J. Barnes. 2003. The fitness-fatigue model revisited: Implications for planning short- and long-term training. *Strength Cond J* 25(6):42.

Chleboun, G.S., J.N. Howell, H.L. Baker, T.N. Ballard, J.L. Graham, H.L. Hallman, L.E. Perkins, J.H. Schauss, and R.R. Conatser. 1995. Intermittent pneumatic compression effect on eccentric exercise-induced swelling, stiffness, and strength loss. *Arch Phys Med Rehabil* 76:744–749.

Choi, D., K.J. Cole, B.H. Goodpaster, W.J. Fink, and D.L. Costill. 1994. Effect of passive and active recovery on the resynthesis of muscle glycogen. *Med Sci Sports Exerc* 26:992–996.

Christensen, N.J., and K. Schultz-Larsen. 1994. Resting venous plasma adrenalin in 70-year-old men correlated positively to survival in a population study: The significance of the physical working capacity. *J Intern Med* 235(3):229–232.

Chu, S. 2008. Olfactory conditioning of positive performance in humans. *Chem Senses* 33:65–71.

Cian, C., N. Koulmann, and P.A. Barraud. 2000. Influence of variations in body hydration on mental efficiency: Effect of hyperhydration, heat stress and exercise induced dehydration. *J Psychophysiol* 14:29–36.

Cissik, J., A. Hedrick, and M. Barnes. 2008. Challenges applying the research on periodization. *Strength Cond J* 30(1):45.

Cizza, G., R. Kvetnansky, M. Tartaglia, M. Blackman, G. Chrousos, and P. Gold. 1993. Immobolisation stress rapidly decreases hypothalamic corticotrophin-releasing hormone secretion in vitro in the male 344/N fischer rat. *Life Sci* 53:233–240.

Cizza, G., P. Romagni, A. Lotsikas, G. Lam, N.E. Rosenthal, and G.P. Chrousos. 2005. Plasma leptin in men and women with seasonal affective disorder and in healthy matched controls. *Horm Metab Res* 37:45–48.

Claremont, A., D. Costill, W. Fink, and P. Van Handel. 1976. Heat tolerance following diuretic induced dehydration. *Med Sci Sports* 8:239–243.

Clarkson, P.M., and M.J. Hubal. 2001. Are women less susceptible to exercise-induced muscle damage? *Curr Opin Clin Nutr Metab Care* 4:527–531.

Clements, J.M., D.J. Casa, J.C. Knight, J.M. McClung, A.S. Blake, P.M. Meenen, A.M. Gilmer, and K.A. Caldwell. 2002. Ice-water immersion and cold-water immersion provide similar cooling rates in runners with exercise-induced hyperthermia. *J Athl Train* 37:146–150.

CNOSF unpublished data. CNOSF, Maison du sport Français, 1 Avenue Pierre de Coubertin, 75640 Paris cedex 13.

Coakley, J. 1992. Burnout among adolescents: A personal failure or a social problem? *J Sport Soc Issues* 9:271–285.

Cochrane, D.J. 2004. Alternating hot and cold water immersion for athlete recovery: A review. *Phys Ther Sport* 5(1):26–32.

Cockerill, I.M., A.M. Nevill, and N. Lyons. 1991. Modelling mood states in athletic performance. *J Sports Sci* 9:205–212.

Coffey, V.G., and J.A. Hawley. 2007. The molecular bases of training adaptation. *Sports Med* 37(9):737–763.

Coffey, V., M. Leveritt, and N. Gill. 2004. Effect of recovery modality on 4-hour repeated treadmill running performance and changes in physiological variables. *J Sci Med Sport* 7(1):1–10.

Coggan, A.R., R.J. Spina, D.S. King, M.A. Rogers, M. Brown, P.M. Nemeth, and J.O. Holloszy. 1992. Skeletal muscle adaptations to endurance training in 60- to 70-yr-old men and women. *J Appl Physiol* 72:1780–1786.

Convertino, V.A., L.E. Armstrong, E.L. Coyle, G.W. Mack, M.N. Sawka, L.C. Senay, and W.M. Sherman. 1996. Exercise and fluids replacement. *Med Sci Sports Exerc* 28:i–vii.

Cooke, B., and E. Ernst. 2000. Aromatherapy: A systematic review. *Br J Gen Pract* 50:493–496.

Cormery, B., M. Marcil, and M. Bouvard. 2008. Rule change incidence on physiological characteristics of elite basketball players: A 10-year-period investigation. *Br J Sports Med* 42:25–30.

Costa, P.B., E.D. Ryan, T.J. Herda, A.A. Walter, K.M. Hoge, and J.T. Cramer. 2010. Acute effects of passive stretching on the electromechanical delay and evoked twitch properties. *Eur J Appl Physiol* 108:301–310.

Costill, D.L., J. Daniels, W. Evans, W. Fink, G. Krahenbuhl, and B. Saltin. 1976. Skeletal muscle enzymes and fiber composition in male and female track athletes. *J Appl Physiol* 40:149–154.

Costill, D.L., W.J. Fink, L.H. Getchell, J.L. Ivy, and F.A. Witzmann. 1979. Lipid metabolism in skeletal muscle of endurance-trained males and females. *J Appl Physiol* 47:787–791.

Costill, D.L., M.G. Flynn, J.P. Kirwan, J.A. Houmard, J.B. Mitchell, R. Thomas, and S.H. Park. 1988. Effects of repeated days of intensified training on muscle glycogen and swimming performance. *Med Sci Sports Exerc* 20:249–254.

Costill, D., and B. Saltin. 1974. Factors limiting gastric emptying during rest and exercise. *J Appl Physiol* 37:679–683.

Costill, D.L., W.M. Sherman, W.J. Fink, C. Maresh, M. Witten, and J.M. Miller. 1981. The role of dietary carbohydrates in muscle glycogen resynthesis after strenuous running. *Am J Clin Nut* 34:1831–1836.

Costill, D., and K. Sparks. 1973. Rapid fluid replacement following thermal dehydration. *J Appl Physiol* 34:299–303.

Coté, D.J., W.E. Prentice Jr., D.N. Hooker, and E.W. Shields. 1988. Comparison of three treatment procedures for minimizing ankle sprain swelling. *Phys Ther* 68(7):1072–1076.

Cotter, J.D., G.G. Sleivert, W.S. Roberts, and M.A. Febbraio. 2001. Effect of pre-cooling, with and without thigh cooling, on strain and endurance exercise performance in the heat. *Comp Biochem Physiol A Mol Integr Physiol* 128:667–677.

Coudert, J. 1992. Anaerobic performance at altitude. *Int J Sports Med* 13(1):S82–S85.

Coutts, A.J., L.K. Wallace, and K.M. Slattery. 2007. Monitoring changes in performance, physiology, biochemistry, and psychology during overreaching and recovery in triathletes. *Int J Sports Med* 28:125–134.

Couzan, S. 2006. Le sportif: Un insuffisant veineux potentiel. *Cardio & Sport* 8:7–20.

Craig, J.A., M.B. Cunningham, D.M. Walsh, G.D. Baxter, and J.M. Allen. 1996. Lack of effect of transcutaneous electrical nerve stimulation upon experimentally induced delayed onset muscle soreness in humans. *Pain* 67:285–289.

Cramp, F.L., G.R. McCullough, A.S. Lowe, and D.M. Walsh. 2002. Transcutaneous electric nerve stimulation: The effect of intensity on local and distal cutaneous blood flow and skin temperature in healthy subjects. *Arch Phys Med Rehabil* 83:5–9.

Crane, R.K. 1962. Hypothesis for mechanism of intestinal active transport of sugars. *Fed Proc* 21:891–895.

Criswell, D., S. Powers, J. Lawler, J. Tew, S. Dodd, Y. Ipyiboz, R. Tulley, and K. Wheeler. 1991. Influence of a carbohydrate electrolyte beverage on performance and blood homeostasis during recovery from football. *Int J Sports Nut* 1:178–191.

Cross, K.M., R.W. Wilson, and D.H. Perrin. 1996. Functional performance following an ice immersion to the lower extremity. *J Athl Train* 31:113–116.

Crowe, M.J., D. O'Connor, and D. Rudd. 2007. Cold water recovery reduces anaerobic performance. *Int J Sports Med* 28(12):994–998.

Cumps, E., J. Pockelé, and R. Meeusen. 2004. Online: A uniform injury registration system. Paper presented at the First International Conference on IT and Sport, Cologne, Germany.

Cunningham, J.J. 1997. Is potassium needed in sports drinks for fluid replacement during exercise? *Int J Sports Nutrition* 7:154–161.

Cury, F., and P. Sarrazin. 2001. *Théories de la motivation et pratiques sportives: État des recherches.* Paris: PUF.

Daniels, J.T. 1985. A physiologist's view of running economy. *Med Sci Sports Exerc* 17(3):332–338.

da Silva, J.F., L.G. Guglielmo, and D. Bishop. 2010. Relationship between different measures of aerobic fitness and repeated-sprint ability in elite soccer players. *J Strength Cond Res* 24:2115–2121.

Davies, H. 1975. Cardiovascular effects of the sauna. *Am J Phys Med* 54:178–185.

Davies, V., K.G. Thompson, and S.M. Cooper. 2009. The effects of compression garments on recovery. *J Strength Cond Res* 23:1786–1794.

Dawson, B., M. Cutler, A. Moody, S. Lawrence, C. Goodman, and N. Randall. 1995. Effects of oral creatine loading on single and repeated maximal short sprints. *Aust J Sci Med Sport* 27:56–61.

Dawson, B., M. Fitzsimons, S. Green, C. Goodman, M. Carey, and K. Cole. 1998. Changes in performance, muscle metabolites, enzymes and fibre types after short sprint training. *Eur J Appl Physiol Occup Physiol* 78:163–169.

Dawson, B., C. Goodman, S. Lawrence, D. Preen, T. Polglaze, M. Fitzsimons, and P. Fournier. 1997. Muscle phosphocreatine repletion following single and repeated short sprint efforts. *Scand J Med Sci Sports* 7:206–213.

Dawson, B., S. Gow, S. Modra, D. Bishop, and G. Stewart. 2005. Effects of immediate post-game recovery procedures on muscle soreness, power and flexibility levels over the next 48 hours. *J Sci Med Sport* 8:210–221.

Day, J.A., R.R. Mason, and S.E. Chesrown. 1987. Effect of massage on serum level of beta-endorphin and betalipotropin in healthy adults. *Phys Ther* 67:926–930.

Deci, E.L., and R.M. Ryan. 2002. *Handbook of self determination research.* Rochester: University of Rochester Press.

Décombaz, J. 2004. Proteins and amino acids in post exercise recovery. *Science & Sports* 3:228–233.

de Graaf-Roelfsema, E., H.A. Keizer, E. van Breda, I.D. Wijnberg, and J.H. van der Kolk. 2007. Hormonal responses to acute exercise, training and overtraining. *Vet Quarty* 29(3):82–101.

Delextrat, A., and D. Cohen. 2008. Physiological testing of basketball players: Toward a standard evaluation of anaerobic fitness. *J Strength Cond Res* 22:1066–1072.

De Meersman, R.E. 1993. Heart rate variability and aerobic fitness. *Am Heart J* 125:726–731.

Denegar, C.R., and D.H. Perrin. 1992. Effect of transcutaneous electrical nerve stimulation, cold, and a combination treatment on pain, decreased range of motion, and strength loss associated with delayed onset muscle soreness. *J Athl Train* 27:200–206.

Denegar, C.R., D.H. Perrin, A.D. Rogol, and R.A. Rutt. 1989. Influence of transcutaneous electrical nerve stimulation on pain, range of motion, and serum cortisol concentration in females experiencing delayed onset muscle soreness. *J Orthop Sports Phys Ther* 11:100–103.

Dennis, S.C., and T.D. Noakes. 1999. Advantages of a smaller body mass in humans when distance-running in warm, humid conditions. *Eur J Appl Physiol* 79:280–284.

De Paula, P., and J. Niebauer. 2010. Effects of high altitude training on exercise capacity: Fact or myth? *Sleep Breath* 1234–1237.

De Pauw, K., B. de Geus, B. Roelands, F. Lauwens, J. Verschueren, E. Heyman, and R.R. Meeusen. 2010. The effect of five different recovery methods on repeated cycle performance. *Med Sci Sports Exerc* 43(5):890–897.

De Schutter, M.F., L. Buyse, R. Meeusen, and B. Roelands. 2004. Hormonal responses to a high-intensity training period in Army recruits. *Med Sci Sport Exer* 36:S295.

De Souza, M.J., and N.I. Williams. 2005. Beyond hypoestrogenism in amenorrheic athletes: Energy deficiency as a contributing factor for bone loss. *Curr Sports Med Rep* 4:38–44.

Deutz, R.C., D. Benardot, D.E. Martin, and M.M. Cody. 2000. Relationship between energy deficits and body composition in elite female gymnasts and runners. *Med Sci Sports Exerc* 32:659–668.

De Vries, H.A., P. Beckmann, H. Huber, and L. Dieckmeir. 1968. Electromyographic evaluation of the effects of sauna on the neuromuscular system. *J Sports Med Phys Fitness* 8(2):61–69.

Diego, M.A., N.A. Jones, T. Field, M. Hernandez-Reif, S. Schanberg, C. Kuhn, V. McAdam, R. Galamaga, and M. Galamaga. 1998. Aromatherapy positively affects mood, EEG patterns of alertness and math computations. *Int J Neurosci* 96:217–224.

Dinges, D.F. 1989. Napping patterns and effects in human adults. In *Sleep and alertness: Chronobiological, behavioral, and medical aspects of napping,* ed. R.J.E. Broughton, 171–204. New York: Raven Press.

———. 1992. Adult napping and its effects on ability to function. In *Why we nap: Evolution, chronobiology, and functions of polyphasic and ultrashort sleep,* ed. C.E. Stampi, 118–134. Boston: Birkhäuser.

Doan, B.K., Y.H. Kwon, R.U. Newton, J. Shim, E.M. Popper, R.A. Rogers, L.R. Bolt, M. Robertson, and W.J. Kraemer. 2003. Evaluation of a lower-body compression garment. *J Sports Sci* 21(8):601–610.

Dodd, S., S.K. Powers, T. Callender, and E. Brooks. 1984. Blood lactate disappearance at various intensities of recovery exercise. *J Appl Physiol* 57:1462–1465.

Dollander, M. 2002. [Etiology of adult insomnia]. *Encephale* 28:493–502.

Donnelly, J.E., and B.K. Smith. 2005. Is exercise effective for weight loss with ad libitum diet? Energy balance, compensation, and gender differences. *Exerc Sport Sci Rev* 33:169–174.

Dorado, C., J. Sanchis-Moysi, and J.A. Calbet. 2004. Effects of recovery mode on performance, O_2 uptake, and O_2 deficit during high-intensity intermittent exercise. *Can J Appl Physiol* 29:227–244.

Doyle, J.A., W.M. Sherman, and R.L. Strauss. 1993. Effects of eccentric and concentric exercise on muscle glycogen replenishment. *J Appl Physiol* 74:1848–1855.

Drust, B., G. Atkinson, W. Gregson, D. French, and D. Binningsley. 2003. The effects of massage on intra muscular temperature in the vastus lateralis in humans. *Int J Sports Med* 24:395–399.

Duchateau, J. 1992. Principe de l'électrostimulation musculaire et recrutement des différents types de fibres. *Science et Motricité* 16:18–24.

Duclos, M., J.B. Corcuff, L. Arsac, F. Moreau-Gaudry, M. Rashedi, P. Roger, A. Tabarin, and G. Manier. 1998. Corticotroph axis sensitivity after exercise in endurance-trained athletes. *Clin Endocrinol* 8:493–501.

Duclos, M., J.B. Corcuff, M. Rashedi, V. Fougere, and G. Manier. 1997. Trained versus untrained men: Different immediate post-exercise responses of pituitary-adrenal axis. *Eur J Appl Phys* 75:343–350.

Duclos, M., C. Gouarne, and D. Bonnemaison. 2003. Acute and chronic effects of exercise on tissue sensitivity to glucocorticoids. *J Appl Physiol* 94:869–875.

Duclos, M., M. Minkhar, A. Sarrieau, D. Bonnemaison, G. Manier, and P. Mormede. 1999. Reversibility of endurance training-induced changes on glucocorticoid sensitivity of monocytes by an acute exercise. *Clin Endocrinol* 1:749–756.

Duffield, R. 2008. Cooling interventions for the protection and recovery of exercise performance from exercise-induced heat stress. *Med Sport Sci* 53:89–103.

Duffield, R., J. Cannon, and M. King. 2010. The effects of compression garments on recovery of muscle performance following high-intensity sprint and plyometric exercise. *J Sci Med Sport* 13(1):136–140.

Duffield, R., B. Dawson, D. Bishop, M. Fitzsimons, and S. Lawrence. 2003. Effect of wearing an ice cooling jacket on repeat sprint performance in warm/humid conditions. *Br J Sports Med* 37:164–169.

Duffield, R., B. Dawson, and C. Goodman. 2004. Energy system contribution to 100-m and 200-m track running events. *J Sci Med Sport* 7:302–313.

———. 2005. Energy system contribution to 1500- and 3000-metre track running. *J Sports Sci* 23:993–1002.

Duffield, R., J. Edge, R. Merrells, E. Hawke, M. Barnes, D. Simcock, and N. Gill. 2008. The effects of compression garments on intermittent exercise performance and recovery on consecutive days. *Int J Sports Physiol Perform* 3:454–468.

Duffield, R., R. Green, P. Castle, and N. Maxwell. 2010. Precooling can prevent the reduction of self-paced exercise intensity in the heat. *Med Sci Sports Exerc* 42:577–584.

Duffield, R., M. King, and M. Skein. 2009. Recovery of voluntary and evoked muscle performance following intermittent-sprint exercise in the heat. *Int J Sports Physiol Perf* 4:254–268.

Duffield, R., and F. Marino. 2007. Effects of pre-cooling procedures on intermittent-sprint exercise performance in warm conditions. *Eur J Appl Physiol* 100:727–735.

Duffield, R., and M. Portus. 2007. Comparison of three types of full-body compression garments on throwing and repeat-sprint performance in cricket players. *Br J Sports Med* 41:409–414.

Dufour, M., P. Colné, P. Gouilly, and G. Chemol. 1999. *Massages et massothérapie*. Paris: Éd. Maloine.

Dugué, B., and E. Leppänen. 1999. Adaptation related to cytokines in man: Effect of regular swimming in ice cold water. *Clin Physiol* 2:114–121.

Dugué, B., J. Smolander, T. Westerlund, J. Oksa, R. Nieminen, E. Moilanen, and M. Mikkelsson. 2005. Acute and long-term effects of winter swimming and whole-body cryotherapy on plasma antioxidative capacity in healthy women. *Scand J Clin Lab Invest* 65(5):395–402.

Dumont, M., and C. Beaulieu. 2007. Light exposure in the natural environment: Relevance to mood and sleep disorders. *Sleep Med* 8:557–565.

Dupont, G., and S. Berthoin. 2004. Time spent at a high percentage of $\dot{V}O_2$max for short intermittent runs: Active versus passive recovery. *Can J Appl Physiol* 29:S3–S16.

Dupont, G., W. Moalla, C. Guinhouya, S. Ahmaidi, and S. Berthoin. 2004. Passive versus active recovery during high-intensity intermittent exercises. *Med Sci Sports Exerc* 36:302–308.

Dupont, G., W. Moalla, R. Matran, and S. Berthoin. 2007. Effect of short recovery intensities on the performance during two Wingate tests. *Med Sci Sports Exerc* 39:1170–1176.

Dykstra, J.H., H.M. Hill, M.G. Miller, C.C. Cheatham, T.J. Michael, and R.J. Baker. 2009. Comparisons of cubed ice, crushed ice, and wetted ice on intramuscular and surface temperature changes. *J Athl Train* 44:136–141.

Eastell, R. 2005. Role of oestrogen in the regulation of bone turnover at the menarche. *J Endocrinol* 185:223–234.

Edwards, R.H.T. 1983. Biochemical bases of fatigue in exercise performance: Catastrophe theory of muscular fatigue. In *Biochemistry of exercise*, ed. H.G. Knuttgen, J.A. Vogel, and J. Poortmans, 3–27. Champaign, IL: Human Kinetics.

Einenkel, D. 1977. Improved health of kindergarten children in the Annenberg district due to regular use of an industrial sauna. *Z Arztl Fortbild (Jena)* 71(22):1069–1077.

Enns, D.L., and P.M. Tiidus. 2010. The influence of estrogen on skeletal muscle: Sex matters. *Sports Med* 40:41–58.

Enwemeka, C.S., C. Allen, P. Avila, J. Bina, J. Konrade, and S. Munns. 2002. Soft tissue thermodynamics before, during, and after cold pack therapy. *Med Sci Sports Exerc* 34:45–50.

Ernst, E. 1998. Does post-exercise massage treatment reduce delayed onset muscle soreness? A systematic review. *Br J Sports Med* 32:212–214.

Ernst, E., E. Pecho, P. Wirz, and T. Saradeth. 1990. Regular sauna bathing and the incidence of common colds. *Ann Med* 22:225–227.

Esbjornsson-Liljedahl, M., K. Bodin, and E. Jansson. 2002. Smaller muscle ATP reduction in women than in men by repeated bouts of sprint exercise. *J Appl Physiol* 93:1075–1083.

Esmarck, B., J.L. Anersen, S. Olsen, E.A. Richter, M. Mizuno, and M. Kjaer. 2001. Timing of postexercise protein intake is important for muscle hypertrophy with resistance training in elderly humans. *Am J Physiol Endocrinol Metab* 280:E4340–4348.

Eston, R., and D. Peters. 1999. Effects of cold water immersion on the symptoms of exercise-induced muscle damage. *J Sports Sci* 17(3):231–238.

Ettinger, S.M., D.H. Silber, K.S. Gray, M.B. Smith, Q.X. Yang, A.R. Kunselman, and L.I. Sinoway. 1998. Effects of the ovarian cycle on sympathetic neural outflow during static exercise. *J Appl Physiol* 85:2075–2081.

Evans, T.A., C. Ingersoll, K.L. Knight, and T. Worrell. 1995. Agility following the application of cold therapy. *J Athl Train* 30:231–234.

Fairchild, T.J., A.A. Armstrong, A. Rao, H. Liu, S. Lawrence, and P.A. Fournier. 2003. Glycogen synthesis in muscle fibers during active recovery from intense exercise. *Med Sci Sports Exerc* 35:595–602.

Falk, B., S. Radom-Isaac, J. Hoffmann, Y. Wang, Y. Yarom, A. Magazanik, and Y. Weinstein. 1998. The effect of heat exposure on performance of and recovery from high-intensity, intermittent exercise. *Int J Sports Med* 19:1–6.

Farr, T., C. Nottle, K. Nosaka, and P. Sacco. 2002. The effects of therapeutic massage on delayed onset muscle soreness and muscle function following downhill walking. *J Sci Med Sport* 5:297–306.

Farthing, M.J.G. 1988. History and rationale for oral rehydration and recent developments in formulating an optimal solution. *Drugs* 36(S4):80–90.

Faucon, M. 2009. *Aromathérapie pratique et usuelle.* Paris: Éditions Ellébore.

Favero, J.P., A.W. Midgley, and D.J. Bentley. 2009. Effects of an acute bout of static stretching on 40 m sprint performance: Influence of baseline flexibility. *Res Sports Med* 17:50–60.

Feibel, A., and A. Fast. 1976. Deep heating of joints: A reconsideration. *Arch Phys Med Rehabil* 57(11):513–514.

Fialkowski, K. 1978. Early hominid brain evolution and heat stress: A hypothesis. *Studies Phys Anthropol* 4:87–92.

———. 1986. A mechanism for the origin of the human brain: A hypothesis. *Curr Anthropol* 27:288–290.

Fischer, J., B.L. Van Lunen, J.D. Branch, and J.L. Pirone. 2009. Functional performance following an ice bag application to the hamstrings. *J Strength Cond Res* 23:44–50.

Flueck, M. 2009. Plasticity of the muscle proteome to exercise at altitude. *High Alt Med Bio* 10:183–193.

Flynn, M., D. Costill, J. Hawley, W. Fink, P. Neufer, R. Fieding, and M. Sleeper. 1987. Influence of selected carbohydrate drinks on cycling performance and glycogen use. *Med Sci Sports Exerc* 19:33–40.

Folkard, S. 1990. Circadian performance rhythms: Some practical and theoretical implications. *Philos Trans R Soc Lond B Biol Sci* 327:543–553.

Foster, C. 1998. Monitoring training in athletes with reference to overtraining syndrome. *Med Sci Sports Exerc* 30(7):1164–1168.

Foster, C., E. Daines, L. Hector, A. Snyder, and R. Welsh. 1996. Athletic performance in relation to training load. *Wisc Med J* 95(6):370–374.

Foster, C., J.A. Florhaug, J. Franklin, L. Gottschall, L.A. Hrovatin, S. Parker, P. Doleshal, and C. Dodge. 2001. A new approach to monitoring exercise training. *J Strength Cond Res* 15(1):109–115.

Foster, C., and M. Lehmann. 1997. Overtraining. In *Running injuries,* ed. G.N. Guten, 173–187. London: WB Saunders.

Foster, C., A. Snyder, N. Thompson, and K. Kuettel. 1988. Normalisation of the blood lactate profile. *Int J Sports Med* 9:198–200.

Fox, E.L., D.K. Mathews, W.S. Kaufman, and R.W. Bowers. 1966. Effects of football equipment on thermal balance and energy cost during exercise. *Res Q* 37:332–339.

Franchini, E., R.C. de Moraes Bertuzzi, M.Y. Takito, and M.A. Kiss. 2009. Effects of recovery type after a judo match on blood lactate and performance in specific and non-specific judo tasks. *Eur J Appl Physiol* 107:377–383.

Franchini, E., M. Yuri Takito, F. Yuzo Nakamura, K. Ayumi Matsushigue, and M.A. Peduti Dal'Molin Kiss. 2003. Effects of recovery type after a judo combat on blood lactate removal and on performance in an intermittent anaerobic task. *J Sports Med Phys Fitness* 43:424–431.

Frank, J.B., K. Weihs, E. Minerva, and D.Z. Lieberman. 1998. Women's mental health in primary care. Depression, anxiety, somatization, eating disorders, and substance abuse. *Med Clin North Am* 82:359–389.

Frank, M.G. 2006. The mystery of sleep function: Current perspectives and future directions. *Rev Neurosci* 17:375–392.

French, D.N., K.G. Thompson, S.W. Garland, C.A. Barnes, M.D. Portas, P.E. Hood, and G. Wilkes. 2008. The effects of contrast bathing and compression therapy on muscular performance. *Med Sci Sports Exerc* 40(7):1297–1306.

Freudenberger, H. 1980. *Burnout: The high cost of high achievement.* New York: Doubleday.

Freund, B.J., S.J. Montain, A.J. Young, M.N. Sawka, J.P. DeLuca, K.B. Pandolf, and C.R. Valeri. 1995. Glycerol hyperhydration: Hormonal, renal, and vascular fluid responses. *J Appl Physiol* 79(6):2069–2077.

Fricke, R. 1989. Ganzkörperkältetherapie in einer Kälterkammer mit Temperaturen um −110 °C. *Z Phys Med Baln Med Klin* 18:1–10.

Fridén, J., and R.L. Lieber. 1998. Segmental muscle fiber lesions after repetitive eccentric contractions. *Cell Tissue Res* 293:165–171.

Friedmann, B., and W. Kindermann. 1989. Energy metabolism and regulatory hormones in women and men during endurance exercise. *Eur J Appl Physiol Occup Physiol* 59:1–9.

Froberg, K., and P.K. Pedersen. 1984. Sex differences in endurance capacity and metabolic response to prolonged, heavy exercise. *Eur J Appl Physiol Occup Physiol* 52:446–450.

Fu, F.H., H.W. Cen, and R.G. Eston. 1997. The effects of cryotherapy on muscle damage in rats subjected to endurance training. *Scand J Med Sci Sports* 7:358–362.

Gabriel, D.A., G. Kamen, and G. Frost. 2006. Neural adaptations to resistive exercise: Mechanisms and recommendations for training practices. *Sports Med* 36(2):133–149.

Gabriel, H., and W. Kindermann. 1997. The acute immune response to exercise: What does it mean? *Int J Sports Med* 18:S28–S45.

Gaitanos, G.C., C. Williams, L.H. Boobis, and S. Brooks. 1993. Human muscle metabolism during intermittent maximal exercise. *J Appl Physiol* 75:712–719.

Galloway, S.D.R. 1999. Dehydration, rehydration, and exercise in the heat: Rehydration strategies for athletic competition. *Can J Appl Physiol* 24:188–200.

Galloway, S.D.R., and R.J. Maughan. 1997. Effects of ambient temperature on the capacity to perform prolonged cycle exercise in man. *Med Sci Sports Exerc* 29:1240–1249.

Galloway, S.D., and J.M. Watt. 2004. Massage provision by physiotherapists at major athletics events between 1987 and 1998. *Br J Sports Med* 38:235–236, 237.

Gandevia, S.C. 2001. Spinal and supraspinal factors in human muscle fatigue. *Physiol Rev* 81:1725–1789.

García-Pallarés, J., M. García-Fernández, L. Sánchez-Medina, and M. Izquierdo. 2010. Performance changes in world-class kayakers following two different training periodization models. *Eur J Appl Physiol* 110(1):99–107.

Garet, M., N. Tournaire, F. Roche, R. Laurent, J.R. Lacour, J.C. Barthelemy, and V. Pichot. 2004. Individual interdependence between nocturnal ANS activity and performance in swimmers. *Med Sci Sports Exerc* 36:2112–2118.

Gastin, P.B. 2001. Energy system interaction and relative contribution during maximal exercise. *Sports Med* 31:725–741.

Gauché, E., R. Lepers, G. Rabita, J.M. Leveque, D. Bishop, J. Brisswalter, and C. Hausswirth. 2006. Vitamin and mineral supplementation and neuromuscular recovery after a running race. *Med Sci Sports Exerc* 38:2110–2117.

Germain, M., M. Jobin, and M. Cabanac. 1987. The effect of face fanning during recovery from exercise hyperthermia. *Can J Physiol Pharmacol* 65:87–91.

Gilbart, M.K., D.J. Oglivie-Harris, C. Broadhurst, and M. Clarfield. 1995. Anterior tibial compartment pressures during intermittent sequential pneumatic compression therapy. *Am J Sports Med* 23:769–772.

Gill, N.D., C.M. Beaven, and C. Cook. 2006. Effectiveness of post-match recovery strategies in rugby players. *Br J Sports Med* 40(3):260–263.

Gisolfi, C.V. 1973. Work-heat tolerance derived from interval training. *J Appl Physiol* 33:349–353.

Gisolfi, C.V., and F. Mora. 2000. *The hot brain: Survival, temperature and the human body.* Cambridge, MA: MIT Press.

Gisolfi, C., and S. Robinson. 1969. Relations between physical training, acclimatization, and heat tolerance. *J Appl Physiol* 26:530–534.

Gisolfi, C., S. Robinson, and E.S. Turrell. 1966. Effects of aerobic work performed during recovery from exhausting work. *J Appl Physiol* 21:1767–1772.

Glaister, M. 2005. Multiple sprint work: Physiological responses, mechanisms of fatigue and the influence of aerobic fitness. *Sports Med* 35:757–777.

———. 2008. Multiple-sprint work: Methodological, physiological, and experimental issues. *Int J Sports Physiol Perform* 3:107–112.

Goats, G.C. 1990. Interferential current therapy. *Br J Sports Med* 24:87–92.

———. 1994a. Massage—The scientific basis of an ancient art: Part 1. The techniques. *Br J Sports Med* 28:149–152.

———. 1994b. Massage—The scientific basis of an ancient art: Part 2. Physiological and therapeutic effects. *Br J Sports Med* 28:153–156.

Goh, S.S., P.B. Laursen, B. Dascombe, and K. Nosaka. 2011. Effect of lower body compression garments on submaximal and maximal running performance in cold (10 °C) and hot (32 °C) environments. *Eur J Appl Physiol* 111(5):819–826.

Goldberg, A.L., J.D. Etlinger, D.F. Goldspink, and C. Jablecki. 1975. Mechanism of work-induced hypertrophy of skeletal muscle. *Med Sci Sports Exerc* 7:248–261.

Goldberg, A.L., and R. Odessey. 1972. Oxidation of amino acids by diaphragms from fed and fasted rats. *Am J Physiol* 223:1384–1391.

Goldberg, J., S.J. Sullivan, and D.E. Seaborne. 1992. The effect of two intensities of massage on H–reflex amplitude. *Phys Ther* 72:449–457.

Goldley, A.L., and H. Goodman. 1999. Amino acid transport during work-induced growth of skeletal muscle. *Am J Physiol* 216:1111–1115.

González-Alonso, J., C. Teller, S.L. Anderson, F.B. Jensen, T. Hyldig, and B. Nielsen. 1999. Influence of body temperature on the development of fatigue during prolonged exercise in the heat. *J Appl Physiol* 86:1032–1039.

Goodall, S., and G. Howatson. 2008. The effects of multiple cold water immersions on indices of muscle damage. *J Sports Sci Med* 7(2):235–241.

Goodger, K., T. Gorely, D. Lavallee, and C. Harwood. 2007. Burnout in sport: A systematic review. *Sport Psychol* 21:127–151.

Goodyear, L.J., M.F. Hirshman, P.S. King, E.D. Horton, C.M. Thompson, and E.S. Horton. 1990. Skeletal muscle plasma membrane glucose transport and glucose tranporters after exercise. *J Appl Physiol* 68:193–198.

Gore, C.J., A.G. Hahn, D.B. Watson, K.I. Norton, D.P. Campbell, G.S. Scroop, D.L. Emonson, R.J. Wood, S.

Ly, S. Bellenger, and E. Lawton. 1996. V\od\O₂max and arterial O₂ saturation at sea level and 610 m. *Med Sci Sports Exerc* 27(S5).

Gore, D.C., P.C. Bourdon, S.M. Woolford, and D.G. Pederson. 1993. Involuntary dehydration during cricket. *Int J Sports Med* 14:387–395.

Gore, D.C., S.E. Wolf, A.P. Sanford, D.N. Herndon, and R.R. Wolfe. 2004. Extremity hyperinsulinemia stimulates muscle protein synthesis in severely injured patients. *Am J Physiol Endocrinol Metab* 286:E529–534.

Goubel, F., and G. Lensel. 2003. *Biomécanique: Éléments de mécanique musculaire* (2nd ed.). Paris: Masson.

Gould, D. 1996. Personal motivation gone awry: Burnout in competitive athletes. *Quest* 48:275–289.

Gould, D., and K. Dieffenbach. 2003. Psychological issues in youth sports: Competitive anxiety, overtraining, and burnout. In *Youth sports: Perspectives for a new century*, ed. R.M. Malina and M.A. Clark, 149–170. Champaign, IL: Human Kinetics.

Gould, D., S. Tuffey, E. Udry, and L. Loehr. 1997. Burnout in competitive junior tennis players: III. Individual differences in the burnout experience. *Sport Psychol* 11:257–276.

Gozal, D., L.D. Serpero, L. Kheirandish-Gozal, O.S. Capdevila, A. Khalyfa, and R. Tauman. 2010. Sleep measures and morning plasma TNF-alpha levels in children with sleep-disordered breathing. *Sleep* 33:319–325.

Graham, T.E. 2001. Caffeine and exercise: Metabolism, endurance and performance. *Sports Med* 31:785–807.

Grandjean, A.C., and N.R. Grandjean. 2007. Dehydration and cognitive performance. *J Am Coll Nutr* 26(5):554S–557S.

Grant, W.B., and M.F. Holick. 2005. Benefits and requirements of vitamin D for optimal health: A review. *Altern Med Rev* 10:94–111.

Gray, A.J., and D.G. Jenkins. 2010. Match analysis and the physiological demands of Australian football. *Sports Med* 40:347–360.

Green, D., C. Cheetham, C. Reed, L. Dembo, and G. O'Driscoll. 2002. Assessment of brachial artery blood flow across the cardiac cycle: Retrograde flows during cycle ergometry. *J Appl Physiol* 93:361–368.

Green, J.M., Z. Yang, C.M. Laurent, J.K. Davis, K. Kerr, R.C. Pritchett, and P.A. Bishop. 2007. Session RPE following interval and constant-resistance cycling in hot and cool environments. *Med Sci Sports Exerc* 39(11):2051–2057.

Greenleaf, J.E. 1989. Energy and thermal regulation during bed rest and spaceflight. *J Appl Physiol* 67(2):507–516.

———. 1992. Problem: Thirst, drinking behavior, and involuntary dehydration. *Med Sci Sports Exerc* 24:645–651.

Greenwood, J.D., G.E. Moses, F.M. Bernardino, G.A. Gaesser, and A. Weltman. 2008. Intensity of exercise recovery, blood lactate disappearance, and subsequent swimming performance. *J Sports Sci* 26:29–34.

Gregorowitcz, H., and Z. Zagrobelny. 1998. Whole-body cryotherapy: Indications and contraindications, the procedure and its clinical and physiological effects. *Acta Bio-Optica Informatica Med* 4:119–131.

Greiwe, J.S., K.S. Staffey, D.R. Melrose, M.D. Narve, R.G. Knowlton. 1998. Effects of dehydration on isometric muscular strength and endurance. *Med Sci Sports Exerc* 30:284–288.

Grinspoon, S., K. Miller, C. Coyle, J. Krempin, C. Armstrong, S. Pitts, D. Herzog, and A. Klibanski. 1999. Severity of osteopenia in estrogen-deficient women with anorexia nervosa and hypothalamic amenorrhea. *J Clin Endocrinol Metab* 84:2049–2055.

Grucza, R. 1990. Efficiency of thermoregulatory system in man under endogenous and exogenous heat loads. *Acta Physiol Pol* 41:123–145.

Grucza, R., M. Szczypaczewska, and S. Kozlowski. 1987. Thermoregulation in hyperhydrated men during physical exercise. *Eur J Appl Physiol Occup Physiol* 56(5):603–607.

Guézennec, C.Y. 1995. Oxidation rates, complex carbohydrates and exercise. Practical recommendations. *Sports Med* 19:365–372.

Guézennec, C.Y., B. Serrurier, D. Merino, and J.M. Clere. 1986. Effect of hypoxia on heart glycogen utilization during exercise. *Aviat Space Environ Med* 57(8):754–758.

Guissard, N., and J. Duchateau. 2006. Neural aspects of muscle stretching. *Exerc Sport Sci Rev* 34:154–158.

Gulick, D.T., and I.F. Kimura. 1996. Delayed onset muscle soreness: What is it and how do we treat it? *J Sport Rehab* 5:234–243.

Gulick, D.T., I.F. Kimura, M. Sitler, A. Paolone, and J.D. Kelly. 1996. Various treatment techniques on signs and symptoms of delayed onset muscle soreness. *J Athl Train* 31:145–152.

Gurjão, A.L., R. Gonçalves, R.F. de Moura, and S. Gobbi. 2009. Acute effect of static stretching on rate of force development and maximal voluntary contraction in older women. *J Strength Cond Res* 23:2149–2154.

Gustafsson, H., G. Kenttä, P. Hassmén, C. Lundqvist, and N. Durand-Bush. 2007. The process of burnout: A multiple case study of three elite endurance athletes. *Int J Sport Psychol* 38:388–416.

Gutiérrez, A., J.L. Mesa, J.R. Ruiz, L.J. Chirosa, and M.J. Castillo. 2003. Sauna-induced rapid weight loss decreases explosive power in women but not in men. *Int J Sports Med* 24(7):518–522.

Haff, G.G. 2004. Roundtable discussion: Periodization of training—part 1. *Strength Cond J* 26(2):50–69.

Hakkinen, K. 1993. Neuromuscular fatigue and recovery in male and female athletes during heavy resistance exercise. *Int J Sports Med* 14:53–59.

Hall, H.K., I.W. Cawthra, and A.W. Kerr. 1997. Burnout: Motivation gone awry or a disaster waiting to happen? In *Innovations in sport psychology: Linking theory and practice. Proceedings of the 9th ISSP World Congress in Sport Psychology, Vol. 1*, ed. R. Lidor and M. Bar-Eli, 306–308. Netanya, Israel: Ministry of Education, Culture and Sport.

Halliburton, W.D. 1915. The death temperature of nerve. *Q J Exp Physiol* 9:193–198.

Halson, S. 2008. Nutrition, sleep and recovery. *Eur J Sport Sci* 8:199–126.

Halson, S.L., M.W. Bridge, R. Meeusen, B. Busschaert, M. Gleeson, D.A. Jones, and A.E. Jeukendrup. 2002. Time course of performance changes and fatigue markers during intensified training in trained cyclists. *J Appl Physiol* 93(3):947–956.

Halson, S.L., and A.E. Jeukendrup. 2004. Does overtraining exist? An analysis of overreaching and overtraining research. *Sports Med* 34(14):967–981.

Halson, S., D.T. Martin, A.S. Gardner, K. Fallon, and J. Gulbin. 2006. Persistent fatigue in a female sprint cyclist after a talent-transfer initiative. *Int J Sports Physiol Perform* 1:65–69.

Halson, S.L., M.J. Quod, D.T. Martin, A.S. Gardner, T.R. Ebert, and P.B. Laursen. 2008. Physiological responses to cold water immersion following cycling in the heat. *Int J Sports Physiol Perform* 3(3):331–346.

Harfouche, J.N., S. Theys, P. Hanson, J.C. Schoevaerdts, and X. Sturbois. 2008. Venous tonus enhancement after a short cycle of intermittent pneumatic compression. *Phlebology* 23:58–63.

Harridge, S.D., R. Bottinelli, M. Canepari, M. Pellegrino, C. Reggiani, M. Esbjornsson, P.D. Balsom, and B. Saltin. 1998. Sprint training, in vitro and in vivo muscle function, and myosin heavy chain expression. *J Appl Physiol* 84:442–449.

Harris, R.C., R.H. Edwards, E. Hultman, L.O. Nordesjo, B. Nylind, and K. Sahlin. 1976. The time course of phosphorylcreatine resynthesis during recovery of the quadriceps muscle in man. *Pflugers Arch* 367:137–142.

Harrison, Y., and J.A. Horne. 2000. The impact of sleep deprivation on decision making: A review. *J Exp Psychol Appl* 6:236–249.

Hartmann, E. 1982. Effects of L-tryptophan on sleepiness and on sleep. *J Psychiatr Res* 17:107–113.

Hartmann, E., and C.L. Spinweber. 1979. Sleep induced by L-tryptophan. Effect of dosages within the normal dietary intake. *J Nerv Ment Dis* 167:497–499.

Hasan, J., M.J. Karvonen, and P. Piironen. 1966. Special review. I. Physiological effects of extreme heat as studied in the Finnish "sauna" bath. *Am J Phys Med* 45(6):296–314.

Hasegawa, H., T. Takatori, T. Komura, and M. Yamasaki. 2005. Wearing a cooling jacket during exercise reduces thermal strain and improves endurance exercise performance in a warm environment. *J Strength Cond Res* 19:122–128.

Haseler, L.J., M.C. Hogan, and R.S. Richardson. 1999. Skeletal muscle phosphocreatine recovery in exercise-trained humans is dependent on O_2 availability. *J Appl Physiol* 86:2013–2018.

Hassmen, P., N. Koivula, and T. Hansson. 1998. Precompetitive mood states and performance of elite male golfers: Do trait characteristics make a difference? *Percept Mot Skills* 86:1443–1457.

Hausswirth, C., A.X. Bigard, R. Lepers, M. Berthelot, and C.Y. Guézennec. 1995. Sodium citrate ingestion and muscle performance in acute hypobaric hypoxia. *Eur J Appl Physiol* 71:362–368.

Havenith, G. 2001. Human surface to mass ratio and body core temperature in exercise heat stress: A concept revisited. *J Therm Biol* 26:387–393.

Hawkins, L. 1992. Seasonal affective disorders: The effects of light on human behaviour. *Endeavour* 16:122–127.

Hawley, J.A. 2009. Molecular responses to strength and endurance training: Are they incompatible? *Appl Physiol Nutr Metab* 34(3):355–361.

Hayashi, M., S. Ito, and T. Hori. 1999. The effects of a 20-min nap at noon on sleepiness, performance and EEG activity. *Int J Psychophysiol* 32:173–180.

Haze, S., K. Sakai, and Y. Gozu. 2002. Effects of fragrance inhalation on sympathetic activity in normal adults. *Jpn J Pharmacol* 90:247–253.

Hedelin, R., U. Wiklund, P. Bjerle, and K. Henriksson-Larsen. 2000. Cardiac autonomic imbalance in an overtrained athlete. *Med Sci Sport Exer* 32(9):1531–1533.

Hedrick, A. 1999. Soccer-specific conditioning. *Strength Cond J* 21:17–21.

Hellard, P., M. Avalos, G. Millet, L. Lacoste, F. Barale, and J.C. Chatard. 2005. Modelling the residual effects and threshold saturation of training: A case study of Olympic swimmers. *J Strength Cond Res* 19(1):67–75.

Hellenbrandt, F., and S. Houtz. 1956. Mechanism of muscle training in man: Experimental demonstration of the overload principle. *Phys Ther Rev* 36:371–383.

Hemmings, B., M. Smith, J. Graydon, and R. Dyson. 2000. Effects of massage on physiological restoration, perceived recovery, and repeated sports performance. *Br J Sports Med* 34:109–114, 115.

Henderson, G.C., J.A. Fattor, M.A. Horning, N. Faghihnia, M.L. Johnson, M. Luke-Zeitoun, and G.A. Brooks. 2008. Glucoregulation is more precise in women than in men during postexercise recovery. *Am J Clin Nutr* 87:1686–1694.

Henderson, G.C., J.A. Fattor, M.A. Horning, N. Faghihnia, M.L. Johnson, T.L. Mau, M. Luke-Zeitoun, and G.A. Brooks. 2007. Lipolysis and fatty acid metabolism in men and women during the postexercise recovery period. *J Physiol* 584:963–981.

Herbert, R.D., and M. de Noronha. 2007. Stretching to prevent or reduce muscle soreness after exercise. *Cochrane Database Syst Rev* 17:1–24.

Hermansen, L., and I. Stensvold. 1972. Production and removal of lactate during exercise in man. *Acta Physiol Scand* 86:191–201.

Hessemer, V., and K. Brück. 1985a. Influence of menstrual cycle on shivering, skin blood flow, and sweating responses measured at night. *J Appl Physiol* 59:1902–1910.

———. 1985b. Influence of menstrual cycle on thermoregulatory, metabolic, and heart rate responses to exercise at night. *J Appl Physiol* 59:1911–1917.

Hessemer, V., D. Langusch, K. Brück, R.H. Bödeker, and T. Breidenbach. 1984. Effect of slightly lowered body temperatures on endurance performance in humans. *J Appl Physiol* 57:1731–1737.

Heuberger, E., T. Hongratanaworakit, C. Bohm, R. Weber, G. Buchbauer. 2001. Effects of chiral fragrances on human autonomic nervous system parameters and self-evaluation. *Chem Senses* 26:281–292.

Heyman, E., B. de Geus, I. Mertens, and R. Meeusen. 2009. Effects of four recovery methods on repeated maximal rock climbing performance. *Med Sci Sports Exerc* 41(6):1303–1310.

Higgins, T.R., I.T. Heazlewood, and M. Climstein. 2011. A random control trial of contrast baths and ice baths for recovery during competition in U/20 Rugby Union. *J Strength Cond Res* 25(4):1046–1051.

Higgins, T., G.A. Naughton, and D. Burgess. 2009. Effects of wearing compression garments on physiological and performance measures in a simulated game-specific circuit for netball. *J Sci Med Sport* 12:223–226.

High, D.M., E.T. Howley, and B.D. Franks. 1989. The effects of static stretching and warm-up on prevention of delayed-onset muscle soreness. *Res Q Exerc Sport* 6:357–361.

Hilbert, J.E., G.A. Sforzo, and T. Swensen. 2003. The effects of massage on delayed onset muscle soreness. *Br J Sports Med* 37:72–75.

Hindmarch, I., U. Rigney, N. Stanley, P. Quinlan, J. Rycroft, and J. Lane. 2000. A naturalistic investigation of the effects of day-long consumption of tea, coffee and water on alertness, sleep onset and sleep quality. *Psychopharmacology* 149:203–216.

Hinds, T., I. McEwan, J. Perkes, E. Dawson, D. Ball, and K. George. 2004. Effects of massage on limb and skin blood flow after quadriceps exercise. *Med Sci Sports Exerc* 36:1308–1313.

Hing, W.A., S.G. White, A. Bouaaphone, and P. Lee. 2008. Contrast therapy: A systematic review. *Phys Ther Sport* 9(3):148–161.

Hinton, P.S., T.C. Sanford, M.M. Davidson, O.F. Yakushko, and N.C. Beck. 2004. Nutrient intakes and dietary behaviors of male and female collegiate athletes. *Int J Sport Nutr Exerc Metab* 14:389–405.

Hirvonen, J., A. Nummela, H. Rusko, S. Rehunen, and M. Harkonen. 1992. Fatigue and changes of ATP, creatine phosphate, and lactate during the 400-m sprint. *Can J Sport Sci* 17:141–144.

Hirvonen, J., S. Rehunen, H. Rusko, and M. Harkonen. 1987. Breakdown of high-energy phosphate compounds and lactate accumulation during short supramaximal exercise. *Eur J Appl Physiol Occup Physiol* 56:253–259.

Hitchins, S., D.T. Martin, L. Burke, K. Yates, K. Fallon, A. Hahn, and G.P. Dobson. 1999. Glycerol hyperhydration improves cycle time trial performance in hot humid conditions. *J Appl Physiol* 80:494–501.

Hocutt, J.E. Jr., R. Jaffe, C.R. Rylander, and J.K. Beebe. 1982. Cryotherapy in ankle sprains. *Am J Sports Med* 10:316–319.

Hoffman, J., and M. Falvo. 2004. Protein: Which is best? *J Sport Sci Med* 3:118–130.

Hoibian, S. 2007. Conditions de vie et aspirations des Français. *Collection des Rapports* 279. Paris: CREDOC.

Holcomb, W.R. 1997. A practical guide to electrical therapy. *J Sport Rehab* 6:272–282.

Holloszy, J., and G. Hansen. 1996. Regulation of glucose transport into skeletal muscle. *Rev Physiol Bioch Pharm* 128:99–103.

Holt, N.L. 2007. Introduction: Positive youth development through sport. In *Positive youth development through sport*, ed. N.L. Holt, 1–5. London: Routledge.

Honda, K., and S. Inoue. 1998. Sleep-enhancing effects of far infrared radiations in rats. *Int J Biometeorol* 32:92–94.

Hooper, S., and L. Mackinnon. 1995. Monitoring overtraining in athletes. Recommendations. *Sports Med* 20(5):321–327.

Hooper, S., L. Mackinnon, and S. Hanrahan. 1997. Mood states as an indication of staleness and recovery. *Int J Sport Psychol* 28:1–12.

Hooper, S.L., L.T. Mackinnon, A. Howard, R.D. Gordon, and A.W. Bachmann. 1995. Markers for monitoring overtraining and recovery. *Med Sci Sport Exer* 27(1):106–112.

Hopkins, W. 1991. Quantification of training in competitive sports. Methods and applications. *Sports Med* 12:161–183.

Hoppeler, H., M. Vogt, W.R. Weibel, and M. Flück. 2008. Special review series: Biogenesis and physiological adaptation of mitochondria response of skeletal muscle mitochondria to hypoxia. *Exp Physiol* 88:109–119.

Horne, J.A., and A.N. Pettitt. 1984. Sleep deprivation and the physiological response to exercise under steady-state conditions in untrained subjects. *Sleep* 7:168–179.

Horne, J.A., and L.A. Reyner. 1996. Counteracting driver sleepiness: Effects of napping, caffeine, and placebo. *Psychophysiology* 33:306–309.

Hornery, D.J., S. Papalia, I. Mujika, and A. Hahn. 2005. Physiological and performance benefits of halftime cooling. *J Sci Med Sport* 8:15–25.

Horton, T.J., M.J. Pagliassotti, K. Hobbs, and J.O. Hill. 1998. Fuel metabolism in men and women during and after long-duration exercise. *J Appl Physiol* 85:1823–1832.

Hoshikawa, M., S. Uchida, T. Sugo, Y. Kumai, Y. Hanai, and T. Kawahara. 2007. Changes in sleep quality of athletes under normobaric hypoxia equivalent to 2,000-m altitude: A polysomnographic study. *J Appl Physiol* 103:2005–2011.

Houghton, L.A., B. Dawson, and S.K. Maloney. 2009. Effects of wearing compression garments on thermoregulation during simulated team sport activity in temperate environmental conditions. *J Sci Med Sport* 12:303–309.

Howarth, K.R., N.A. Moreau, S.M. Philips, and M.J. Gibala. 2009. Coingestion of protein with carbohydrate during recovery from endurance exercise stimulates skeletal muscle protein synthesis in humans. *J Appl Physiol* 106:1394–1402.

Howat, P.M., M.L. Carbo, G.Q. Mills, and P. Wozniak. 1989. The influence of diet, body fat, menstrual cycling, and activity upon the bone density of females. *J Am Diet Assoc* 89:1305–1307.

Howatson, G., D. Gaze, and K.A. van Someren. 2005. The efficacy of ice massage in the treatment of exercise-induced muscle damage. *Scand J Med Sci Sports* 15:416–422.

Howatson, G., S. Goodall, and K.A. van Someren. 2009. The influence of cold water immersions on adaptation following a single bout of damaging exercise. *Eur J Appl Physiol* 105(4):615–621.

Howatson, G., and K.A. van Someren. 2003. Ice massage. Effects on exercise-induced muscle damage. *J Sports Med Phys Fitness* 43(4):500–505.

———. 2008. The prevention and treatment of exercise-induced muscle damage. *Sports Med* 38:483–503.

Hubbard, R.W., B.L. Sandick, W.T. Matthew, R.P. Francesconi, J.B. Sampson, M.J. Durkot, O. Maller, and D.B. Engell. 1984. Voluntary dehydration and alliesthesia for water. *J Appl Physiol* 57:866–873.

Hughes, G.S. Jr., P.R. Lichstein, D. Whitlock, and C. Harker. 1984. Response of plasma beta-endorphins to transcutaneous electrical nerve stimulation in healthy subjects. *Phys Ther* 64:1062–1066.

Hultman, E., and H. Sjoholm. 1983. Energy metabolism and contraction force of human skeletal muscle in situ during electrical stimulation. *J Physiol* 345:525–532.

Hunt, J., and J. Pathak. 1960. The osmotic effects of some simple molecules and ions on gastric emptying. *J Physiol (London)* 154:254–257.

Hunter, A., Y. Albertus-Kajee, and A. St Clair Gibson. 2011. The effect of exercise induced hyperthermia on muscle fibre conduction velocity during sustained isometric contraction. *J Electromyogr Kinesiol* 21:834–840.

Hunter, J.R., B.J. O'Brien, M.G. Mooney, J. Berry, W.B. Young, and N. Down. 2011. Repeated sprint training improves intermittent peak running speed in team-sport athletes. *J Strength Cond Res* 25(5):1318–1325.

Hynynen, E., A. Uusitalo, N. Konttinen, and H. Rusko. 2006. Heart rate variability during night sleep and after awakening in overtrained athletes. *Med Sci Sports Exerc* 38:313–317.

Iellamo, F., J.M. Legramante, F. Pigozzi, A. Spataro, G. Norbiato, D. Lucini, and M. Pagani. 2002. Conversion from vagal to sympathetic predominance with strenuous training in high-performance world class athletes. *Circulation* 105:2719–2724.

Ikegami, K., S. Ogyu, Y. Arakomo, K. Suzuki, K. Mafune, H. Hiro, and S. Nagata. 2009. Recovery of cognitive performance and fatigue after one night of sleep deprivation. *J Occup Health* 51:412–422.

Ilmberger, J., E. Heuberger, C. Mahrhofer, H. Dessovic, D. Kowarik, and G. Buchbauer. 2001. The influence of essential oils on human attention. I: Alertness. *Chem Senses* 26:239–245.

Ingram, J., B. Dawson, C. Goodman, K. Wallman, and J. Beilby. 2009. Effect of water immersion methods on post-exercise recovery from simulated team sport exercise. *J Sci Med Sport* 12(3):417–421.

Isabell, W.K., E. Durrant, W. Myrer, and S. Anderson. 1992. The effects of ice massage, ice massage with exercise, and exercise on the prevention and treatment of delayed onset muscle soreness. *J Athl Train* 27:208–217.

Isbell, L.A., and T.P. Young. 1996. The evolution of bipedalism in hominids and reduced group size in chimpanzees: Alternative responses to decreasing resource availability. *J Hum Evol* 30:389–397.

Issurin, V.B. 2010. New horizons for the methodology and physiology of training periodization. *Sports Med* 40(3):189–206.

Ivins, D. 2006. Acute ankle sprain: An update. *Am Fam Physician* 74:1714–1720.

Ivy, J.L., H.W. Goforth Jr., B.M. Damon, T.R. McCauley, E.C. Parsons, and T.B. Price. 2002. Early postexercise muscle glycogen recovery is enhanced with a carbohydrate–protein supplement. *J Appl Physiol* 93:1337–1344.

Ivy, J.L., A.L. Katz, C.L. Cutler, W.M. Sherman, and E.F. Coyle. 1988. Muscle glycogen synthesis after exercise: Effect of time of carbohydrate ingestion. *J Appl Physiol* 64:1480–1485.

Ivy, J.L., and C.H. Kuo. 1998. Regulation of GLUT-4 protein and glycogen synthase during muscle glycogen synthesis after exercise. *Acta Physiol Scand* 162:295–304.

Ivy, J.L., M.C. Lee, J.T. Brozinick, and M. Reed. 1988. Muscle glycogen storage after different amounts of carbohydrate ingestion. *J App Physiol* 65:2018–2023.

Jacobs, I., A. Anderberg, R. Schele, and H. Lithell. 1983. Muscle glycogen in soldiers on different diets during military field manoeuvres. *Aviat Space Environ Med* 54:898–900.

Jacobs, I., M. Esbjornsson, C. Sylven, I. Holm, and E. Jansson. 1987. Sprint training effects on muscle myoglobin, enzymes, fiber types, and blood lactate. *Med Sci Sports Exerc* 19:368–374.

Jakeman, J.R., C. Byrne, and R.G. Eston. 2010. Lower limb compression garment improves recovery from exercise-induced muscle damage in young, active females. *Eur J Appl Physiol* 109:1137–1144.

Jakeman, J.R., R. Macrae, and R. Eston. 2009. A single 10-min bout of cold-water immersion therapy after strenuous plyometric exercise has no beneficial effect on recovery from the symptoms of exercise-induced muscle damage. *Ergonomics* 52(4):456–460.

Janeira, M.A., and J. Maia. 1998. Game intensity in basketball: An interactionist view linking time-motion analysis, lactate concentration and heart rate. *Coaching Sport Sci J* 3:26–30.

Janský, L., H. Janáková, B. Ulicný, P. Srámek, V. Hosek, J. Heller, and J. Parízková. 1996. Changes in thermal homeostasis in humans due to repeated cold water immersions. *Pflugers Arch* 432(3):368–372.

Janssen, G.M., H.R. Scholte, M.H. Vaandrager-Verduin, and J.D. Ross. 1989. Muscle carnitine level in endurance training and running a marathon. *Int J Sports Med* 10(3):S153–S155.

Janwantanakul, P. 2009. The effect of quantity of ice and size of contact area on ice pack/skin interface temperature. *Physiotherapy* 95:120–125.

Jemni, M., W.A. Sands, F. Friemel, and P. Delamarche. 2003. Effect of active and passive recovery on blood lactate and performance during simulated competition in high level gymnasts. *Can J Appl Physiol* 28:240–256.

Jenkins, D.J.A., D. Cuff, T.M.S. Wolever, D. Knowland, L. Thompson, Z. Cohen, and E.J. Propikchuk. 1987. Digestibility of carbohydrate foods in an ileostomate: Relationship to dietary fibre, in vitro digestibility, and glycemic responses. *Am J Gastro* 82:709–717.

Jenkins, R.R. 1988. Free radical chemistry, relationship to exercise. *Sports Med* 5:156–170.

Jentjens, R.L., and A.E. Jeukendrup. 2003. Determinants of post-exercise glycogen synthesis during short-term recovery. *Sports Med* 33:117–144.

Jentjens, R.L., L.J.C. van Loon, C.H. Mann, A.J.M. Wagenmakers, and A.E. Jeukendrup. 2001. Addition of protein and amino acids to carbohydrates does not enhance postexercise muscle glycogen synthesis. *J Appl Physiol* 91:839–846.

Jessen, C. 1998. Brain cooling: An economy mode of temperature regulation in artiodactyls. *News Physiol Sci* 13:281–286.

———. 2001. Selective brain cooling in mammals and birds. *Jpn J Physiol* 51:291–301.

Jessen, C., H.P. Laburn, M.H. Knight, G. Kuhnen, K. Goelst, and D. Mitchell. 1994. Blood and brain temperatures of free-ranging black wildebeest in their natural environment. *Am J Physiol Regul Integr Comp Physiol* 267:R1528–R1536.

Jeukendrup, A.E., M.K. Hesselink, A.C. Snyder, H. Kuipers, and H.A. Keizer. 1992. Physiological changes in male competitive cyclists after two weeks of intensified training. *Int J Sports Med* 13:534–541.

Jezová, D., B.B. Johansson, Z. Oprsalová, and M. Vigas. 1989. Changes in blood-brain barrier function modify the neuroendocrine response to circulating substances. *Neuroendocrinology* 49(4):428–433.

Jezová, D., M. Vigas, P. Tatar, J. Jurcovicova, and M. Palat. 1985. Rise in plasma β-endorphin and ACTH in response to hyperthermia in sauna. *Horm Metab Res* 17(12):693–694.

Johansson, P.H., L. Lindström, G. Sundelin, and B. Lindström. 1999. The effects of preexercise stretching on muscular soreness, tenderness and force loss following heavy eccentric exercise. *Scand J Med Sci Spor* 9:219–225.

Johnson, D.C., C.T. Burt, W.C. Perng, and B.M. Hitzig. 1993. Effects of temperature on muscle pH and phosphate metabolites in newts and lungless salamanders. *Am J Physiol* 265:R1162–R1167.

Johnson, M.I., and G. Tabasam. 2003. An investigation into the analgesic effects of interferential currents and transcutaneous electrical nerve stimulation on experimentally induced ischemic pain in otherwise pain-free volunteers. *Phys Ther* 83:208–223.

Jones, N.L., N. McCartney, T. Graham, L.L. Spriet, J.M. Kowalchuk, G.J. Heigenhauser, and J.R. Sutton. 1985. Muscle performance and metabolism in maximal isokinetic cycling at slow and fast speeds. *J Appl Physiol* 59:132–136.

Jones, N.L., J.R. Sutton, R. Taylor, and C.J. Toews. 1977. Effect of pH on cardiorespiratory and metabolic responses to exercise. *J Appl Physiol* 43:959–964.

Josse, A.R., J.E. Tang, M.A. Tarnopolsky, and S.M. Phillips. 2010. Body composition and strength changes in women with milk and resistance exercise. *Med Sci Sports Exerc* 42:1122–1130.

Joszi, A.C., T.A. Trappe, R.D. Starling, B. Goodpaster, S.W. Trappe, W.J. Fink, and D.L. Costill. 1996. The influence of starch structure on glycogen resynthesis and subsequent cycling performance. *Int J Sports Med* 17:373–378.

Jouvet, M. 1999. Sleep and serotonin: An unfinished story. *Neuropsychopharmacol* 21:S24–S27.

Joyner, M.J., and E.F. Coyle. 2008. Endurance exercise performance: The physiology of champions. *J Physiol* 586:35–44.

Juel, C. 1998. Muscle pH regulation: Role of training. *Acta Physiol Scand* 162:359–366.

Jung, H. 1986. A generalised concept for cell killing by heat. *Radiat Res* 106:56–72.

Jurimae, J., J. Maestu, P. Purge, and T. Jurimae. 2004. Changes in stress and recovery after heavy training in rowers. *J Sci Med Sport* 7:335–339.

Kaciuba-Uscilko, H., and R. Grucza. 2001. Gender differences in thermoregulation. *Curr Opin Clin Nutr Metab Care* 4:533–536.

Kahn, S.R., L. Azoulay, A. Hirsch, M. Haber, C. Strulovitch, and I. Shrier. 2003. Effect of graduated elastic compression stockings on leg symptoms and signs during exercise in patients with deep venous thrombosis: A randomized cross-over trial. *J Thromb Haemost* 1(3):494–499.

Kameda, T., H. Mano, T. Yuasa, Y. Mori, K. Miyazawa, M. Shiokawa, Y. Nakamaru, E. Hiroi, K. Hiura, A. Kameda, N.N. Yang, Y. Hakeda, and M. Kumegawa. 1997. Estrogen inhibits bone resorption by directly inducing apoptosis of the bone-resorbing osteoclasts. *J Exp Med* 186:489–495.

Karacan, I., J.I. Thornby, M. Anch, G.H. Booth, R.L. Williams, and P.J. Salis. 1976. Dose-related sleep disturbances induced by coffee and caffeine. *Clin Pharmacol Ther* 20:682–689.

Karlsson, H.K., P.A. Nilsson, and J. Nilsson. 2004. Branched chain amino acids increase p70s6 kinase phosphorylation in human skeletal muscle after resistance exercise. *Am J Physiol Endocrinol Metab* 287:E1–E7.

Kasperek, G.J., and R.D. Snider. 1987. Effect of exercise intensity and starvation on the activation of branched chain keto acid dehydrogenase by exercise. *Am J Physiol* 252:E33–E37.

Katayama, K., G. Kasuchije, K. Ishida, and F. Ogita. 2010. Substrate utilization during exercise and recovery at moderate altitude. *Metabolism* 59:959–966.

Kauppinen, K. 1989. Sauna, shower, and ice water immersion. Physiological responses to brief exposures to heat, cool, and cold. Part I: Body fluid balance. *Arct Med Res* 48:55–56.

Kauppinen, K., and I. Vuori. 1986. Man in the sauna. *Ann Clin Res* 18(4):173–185.

Kay, D., F.E. Marino, J. Cannon, A. St Clair Gibson, M.I. Lambert, and T.D. Noakes. 2001. Evidence for neuromuscular fatigue during high-intensity cycling in warm, humid conditions. *Eur J Appl Physiol* 84:115–121.

Kechijian, D. 2011. Optimizing nutrition for performance at altitude: A literature review. *J Special Op Med* 11:12–17.

Keir, K.A., and G.C. Goats. 1991. Introduction to manipulation. *Br J Sports Med* 25:221–226.

Kellmann, M. 2002. *Enhancing recovery: Preventing underperformance in athletes.* Champaign, IL: Human Kinetics.

———. 2009. Is recovery important? Oral presentation at the 12th ISSP World Congress, Marrakesh, Morocco. www.issponline.org/documents/001.rar.

———. 2010. Preventing overtraining in athletes in high-intensity sports and stress/recovery monitoring. *Scan J Med Sci Sports* 20(2):S95–S102.

Kellmann, M., and K.W. Kallus. 2001. *Recovery stress questionnaire for athletes: User manual.* Champaign, IL: Human Kinetics.

Kelly, V.G., and A.J. Coutts. 2007. Planning and monitoring training loads during the competition phase in team sports. *Strength Cond J* 29(4):32.

Kemmler, W., S. von Stengel, C. Kockritz, J. Mayhew, A. Wassermann, and J. Zapf. 2009. Effect of compression stockings on running performance in men runners. *J Strength Cond Res* 23:101–105.

Kennedy, D.O., S. Pace, C. Haskell, E.J. Okello, A. Milne, and A.B. Scholey. 2006. Effects of cholinesterase inhibiting sage (Salvia officinalis) on mood, anxiety and performance on a psychological stressor battery. *Neuropsychopharmacology* 31:845–852.

Kenny, G.P., and O. Jay. 2007. Sex differences in postexercise esophageal and muscle tissue temperature response. *Am J Physiol Regul Integr Comp Physiol* 292:R1632–R1640.

Kenny, G.P., J.E. Murrin, W.S. Journeay, and F.D. Reardon. 2006. Differences in the postexercise threshold for cutaneous active vasodilation between men and women. *Am J Physiol Regul Integr Comp Physiol* 290:R172–R179.

Kentta, G., and P. Hassmen. 1998. Overtraining and recovery. A conceptual model. *Sports Med* 26(1):1–16.

Kerksick, C., T. Harvey, J.L. Ivy, and J. Antonio. 2008. International society of sports nutrition position: Nutrient timing. *J Int Soc Sports Nut* 12:5–17.

Kiens, B., and E.A. Richter. 1996. Types of carbohydrate in an ordinary diet affect insulin action and muscle substrates in humans. *Am J Clin Nut* 63:47–53.

Kim, D.J., H.P. Lee, M.S. Kim, Y.J. Park, H.J. Go, K.S. Kim, S.P. Lee, J.H. Chae, and C.T. Lee. 2001. The effect of total sleep deprivation on cognitive functions in normal adult male subjects. *Int J Neurosci* 109:127–137.

Kimura, I.F., G.T. Thompson, and D.T. Gulick. 1997. The effect of cryotherapy on eccentric plantar flexion peak torque and endurance. *J Athl Train* 32:124–126.

King, M., and R. Duffield. 2009. The effects of recovery interventions on consecutive days of intermittent sprint exercise. *J Strength Cond Res* 23(6):1795–1802.

Kinugasa, T., and A.E. Kilding. 2009. A comparison of post-match recovery strategies in youth soccer players. *J Strength Cond Res* 23(5):1402–1407.

Knaflitz, M., R. Merletti, and C.J. De Luca. 1990. Inference of motor unit recruitment order in voluntary and electrically elicited contractions. *J Appl Physiol* 68:1657–1667.

Knicker, A.J., I. Renshaw, A.R. Oldham, and S.P. Cairns. 2011. Interactive processes link the multiple symptoms of fatigue in sport competition. *Sports Med* 41:307–328.

Kohsaka, M., N. Fukuda, H. Honma, R. Kobayashi, S. Sakakibara, E. Koyama, T. Nakano, and H. Matsubara. 1999. Effects of moderately bright light on subjective evaluations in healthy elderly women. *Psychiatry Clin Neurosci* 53:239–241.

Kolka, M.A., and L.A. Stephenson. 1988. Exercise thermoregulation after prolonged wakefulness. *J Appl Physiol* 64:1575–1579.

Koopman, R., A.J.M. Wagenmakers, R.J.F. Manders, H.G. Zorenc, J.M.G. Senden, M. Gorselink, H.A. Keizer, and L.J.C. van Loon. 2005. Combined ingestion of protein and free leucine with carbohydrate increases postexercise muscle protein synthesis in vivo in male subjects. *Am J Physiol Endocrinol Metab* 288:E645–E653.

Kopp-Woodroffe, S.A., M.M. Manore, C.A. Dueck, J.S. Skinner, and K.S. Matt. 1999. Energy and nutrient status of amenorrheic athletes participating in a diet and exercise training intervention program. *Int J Sport Nutr* 9:70–88.

Koulmann, N., C. Jimenez, and D. Regal. 2000. Use of BIA to estimate body fluid compartments after acute variations of the body hydration level. *Med Sci Sports Exerc* 32:857–864.

Kovacs, E.M., R.M. Schahl, J.N. Sender, and F. Browns. 2002. Effect of high and low rates of fluids intake on postexercise rehydration. *Int J Sports Nut Exer Metab* 12:14–23.

Kraemer, W.J., J.A. Bush, J.A. Bauer, N.T. Triplett-McBride, N.J. Paxton, A. Clemson, L.P. Koziris, L.C. Mangino, A.C. Fry, and R.U. Newton. 1996. Influence of compression garments on vertical jump performance in NCAA division I volleyball players. *J Stregth Cond Res* 10:180–183.

Kraemer, W.J., J.A. Bush, R.B. Wickham, C.R. Denegar, A.L. Gomez, L.A. Gotshalk, N.D. Duncan, J.S. Volek, M. Putukian, and W.J. Sebastianelli. 2001. Influence of compression therapy on symptoms following soft tissue injury from maximal eccentric exercise. *J Orthop Sports Phys Ther* 31(6):282–290.

Kraemer, W.J., S.J. Fleck, and W.J. Evans. 1996. Strength and power training: Physiological mechanisms of adaptation. *Exerc Sport Sci Rev* 24:363–397.

Kreider, R. 1998. Central fatigue hypothesis and overtraining. In *Overtraining in sport*, ed. R. Kreider, A.C. Fry, M. O'Toole, 309–334. Champaign, IL: Human Kinetics.

Krueger, J.M., and J.A. Madje. 1990. Sleep as a host defense: Its regulation by microbial products and cytokines. *Clin Immunol Immunopathol* 57:188–199.

Krustrup, P., M. Mohr, A. Steensberg, J. Bencke, M. Kjaer, and J. Bangsbo. 2006. Muscle and blood metabolites during a soccer game: Implications for sprint performance. *Med Sci Sports Exerc* 38:1165–1174.

Kubesch, S., V. Bretschneider, R. Freudenmann, N. Weidenhammer, M. Lehmann, M. Spitzer, and G. Gron. 2003. Aerobic endurance exercise improves executive functions in depressed patients. *J Clin Psychiat* 64(9):1005–1012.

Kuipers, H. 1996. How much is too much? Performance aspects of overtraining. *Res Q Exer Sport* 67(3):S65–S69.

Kuipers, H., W.H. Saris, F. Brouns, H.A. Keizer, and C. ten Bosch. 1989. Glycogen synthesis during exercise and rest with carbohydrate feeding in males and females. *Int J Sports Med* 10(1):S63–S67.

Kukkonen-Harjula, K., and K. Kauppinen. 1988. How the sauna affects the endocrine system. *Ann Clin Res* 20(4):262–266.

———. 2006. Health effects and risks of sauna bathing. *Int J Circumpolar Health* 65(3):195–205.

Kukkonen-Harjula, K., P. Oja, K. Laustiola, I. Vuori, J. Jolkkonen, S. Siitonen, and H. Vapaatalo. 1989. Haemodynamic and hormonal responses to heat exposure in a Finnish sauna bath. *Eur J Appl Physiol Occup Physiol* 58(5):543–550.

Kuo, C.C., J.A. Fattor, G.C. Henderson, and G.A. Brooks. 2005. Lipid oxidation in fit young adults during postexercise recovery. *J Appl Physiol* 99:349–356.

Kuriyama, H., S. Watanabe, T. Nakaya, I. Shigemori, M. Kita, N. Yoshida, D. Masaki, T. Tadai, K. Ozasa, K. Fukui, and J. Imanishi. 2005. Immunological and psychological benefits of aromatherapy massage. *Evid Based Complement Alternat Med* 2:179–184.

Kusaka, Y., H. Kondou, and K. Morimoto. 1992. Healthy lifestyles are associated with higher natural killer cell activity. *Prev Med* 21:602–615.

Lachuer, J., I. Delton, M. Buda, and M. Tappaz. 1994. The habituation of brainstem catecholaminergic groups to chronic daily restraint stress is stress specific like that of the hypothalamo-pituitary-adrenal axis. *Brain Res* 638:196–202.

Lahmeyer, H.W. 1991. Seasonal affective disorders. *Psychiatr Med* 9:105–114.

Lahti, T.A., S. Leppamaki, J. Lonnqvist, and T. Partonen. 2008. Transitions into and out of daylight saving time compromise sleep and the rest-activity cycles. *BMC Physiol* 8:3.

Laitinen, L.A., and A. Laitinen. 1988. Mucosal inflammation and bronchial hyperreactivity. *Eur Respir J* 1(5):488–489.

Lakomy, J., and D.T. Haydon. 2004. The effects of enforced, rapid deceleration on performance in a multiple sprint test. *J Strength Cond Res* 18:579–583.

Lam, R., and A. Levitt. 1999. *Canadian consensus guidelines for the treatment of seasonal affective disorder.* Vancouver, BC: Clinical and Academic Press.

Lamb, D.R., and G.R. Brodowicz. 1986. Optimal use of fluids of varying formulations to minimize exercise-induced disturbances in homeostasis. *Sports Med* 3:247–274.

Lambert, M.I., and J. Borresen. 2006. A theoretical basis of monitoring fatigue: A practical approach for coaches. *Int J Sport Sci Coach* 1(4):371–388.

Lambert, M.I., and L.R. Keytel. 2000. Training habits of top runners in different age groups in a 56 km race. *S Afr J Sports Med* 7:27–32.

Lambert, M.I., Z.H. Mbambo, and A. St Clair Gibson. 1998. Heart rate during training and competition for long-distance running. *J Sports Sci* 16:85–90.

Lambert, M.I., W. Viljoen, A. Bosch, A.J. Pearce, and M. Sayers. 2008. General principles of training. In *Olympic textbook of medicine in sport*, ed. M.P. Schwellnus, 1–48. Chichester, UK: Blackwell.

Lamberts, R.P., and M.I. Lambert. 2009. Day-to-day variation in heart rate at different levels of submaximal exertion: Implications for monitoring training. *J Strength Cond Res* 23(3):1005–1010.

Lamberts, R.P., G.J. Rietjens, H.H. Tijdink, T.D. Noakes, and M.I. Lambert. 2010. Measuring submaximal performance parameters to monitor fatigue and predict cycling performance: A case study of a world-class cyclo-cross cyclist. *Eur J Appl Physiol* 108:183–190.

Lamont, L.S., P.W. Lemon, and B.C. Bruot. 1987. Menstrual cycle and exercise effects on protein catabolism. *Med Sci Sports Exerc* 19:106–110.

Lane, K.N., and H.A. Wenger. 2004. Effect of selected recovery conditions on performance of repeated bouts of intermittent cycling separated by 24 hours. *J Strength Cond Res* 18(4):855–860.

Lange, T., S. Dimitrov, and J. Born. 2010. Effects of sleep and circadian rhythm on the human immune system. *Ann N Y Acad Sci* 1193:48–59.

Lardry, J.M. 2007. Les principales huiles essentielles utilisées en massage. *Kinesither Rev* 61:24–29.

Lardry, J.M., and V. Haberkorn. 2007a. L'aromathérapie et les huiles essentielles. *Kinesither Rev* 61:14–17.

———. 2007b. Les huiles essentielles: Principes d'utilisation. *Kinesither Rev* 61:18–23.

Larson-Meyer, D.E., O.N. Borkhsenious, J.C. Gullett, R.R. Russell, M.C. Devries, S.R. Smith, and E. Ravussin. 2008. Effect of dietary fat on serum and intramyocellular lipids and running performance. *Med Sci Sports Exerc* 40:892–902.

Larson-Meyer, D.E., B.R. Newcomer, and G.R. Hunter. 2002. Influence of endurance running and recovery diet on intramyocellular lipid content in women: A 1H NMR study. *Am J Physiol Endocrinol Metab* 282:E95–E106.

Lashley, F.R. 2004. Measuring sleep. In *Instruments for clinical health-care research,* ed. M. Frank-Stromberg and S.J. Olsen, 715. Sudbury, MA: Jones and Bartlett.

Lattier, G., G.Y. Millet, A. Martin, and V. Martin. 2004. Fatigue and recovery after high-intensity exercise. Part II: Recovery interventions. *Int J Sports Med* 25:509–515.

Laughlin, G.A., and S.S. Yen. 1996. Nutritional and endocrine-metabolic aberrations in amenorrheic athletes. *J Clin Endocrinol Metab* 81:4301–4309.

Laurent, D., B. Authier, J.F. Lebas, and A. Rossi. 1992. Effect of prior exercise in Pi/PC ratio and intracellular pH during a standardized exercise. A study on human muscle using [31P] NMR. *Acta Physiol Scand* 144:31–38.

Laureys, S., P. Peigneux, F. Perrin, and P. Maquet. 2002. Sleep and motor skill learning. *Neuron* 35:5–7.

Laursen, P.B. 2010. Training for intense exercise performance: High-intensity or high-volume training? *Scand J Med Sci Sports* 20(2):S1–S10.

Laursen, P.B., and D.G. Jenkins. 2002. The scientific basis for high-intensity interval training: Optimising training programmes and maximising performance in highly trained endurance athletes. *Sports Med* 32(1):53–73.

Lavoie, M. 2002. La photothérapie dans le traitement et la prévention du trouble affectif saisonnier. *Le Médecin du Québec* 3(37).

Leal, E.C. Jr., V. de Godoi, J.L. Mancalossi, R.P. Rossi, T. De Marchi, M. Parente, D. Grosselli, R.A. Generosi, M. Basso, L. Frigo, S.S. Tomazoni, J.M. Bjordal, and R.A.B. Lopes-Martins. 2010. Comparison between cold water immersion therapy (CWIT) and light emitting diode therapy (LEDT) in short-term skeletal muscle recovery after high-intensity exercise in athletes—preliminary results. *Lasers Med Sci* 26(4):493–501.

Leddy, J., P. Horvath, J. Rowland, and D. Pendergast. 1997. Effect of a high or a low fat diet on cardiovascular risk factors in male and female runners. *Med Sci Sports Exerc* 29:17–25.

Lee, D.T., and E.M. Haymes. 1995. Exercise duration and thermoregulatory responses after whole body precoooling. *J Appl Physiol* 79:1971–1976.

Lee-Chiong, T. 2006. *Sleep: A comprehensive handbook.* Hoboken, NJ: John Wiley & Sons.

Leeder, J., C. Gissane, K. van Someren, W. Gregson, and G. Howatson. 2012. Cold water immersion and recovery from strenuous exercise: A meta-analysis. *Br J Sport Med* 46:233–240.

Lehmann, J.F., and B.J. Delateur. 1990. Cryotherapy. In *Therapeutic heat and cold* (4th ed.), ed. J.F. Lehmann, 590. Baltimore: Williams & Wilkins.

Lehmann, M., C. Foster, U. Gastmann, H. Keizer, and J. Steinacker. 1999. Definitions, types, symptoms, findings, underlying mechanisms, and frequency of overtraining and overtraining syndrome. In *Overload, performance incompetence, and regeneration in sport,* ed. M. Lehmann, C. Foster, U. Gastmann, H. Keizer, and J. Steinacker, 1–6. New York: Kluwer Academic/Plenum.

Lehmann, M., C. Foster, and J. Keul. 1993. Overtraining in endurance athletes: A brief review. *Med Sci Sport Exer* 25(7):854–862.

Lehmann, M., U. Gastmann, S. Baur, Y. Liu, W. Lormes, A. Opitz-Gress, S. Reissnecker, C. Simsch, and J.M. Steinacker. 1999. Selected parameters and mechanisms of peripheral and central fatigue and regeneration in overtrained athletes. In *Overload, performance incompetence, and regeneration in sport,* ed. M. Lehmann, C. Foster, U. Gastmann, H. Keizer, and J. Steinacker, 7–25. New York: Kluwer Academic/Plenum.

Leick, L., P. Plomgaard, L. Gronlokke, F. Al-Abaiji, J.F. Wojtaszewski, and H. Pilegaard. 2010. Endurance exercise induces mRNA expression of oxidative enzymes in human skeletal muscle late in recovery. *Scand J Med Sci Sports* 20(4):593–599.

Leiper, J.B., and R.J. Maughan. 1986. Absorption of water and electrolytes from hypotonic, isotonic and hypertonic solutions. *J Physiol* 373:90.

Lemon, P.W.R. 1991. Effects of exercise on protein requirements. *J Sports Sci* 9:53–70.

————. 1997. Dietary protein requirements in athletes. *J Nutr Biochem* 87:982–992.

Lemon, P.W.R., and J.P. Mullin. 1980. Effect of initial muscle glycogen levels on protein catabolism during exercise. *J Appl Physiol* 48:624–629.

Lemyre, P.N. 2005. *Determinants of burnout in elite athletes: A multidimensional perspective.* Oslo, Norway: Norwegian University of Sport Sciences.

Lemyre, P.N., H.K. Hall, and G.C. Roberts. 2008. A social cognitive approach to burnout in elite athletes. *Scand J Med Sci Sports* 18:221–234.

Lemyre, P.N., G.C. Roberts, and J. Stray-Gundersen. 2007. Motivation, overtraining and burnout: Can self-determined motivation predict overtraining and burnout in elite athletes. *Eur J Sport Sci* 7:115–132.

Lemyre, P.N., D.C. Treasure, and G.C. Roberts. 2006. Influence of variability in motivation and affect on elite athlete burnout. *J Sport Exercise Psy* 28:32–48.

Leppäluoto, J. 1988. Human thermoregulation in sauna. *Ann Clin Res* 20:240–243.

Leppäluoto, J., T. Ranta, U. Laisi, J. Partanen, P. Virkkunen, and H. Lybeck. 1975. Strong heat exposure and adenohypophyseal hormone secretion in man. *Horm Metab Res* 7:439–440.

Leppäluoto, J., P. Tapanainen, and M. Knip. 1987. Heat exposure elevates plasma immunoreactive growth hormone releasing hormone levels in man. *J Clin Endocr Metab* 97:21–29.

Leppäluoto, J., T. Westerlund, P. Huttunen, J. Oksa, J. Smolander, B. Dugué, and M. Mikkelsson. 2008. Effects of long-term whole-body cold exposures on plasma concentrations of ACTH, beta-endorphin, cortisol, catecholamines and cytokines in healthy females. *Scand J Clin Lab Inv* 68(2):145–153.

Levenhagen, D.K., J.D. Gresham, M.G. Carlson, D.J. Maron, M.J. Borel, and P.J. Flakoll. 2001. Post exercise nutrient intake timing is critical to recovery of leg glucose and protein homeostasis. *Am J Physiol* 280:E982–E993.

Lieber, R.L., L.E. Thornell, and J. Fridén. 1996. Muscle cytoskeletal disruption occurs within the first 15 min of cyclic eccentric contraction. *J Appl Physiol* 80:278–284.

Lin, C.C., C.F. Chang, M.Y. Lai, T.W. Chen, P.C. Lee, and W.C. Yang. 2007. Far-infrared therapy: A novel treatment to improve access blood flow and unassisted patency of arteriovenous fistula in hemodialysis patients. *J Am Soc Nephrol* 18(3):985–992.

Lin, Z.P., and J.J. Hsu. 2007. Aromatherapy massage. *Sport Digest* 15.

————. 2008. Aromatherapy in sports medicine practice with elite athletes. *Sport Digest* 16.

Linnamo, V., K. Hakkinen, and P.V. Komi. 1998. Neuromuscular fatigue and recovery in maximal compared to explosive strength loading. *Eur J Appl Physiol Occup Physiol* 77:176–181.

Linossier, M.T., C. Denis, D. Dormois, A. Geyssant, and J.R. Lacour. 1993. Ergometric and metabolic adaptation to a 5-s sprint training programme. *Eur J Appl Physiol Occup Physiol* 67:408–414.

Lonsdale, C., K. Hodge, and S.A. Jackson. 2007. Athlete engagement: II. Development and initial validation of the athlete engagement questionnaire. *Int J Sport Psychol* 38:471–492.

Lonsdale, C., K. Hodge, and T.D. Raedeke. 2007. Athlete engagement: I. A qualitative investigation of relevance of dimensions. *Int J Sport Psychol* 38:451–470.

Loucks, A.B. 2003. Energy availability, not body fatness, regulates reproductive function in women. *Exerc Sport Sci Rev* 31:144–148.

Loy, S.F., J.J. Hoffmann, and G.J. Holland. 1995. Benefits and practical use of cross-training in sports. *Sports Med* 19:1–8.

Lubkowska, A., M. Chudecka, A. Klimek, Z. Sgygula, and B. Fraczek. 2008. Acute effect of a single whole-body cryostimulation or prooxidant-antioxidant balance in blood of healthy, young men. *J Therm Biol* 33:464–467.

Lucey, D.R., M. Clerici, and G.M. Shearer. 1996. Type 1 and type 2 cytokine dysregulation in human infectious, neoplastic, and inflammatory diseases. *Clin Microbiol Rev* 9:532–562.

Lund, H., P. Vestergaard-Poulsen, I.L. Kanstrup, and P. Sejrsen. 1998. The effect of passive stretching on delayed onset muscle soreness, and other detrimental effects following eccentric exercise. *Scand J Med Sci Spor* 8:216–221.

Lurie, S.J., B. Gawinski, D. Pierce, and S.J. Rousseau. 2006. Seasonal affective disorder. *Am Fam Physician* 74:1521–1524.

Lynch, G., and R. Granger. 2008. *Big brain: The origins and future of human intelligence.* New York: Macmillan.

Lyons, T.P., M.L. Riedesel, L.E. Meulin, and T.W. Chick. 1990. Effects of glycerol-induced hyperhydration prior to exercise in the heat on sweating and core temperature. *Med Sci Sports Exerc* 22(4):477–483.

MacAuley, D.C. 2001. Ice therapy: How good is the evidence? *Int J Sports Med* 22:379–384.

MacDougall, J.D., A.L. Hicks, J.R. MacDonald, R.S. McKelvie, H.J. Green, and K.M. Smith. 1998. Muscle performance and enzymatic adaptations to sprint interval training. *J Appl Physiol* 84:2138–2142.

MacIntyre, D.L., W.D. Reid, D.M. Lyster, and D.C. McKenzie. 2000. Different effects of strenuous eccentric exercise on the accumulation of neutrophils in muscle in women and men. *Eur J Appl Physiol* 81:47–53.

Magness, J.L., T.R. Garrett, and D.J. Erickson. 1970. Swelling of the upper extremity during whirlpool baths. *Arch Phys Med Rehabil* 51(5):297–299.

Magnusson, S.P., P. Aagard, E. Simonsen, and F. Bojsen-Møller. 1998. A biomechanical evaluation of cyclic and static stretch in human skeletal muscle. *Int J Sports Med* 19:310–316.

Magnusson, S.P., E.B. Simonsen, P. Dyhre-Poulsen, P. Aagaard, T. Mohr, and M. Kjaer. 1996. Viscoelastic stress relaxation during static stretch in human skeletal muscle in the absence of EMG activity. *Scand J Med Sci Sports* 6:323–328.

Maïsetti, O., J. Sastre, J. Lecompte, and P. Portero. 2007. Differential effects of an acute bout of passive stretching on maximal voluntary torque and the rate of torque development of the calf muscle-tendon unit. *Isokinet Exerc Sci* 15:11–17.

Maitre, S., C. Hautier, H. Toumi, G. Poumarat, and N. Fellmann. 2001. Exercice répété sur presse inclinée. Influence de la récupération par électrostimulation sur la fatigue et la lactatémie sanguine. Paper presented at the 3rd Conference on Biology and Muscular Exercise, Clermont-Ferrand, France.

Manore, M.M. 1999. Nutritional needs of the female athlete. *Clin Sports Med* 18:549–563.

———. 2002. Dietary recommendations and athletic menstrual dysfunction. *Sports Med* 32:887–901.

———. 2005. Exercise and the Institute of Medicine recommendations for nutrition. *Curr Sports Med Rep* 4:193–198.

Marcus, R., C. Cann, P. Madvig, J. Minkoff, M. Goddard, M. Bayer, M. Martin, L. Gaudiani, W. Haskell, and H. Genant. 1985. Menstrual function and bone mass in elite women distance runners. Endocrine and metabolic features. *Ann Intern Med* 102:158–163.

Marieb, E.N. 1999. *Anatomie et physiologie humaines* (4th ed.). Bruxelles: De Boeck University.

Marino, F.E. 2002. Methods, advantages, and limitations of body cooling for exercise performance. *Br J Sports Med* 36:89–94.

———. 2004. Anticipatory regulation and avoidance of catastrophe during exercise-induced hyperthermia. *Comp Biochem Physiol Part B Biochem Mol Biol* 139:561–569.

———. 2011. The critical limiting temperature and selective brain cooling: Neuroprotection during exercise? *Int J Hyperth* 27:582–590.

Marino, F.E., M.I. Lambert, and T.D. Noakes. 2004. Superior performance of African runners in warm humid but not in cool environmental conditions. *J Appl Physiol* 96:124–130.

Marino, F.E., Z. Mbambo, E. Kortekaas, G. Wilson, M.I. Lambert, T.D. Noakes, and S.C. Dennis. 2000. Advantages of smaller body mass during distance running in warm, humid environments. *Pflügers Arch* 441:359–367.

Marino, F.E., J.M. Sockler, and J.M. Fry. 1998. Thermoregulatory, metabolic and sympathoadrenal responses to repeated brief exposure to cold. *Scand J Clin Lab Investig* 58:537–546.

Markus, C.R., L.M. Jonkman, J.H. Lammers, N.E. Deutz, M.H. Messer, and N. Rigtering. 2005. Evening intake of alpha-lactalbumin increases plasma tryptophan availability and improves morning alertness and brain measures of attention. *Am J Clin Nutr* 81:1026–1033.

Marsh, S.A., and D.G. Jenkins. 2002. Physiological responses to the menstrual cycle: Implications for the development of heat illness in female athletes. *Sports Med* 32:601–614.

Martin, A. 2001. *Apports nutritionnels conseillés pour la population française* (3rd ed.). Paris: Tec & Doc.

Martin, B.J., P.R. Bender, and H. Chen. 1986. Stress hormonal response to exercise after sleep loss. *Eur J Appl Physiol Occup Physiol* 55:210–214.

Martin, B.J., and H.I. Chen. 1984. Sleep loss and the sympathoadrenal response to exercise. *Med Sci Sports Exerc* 16:56–59.

Martin, B.J., and G.M. Gaddis. 1981. Exercise after sleep deprivation. *Med Sci Sports Exerc* 13:220–223.

Martin, B., and R. Haney. 1982. Self-selected exercise intensity is unchanged by sleep loss. *Eur J Appl Physiol Occup Physiol* 49:79–86.

Martin, P., F.E. Marino, J. Rattey, D. Kay, and J. Cannon. 2005. Reduced voluntary activation of skeletal muscle during shortening and lengthening contractions in whole body hyperthermia. *Exp Physiol* 90:225–236.

Martin, V., G.Y. Millet, G. Lattier, and L. Perrod. 2004. Effects of recovery modes after knee extensor muscles eccentric contractions. *Med Sci Sports Exerc* 36:1907–1915.

Masago, R., T. Matsuda, Y. Kikuchi, Y. Miyazaki, K. Iwanaga, H. Harada, and T. Katsuura. 2000. Effects of inhalation of essential oils on EEG activity and sensory evaluation. *J Physiol Anthropol Appl Human Sci* 19:35–42.

Masuda, A., T. Kihara, T. Fukudome, T. Shinsato, S. Minagoe, and C. Tei. 2005. The effects of repeated thermal therapy for two patients with chronic fatigue syndrome. *J Psychosom Res* 58(4):383–387.

Masuda, A., Y. Koga, M. Hattanmaru, S. Minagoe, and C. Tei. 2005. The effects of repeated thermal therapy for patients with chronic pain. *Psychother Psychosom* 74(5):288–294.

Mathews, D.K., E.L. Fox, and D. Tanzi. 1969, May. Physiological responses during exercise and recovery in a football uniform. *J Appl Physiol* 26(5):611–615.

Maton, B., G. Thiney, S. Dang, S. Tra, S. Bassez, P. Wicart, and A. Ouchene. 2006. Human muscle fatigue and elastic compressive stockings. *Eur J Appl Physiol* 97:432–442.

Matzen, L.E., B.B. Andersen, B.G. Jensen, H.J. Gjessing, S.H. Sindrup, and J. Kvetny. 1990. Different short-term effect of protein and CH intake on TSH, growth hormone, insulin, C-peptide, and glucagon in humans. *Scand J Clin Lab Invest* 50:801–805.

Maughan, R.J., J.B. Leiper, and S.M. Shirreffs. 1997. Factors influencing the restoration of fluid and electrolyte balance after exercise in the heat. *Br J Sports Med* 31:175–182.

Maughan, R.J., and S.M. Shirreffs. 1997. Recovery from prolonged exercise: Restoration of water and electrolyte balance. *J Sports Sci* 15:297–303.

———. 2004. Exercise in the heat: Challenges and opportunities. *J Sports Sci* 22:917–927.

Maxwell, S., S. Kohn, A. Watson, and R.J. Balnave. 1988. Is stretching effective in the prevention of or amelioration of delayed-onset muscle soreness? In *The athlete maximising participation and minimising risk,* ed. M Torode, 109–118. Sydney: Cumberland College of Health Sciences.

Mayrovitz, H.N. 2007. Interface pressures produced by two different types of lymphedema therapy devices. *Phys Ther* 87:1379–1388.

McAinch, A.J., M.A. Febbraio, J.M. Parkin, S. Zhao, K. Tangalakis, L. Stojanovska, and M.F. Carey. 2004. Effect of active versus passive recovery on metabolism and performance during subsequent exercise. *Int J Sport Nutr Exerc Metab* 14:185–196.

McArdle, W.D., F.I. Katch, and V.C. Katch. 2004. *Nutrition et performances sportives.* Bruxelles: De Boeck.

McCaffrey, R., D.J. Thomas, and A.O. Kinzelman. 2009. The effects of lavender and rosemary essential oils on test taking anxiety among graduate nursing students. *Holist Nurs Pract* 23:88–93.

McCleland, G.B., P.W. Hochachka, and J.M. Weber. 1998. Carbohydrate utilization during exercise after high altitude acclimatization: A new perspective. *Proc Natl Acad Sci* 95:10288–10293.

McCoy, M., J. Proietto, and M. Hargreaves. 1996. Skeletal muscle GLUT-4 and postexercise muscle glycogen storage in humans. *J Appl Physiol* 80:411–415.

McGlynn, G.H., N.T. Laughlin, and V. Rowe. 1979. Effect of electromyographic feedback and static stretching on artificially induced muscle soreness. *Amer J Physical Med* 58:139–148.

McInnes, S.E., J.S. Carlson, C.J. Jones, and M.J. McKenna. 1995. The physiological load imposed on basketball players during competition. *J Sport Sci* 3:387–397.

McKenna, M.J., T.A. Schmidt, M. Hargreaves, L. Cameron, S.L. Skinner, and K. Kjeldsen. 1993. Sprint training increases human skeletal muscle $Na^{(+)}$-$K^{(+)}$-ATPase concentration and improves K^+ regulation. *J Appl Physiol* 75:173–180.

McKenzie, S., S.M. Phillips, S.L. Carter, S. Lowther, M.J. Gibala, and M.A. Tarnopolsky. 2000. Endurance exercise training attenuates leucine oxidation and BCOAD activation during exercise in humans. *Am J Physiol Endocrinol Metab* 278:E580–E587.

McMahon, S., and D. Jenkins. 2002. Factors affecting the rate of phosphocreatine resynthesis following intense exercise. *Sports Med* 32:761–784.

McNair, P.J., E.W. Dombroski, D.J. Hewson, and S.N. Stanley. 2001. Stretching at the ankle joint: Viscoelastic responses to holds and continuous passive motion. *Med Sci Sports Exerc* 33:354–358.

McNaughton, L., K. Backx, G. Palmer, and N. Strange. 1999. Effects of chronic bicarbonate ingestion on the performance of high-intensity work. *Eur J Appl Physiol* 80:333–336.

McNaughton, L., and R. Cedaro. 1992. Sodium citrate ingestion and its effects on maximal anaerobic exercise of different durations. *Eur J Appl Physiol* 64:36–41.

Medbo, J.I., P. Gramvick, and E. Jebens. 1999. Aerobic and anaerobic energy release during 10 and 30 s bicycle sprints. *Acta Kinesiol Univ Tartuensis* 4:122–146.

Medbo, J.I., and I. Tabata. 1993. Anaerobic energy release in working muscle during 30 s to 3 min of exhausting bicycling. *J Appl Physiol* 75:1654–1660.

Meeusen, R. 1999. Overtraining and the central nervous system, the missing link? In *Overload, performance incompetence, and regeneration in sport,* ed. M. Lehmann, C. Foster, U. Gastmann, H. Keizer, and J. Steinacker, 187–202. New York: Kluwer Academic/Plenum.

Meeusen, R., M. Duclos, C. Foster, A. Fry, M. Gleeson, D. Nieman, J. Raglin, G. Rietjens, J. Steinacker, and A. Urhausen. 2013a. Prevention, diagnosis and treatment of the overtraining syndrome: Joint consensus statement of the European College of Sport Science (ECSS) and the American College of Sports Medicine (ACSM), *European Journal of Sport Science,* DOI:10.1080/1746 1391.2012.730061.

Meeusen, R., M. Duclos, C. Foster, A. Fry, M. Gleeson, D. Nieman, J. Raglin, G. Rietjens, J. Steinacker, and A. Urhausen. 2013b. Prevention, diagnosis and treatment of the overtraining syndrome: Joint consensus statement of the European College of Sport Science (ECSS) and the American College of Sports Medicine (ACSM), *Med Sci Sports Exer* 45(1):186–205.

Meeusen, R., M. Duclos, M. Gleeson, G.J. Rietjens, J.M. Steinacker, and A. Urhausen. 2006. Prevention, diagnosis and treatment of the overtraining syndrome. *Eur J Sport Sci* 6(1):1–14.

Meeusen, R., E. Nederhof, L. Buyse, B. Roelands, G. de Schutter, and M.F. Piacentini. 2010. Diagnosing overtraining in athletes using the two bout exercise protocol. *Br J Sports Med* 44(9):642–648.

Meeusen, R., M.F. Piacentini, B. Busschaert, L. Buyse, G. de Schutter, and J. Stray-Gundersen. 2004. Hormonal responses in athletes: The use of a two bout exercise protocol to detect subtle differences in (over)training status. *Eur J Appl Physiol* 91:140–146.

Melanson, E.L., T.A. Sharp, H.M. Seagle, T.J. Horton, W.T. Donahoo, G.K. Grunwald, J.T. Hamilton, and J.O. Hill. 2002. Effect of exercise intensity on 24-h energy expenditure and nutrient oxidation. *J Appl Physiol* 92:1045–1052.

Melzack, R., and P.D. Wall. 1965. Pain mechanisms: A new theory. *Science* 150:971–979.

Mendelssohn, M. 1895. Ueber den thermotropismus einzelliger organismen. *Pflügers Arch* 60:1–27.

Meney, I., J. Waterhouse, G. Atkinson, T. Reilly, and D. Davenne. 1998. The effect of one night's sleep deprivation on temperature, mood, and physical performance in subjects with different amounts of habitual physical activity. *Chronobiol Int* 15:349–363.

Menzies, P., C. Menzies, L. McIntyre, P. Paterson, J. Wilson, and O.J. Kemi. 2010. Blood lactate clearance during active recovery after an intense running bout depends on the intensity of the active recovery. *J Sports Sci* 28:975–982.

Merrick, M.A., J.M. Rankin, F.A. Andres, and C.L. Hinman. 1999. A preliminary examination of cryotherapy and secondary injury in skeletal muscle. *Med Sci Sports Exerc* 31:1516–1521.

Metzger, D., C. Zwingmann, W. Protz, and W.H. Jackel. 2000. Whole-body cryotherapy in rehabilitation of patients with rheumatoid diseases: Pilot study. *Rehabilitation (Stuttg)* 39:93–100.

Micklewright, D., M. Griffin, V. Gladwell, and R. Beneke. 2005. Mood state response to massage and subsequent exercise performance. *Sport Psychol* 19:234–250.

Mika, A., P. Mika, B. Fernhall, and V.B. Unnithan. 2007. Comparison of recovery strategies on muscle performance after fatiguing exercise. *Am J Phys Med Rehabil* 86:474–481.

Miller, B.F., K.G. Gruben, and B.J. Morgan. 2000. Circulatory responses to voluntary and electrically induced muscle contractions in humans. *Phys Ther* 80:53–60.

Miller, S.L., W.J. Schopf, and A. Lazcano. 1997. Oparin's *Origin of Life*: Sixty years later. *J Mol Evol* 44:351–353.

Millet, G., S. Perrey, C. Divert, and M. Foissac. 2006. The role of engineering in fatigue reduction during human locomotion: A review. *Sports Eng* 19(4):209–220.

Milner-Brown, H.S., R.B. Stein, and R. Yemm. 1973. The orderly recruitment of human motor units during voluntary isometric contractions. *J Physiol* 230:359–370.

Minett, G.M., R. Duffield, F.E. Marino, and M. Portus. 2011. Volume-dependent response of pre-cooling for intermittent-sprint exercise in the heat. *Med Sci Sports Exerc* 43:1760–1769.

Mitchell, D., S.K. Maloney, C. Jessen, H.P. Laburn, P.R. Kamerman, G. Mitchell, and A. Fuller. 2002. Adapative heterothermy and selective brain cooling in arid-zone mammals. *Comp Biochem Physiol Part B Biochem Mol Biol* 131:571–585.

Mitchell-Taverner, C. 2005. *Field hockey techniques and tactics.* Champaign, IL: Human Kinetics.

Mittendorfer, B., J.F. Horowitz, and S. Klein. 2001. Gender differences in lipid and glucose kinetics during short-term fasting. *Am J Physiol Endocrinol Metab* 281:E1333–E1339.

Mizuno, T., Y. Takanashi, K. Yoshizaki, and M. Kondo. 1994. Fatigue and recovery of phosphorus metabolites and pH during stimulation of rat skeletal muscle: An evoked electromyography and in vivo 31P-nuclear magnetic resonance spectroscopy study. *Eur J Appl Physiol Occup Physiol* 69:102–109.

Moesch, K., A.M. Elbe, M.L.T. Hauge, and J.M. Wikman. 2011. Late specialization: The key to success in centimeters, grams, or seconds (cgs) sports. *Scand J Med Sci Sports* 21(6):282–290.

Monedero, J., and B. Donne. 2000. Effect of recovery interventions on lactate removal and subsequent performance. *Int J Sports Med* 21:593–597.

Montgomery, P.G., D.B. Pyne, A.J. Cox, W.G. Hopkins, C.L. Minahan, and P.H. Hunt. 2008. Muscle damage, inflammation, and recovery interventions during a 3-day basketball tournament. *J Sports Sci Med* 8(5):241–250.

Montgomery, P.G., D.B. Pyne, W.G. Hopkins, J.C. Dorman, K. Cook, and C.L. Minahan. 2008. The effect of recovery strategies on physical performance and cumulative fatigue in competitive basketball. *J Sports Sci* 26(11):1135–1145.

Moraska, A. 2007. Therapist education impacts the massage effect on postrace muscle recovery. *Med Sci Sport Exer* 39:34–37.

Morelli, M., C.E. Chapman, and S.J. Sullivan. 1999. Do cutaneous receptors contribute to the changes in the amplitude of the H-reflex during massage? *Electromyogr Clin Neurophysiol* 39:441–447.

Morgan, W.P., D. Brown, J. Raglin, P. O'Connor, and K. Ellickson. 1987. Psychological monitoring of overtraining and staleness. *Br J Sports Med* 21:107–114.

Morgan, W., D. Costill, M. Flynn, J. Raglin, and P. O'Connor. 1988. Mood disturbance following increased training in swimmers. *Med Sci Sport Exer* 20:408–414.

Mori, H., H. Ohsawa, T.H. Tanaka, E. Taniwaki, G. Leisman, and K. Nishijo. 2004. Effect of massage on blood flow and muscle fatigue following isometric lumbar exercise. *Med Sci Monit* 10:CR173–178.

Morris, R.J., N.D. Pugh, D.P. Coleman, and J.P. Woodcock. 2003. Spatial-temporal display of colour-flow ultrasound for the assessment of the effects of intermittent pneumatic compression in limbs. *Ultrasound Med Biol* 29:1805–1807.

Morrison, S., G.G. Sleivert, and S.C. Cheung. 2004. Passive hyperthermia reduces voluntary activation and isometric force production. *Eur J Appl Physiol* 91:729–736.

Morse, C.I., H. Degens, O.R. Seynnes, C.N. Maganaris, and D.A. Jones. 2008. The acute effect of stretching on the passive stiffness of the human gastrocnemius muscle tendon unit. *J Physiol* 586:97–106.

Morton, R.H. 2001. Modelling training and overtraining. *J Sports Sci* 15(3):335–340.

———. 2007. Contrast water immersion hastens plasma lactate decrease after intense anaerobic exercise. *J Sci Med Sport* 10(6):467–470.

Mougin, F., H. Bourdin, M.L. Simon-Rigaud, N.U. Nguyen, J.P. Kantelip, and D. Davenne. 2001. Hormonal responses to exercise after partial sleep deprivation and after a hypnotic drug-induced sleep. *J Sports Sci* 19:89–97.

Mougin, F., M.L. Simon-Rigaud, D. Davenne, A. Renaud, A. Garnier, J.P. Kantelip, and P. Magnin. 1991. Effects of sleep disturbances on subsequent physical performance. *Eur J Appl Physiol Occup Physiol* 63:77–82.

Mountain, S.J., S.A. Smith, and R.P. Mattot. 1998. Hypohydration effects on skeletal muscle performance and metabolism: A 31P-MRS study. *J Appl Physiol* 84:1889–1894.

Mourot, L., C. Cluzeau, and J. Regnard. 2007. [Physiological assessment of a gaseous cryotherapy device: Thermal effects and changes in cardiovascular autonomic control]. *Ann Readapt Med Phys* 50:209–217.

Mujika, I., and S. Padilla. 2000. Detraining: Loss of training-induced physiological and performance adaptations. Part I. Short-term insufficient training stimulus. *Sports Med* 30:79–87.

———. 2003. Scientific bases for precompetition tapering strategies. *Med Sci Sports Exerc* 35(7):1182–1187.

Muoio, D.M., J.J. Leddy, P.J. Horvath, A.B. Awad, and D.R. Pendergast. 1994. Effect of dietary fat on metabolic adjustments to maximal $\dot{V}O_2$ and endurance in runners. *Med Sci Sports Exerc* 26:81–88.

Murphy, R.J., and W.F. Ashe. 1965. Prevention of heat illness in football players. *JAMA* 194:650–654.

Myrer, J.W., D.O. Draper, and E. Durrant. 1994. Contrast therapy and intramuscular temperature in the human leg. *J Athl Train* 29:318–322.

Myrer, J.W., G. Measom, E. Durrant, and G.W. Fellingham. 1997. Cold- and hot-pack contrast therapy: Subcutaneous and intramuscular temperature change. *J Athl Train* 32:238–241.

Myrer, J.W., G. Measom, and G.W. Fellingham. 1998. Temperature changes in the human leg during and after two methods of cryotherapy. *J Athl Train* 33:25–29.

———. 2000. Exercise after cryotherapy greatly enhances intramuscular rewarming. *J Athl Train* 35:412–416.

Myrer, J.W., K.A. Myrer, G.J. Measom, G.W. Fellingham, and S.L. Evers. 2001. Muscle temperature is affected by overlying adipose when cryotherapy is administered. *J Athl Train* 36:32–36.

Nadel, E., S. Fortney, and C. Wenger. 1980. Effect of hydration state on circulatory and thermal regulations. *J Appl Physiol: Respir Environ Exercise Physiol* 49:715–721.

Nadel, E.R., G.W. Mack, and H. Nose. 1990. Influence of fluid replacement beverages on body fluid homeostasis during exercise and recovery. In *Perspectives in exercise science and sports medicine, Vol. 3. Fluid homeostasis during exercise*, ed. C.V. Gisolfi and D.R. Lamb, 181–205. Carmel, IN: Benchmark Press.

Nadler, S.F., K. Weingand, and R.J. Kruse. 2004. The physiologic basis and clinical applications of cryotherapy and thermotherapy for the pain practitioner. *Pain Physician* 7:395–399.

Nair, K.S., R.G. Schwartz, and S. Welle. 1992. Leucine as a regulator of whole body and skeletal muscle protein metabolism in humans. *Am J Physiol* 263:E928–E934.

Naitoh, P., C.E. Englund, and D. Ryman. 1982. Restorative power of naps in designing continuous work schedules. *J Hum Ergol (Tokyo)* 11:S259–S278.

Naitoh, P., T.L. Kelly, and H. Babkoff. 1992. Napping, stimulant, and four-choice performance. In *Sleep, arousal, and performance*, ed. R.J. Broughton and R.D. Ogilvie, 198–219. Boston: Birkhäuser.

Nakachi, K., and K. Imai. 1992. Environmental and physiological influences on human natural killer cell activity in relation to good health practices. *Jpn J Cancer Res* 83:798–805.

Naliboff, B.D., and K.H. Tachiki. 1991. Autonomic and skeletal muscle responses to nonelectrical cutaneous stimulation. *Percept Mot Skills* 72:575–584.

Nattiv, A. 2000. Stress fractures and bone health in track and field athletes. *J Sci Med Sport* 3:268–279.

Nemet, D., Y. Meckel, S. Bar-Sela, F. Zaldivar, D.M. Cooper, and A. Eliakim. 2009. Effect of local cold-pack application on systemic anabolic and inflammatory response to sprint-interval training: A prospective comparative trial. *Eur J Appl Physiol* 107:411–417.

Neric, F.B., W.C. Beam, L.E. Brown, and L.D. Wiersma. 2009. Comparison of swim recovery and muscle stimulation on lactate removal after sprint swimming. *J Strength Cond Res* 23:2560–2567.

Nesher, D.L., J.E. Karl, and D.M. Kipnis. 1985. Dissociation of effects of insulin and contraction on glucose transport in rat epitrochlearis muscle. *Am J Physiol* 249:C226–C232.

Nevill, M.E., L.H. Boobis, S. Brooks, and C. Williams. 1989. Effect of training on muscle metabolism during treadmill sprinting. *J Appl Physiol* 67:2376–2382.

Newsholme, E.A. 1986. Application of knowledge of metabolic integration to the problem of metabolic limitations in middle-distance and marathon running. *Acta Physiol Scand* 128:93–97.

Nicholls, J.G. 1989. *The competitive ethos and democratic education*. Cambridge, MA: Harvard University Press.

Nielsen, B. 1984. The effect of dehydration on circulation and temperature regulation during exercise. *J Therm Biol* 9:107–112.

———. 1996. Olympics in Atlanta: A fight against physics. *Med Sci Sports Exerc* 8(6):665–668.

Nielsen, B., J.R.S. Hales, S. Strange, N.J. Christensen, J. Warberg, and B. Saltin. 1993. Human circulatory and thermoregulatory adaptations with heat acclimation and exercise in a hot, dry environment. *J Physiol* 460:467–485.

Nielsen, M. 1938. Die regulation der kopertemperatur bei muskelarbeit. *Skand Arch Physiol* 79:193–230.

Nieman, D.C. 1994. Exercise, infection, and immunity. *Int J Sports Med* 15:S131–141.

Nieman, D.C., D.A. Henson, L.L. Smith, A.C. Utter, D.M. Vinci, J.M. Davis, D.E. Kaminsky, and M. Shute. 2001. Cytokine changes after a marathon race. *J Appl Physiol* 91:109–114.

Nix, G.A., R.M. Ryan, J.B. Manly, and E.L. Deci. 1999. Revitalization through self-regulation: The effects of autonomous and controlled motivation on happiness and vitality. *J Exp Soc Psychol* 35:266–284.

Noakes, T.D., N. Goodwin, and B.L. Rayner. 1985. Water intoxication: A possible complication during endurance exercise. *Med Sci Sports Exerc* 17:370–375.

Noakes, T.D., and A. St Clair Gibson. 2004. Logical limitations to the catastrophe models of fatigue during exercise in humans. *Br J Sports Med* 38:648–649.

Nordez, A., P. Casari, J.P. Mariot, and C. Cornu. 2009. Modeling of the passive mechanical properties of the musculoarticular complex: Acute effects of cyclic and static stretching. *J Biomech* 42:767–773.

Nordez, A., P.J. McNair, P. Casari, and C. Cornu. 2009. The effect of angular velocity and cycle on the dissipative properties of the knee during passive cyclic stretching: A matter of viscosity or solid friction. *Clin Biomech* 24:77–81.

———. 2010. Static and cyclic stretching: Their different effects on the passive torque-angle curve. *J Sci Med Sport* 13:156–160.

Norlander, T., and T. Archer. 2002. Predicting performance in ski and swim championships: Effectiveness of mood, perceived exertion, and dispositional optimism. *Percept Mot Skills* 94:153–164.

Norton, K., and T. Olds. 2001. Morphological evolution of athletes over the 20th century: Causes and consequences. *J Sports Med* 31:763–783.

Nose, H., G.W. Mack, S. Xiangrong, and E.R. Nadel. 1988. Role of osmolality and plasma volume during rehydration in humans. *J Appl Physiol* 17:325–331.

Nunneley, S.A. 1978. Physiological responses of women to thermal stress: A review. *Med Sci Sports* 10:250–255.

Nybo, L., and B. Nielsen. 2001a. Hyperthermia and central fatigue during prolonged exercise in humans. *J Appl Physiol* 91:1055–1060.

———. 2001b. Perceived exertion is associated with an altered brain activity during exercise with progressive hyperthermia. *J Appl Physiol* 91:2017–2023.

Nybo, L., N.H. Secher, and B. Nielsen. 2002. Inadequate heat release from the human brain during prolonged exercise with hyperthermia. *J Physiol* 545:667–704.

Nybo, L., and M.D. White. 2008. Do humans have selective brain cooling? In *Physiological bases of human performance during work and exercise,* ed. N.A.S. Taylor and H. Groeller, 473–479. London: Churchill Livingstone Elsevier.

Obal, F. Jr., and J.M. Krueger. 2004. GHRH and sleep. *Sleep Med Rev* 8:367–377.

O'Brien, C.P., and F. Lyons. 2000. Alcohol and the athlete. *Sports Med* 29:295–300.

O'Brien, W.J., F.M. Rutan, C. Sanborn, and G.E. Omer. 1984. Effect of transcutaneous electrical nerve stimulation on human blood beta-endorphin levels. *Phys Ther* 64:1367–1374.

O'Connor, P. 1997. Overtraining and staleness. In *Physical activity & mental health,* ed. W.P. Morgan, 145–160. Washington, DC: Taylor & Francis.

O'Connor, P., W. Morgan, J. Raglin, C. Barksdale, and N. Kalin. 1989. Mood state and salivary cortisol levels following overtraining in female swimmers. *Psychoneuroendocrino* 14:303–310.

Ogai, R., M. Yamane, T. Matsumoto, and M. Kosaka. 2008. Effects of petrissage massage on fatigue and exercise performance following intensive cycle pedalling. *Br J Sports Med* 42:834–838.

Ojuka, E.O., T.E. Jones, D.H. Han, M. Chen, and J.O. Holloszy. 2003. Raising Ca^{2+} in L6 myotubes mimics effects of exercise on mitochondrial biogenesis in muscle. *FASEB J* 17(6):675–681.

Oksa, J., H. Rintamaki, S. Rissanen, S. Rytky, U. Tolonen, and P.V. Komi. 2000. Stretch- and H-reflexes of the lower leg during whole body cooling and local warming. *Aviat Space Environ Med* 71(2):156–161.

Oliver, S.J., R.J. Costa, S.J. Laing, J.L. Bilzon, and N.P. Walsh. 2009. One night of sleep deprivation decreases treadmill endurance performance. *Eur J Appl Physiol* 107:155–161.

Olschewski, H., and K. Brück. 1988. Thermoregulatory, cardiovascular, and musclar factors related to exercise after precooling. *J Appl Physiol* 64:803–811.

Onen, S.H., A. Alloui, A. Gross, A. Eschallier, and C. Dubray. 2001. The effects of total sleep deprivation, selective sleep interruption and sleep recovery on pain tolerance thresholds in healthy subjects. *J Sleep Res* 10:35–42.

Opp, M.R., L. Kapas, and L.A. Toth. 1992. Cytokine involvement in the regulation of sleep. *Proc Soc Exp Biol Med* 201:16–27.

Ortenblad, N., P.K. Lunde, K. Levin, J.L. Andersen, and P.K. Pedersen. 2000. Enhanced sarcoplasmic reticulum $Ca^{(2+)}$ release following intermittent sprint training. *Am J Physiol Regul Integr Comp Physiol* 279:R152–R160.

Owen, M.D., K.C. Kregel, P.T. Wall, and C.V. Gisolfi. 1986. Effects of ingesting carbohydrate beverages during exercise in the heat. *Med Sci Sports Exerc* 18:568–575.

Pallikarakis, N., B. Jandrain, F. Pirnay, F. Mosora, M. Lacroix, A. Luyckx, and P. Lefevre. 1986. Remarkable metabolic availability of oral glucose during long duration exercise in humans. *J Appl Physiol* 60:1035–1042.

Pallotta, J.A., and P.J. Kennedy. 1968. Responses of plasma insulin and growth hormone to carbohydrate and protein feeding. *Metabolism* 17:901–908.

Pandolf, K. 1998. Time course of heat acclimation and its decay. *Int J Sports Med* 19:S157–S160.

Pandolf, K., M. Sawka, and R. Gonzales. 1988. *Human performance physiology and environmental medicine at terrestrial extremes.* Indianapolis: Benchmark Press.

Paolone, A.M., W.T. Lanigan, R.R. Lewis, and M.J. Goldstein. 1980. Effects of a postexercise sauna bath on ECG pattern and other physiologic variables. *Aviat Space Environ Med* 51(3):224–229.

Parkhouse, W.S., and D.C. McKenzie. 1984. Possible contribution of skeletal muscle buffers to enhanced anaerobic performance: A brief review. *Med Sci Sports Exerc* 16:328–338.

Parkin, J.A.M., M.F. Carey, I.K. Martin, L. Stojanovska, and M.A. Febbraio. 1997. Muscle glycogen storage following prolonged exercise: Effect of timing of ingestion of high glycemic index food. *Med Sci Sports Exerc* 29:220–224.

Parolin, M.L., A. Chesley, M.P. Matsos, L.L. Spriet, N.L. Jones, and G.J. Heigenhauser. 1999. Regulation of skeletal muscle glycogen phosphorylase and PDH during maximal intermittent exercise. *Am J Physiol* 277:E890–E900.

Parouty, J., H. Al Haddad, M. Quod, P.M. Lepretre, S. Ahmaidi, and M. Buchheit. 2010. Effect of cold water immersion on 100-m sprint performance in well-trained swimmers. *Eur J Appl Physiol* 109(3):483–490.

Parra, J., J.A. Cadefau, G. Rodas, N. Amigo, and R. Cusso. 2000. The distribution of rest periods affects performance and adaptations of energy metabolism induced by high-intensity training in human muscle. *Acta Physiol Scand* 169:157–165.

Peake, J., K. Nosaka, and K. Suzuki. 2005. Characterization of inflammatory responses to eccentric exercise in humans. *Exerc Immunol Rev* 11:64–85.

Peiffer, J.J., C.R. Abbiss, K. Nosaka, J.M. Peake, and P.B. Laursen. 2009. Effect of cold-water immersion after exercise in the heat on muscle function, body temperatures, and vessel diameter. *J Sci Med Sport* 12(1):91–96.

Peiffer, J.J., C.R. Abbiss, B.A. Wall, G. Watson, K. Nosaka, and P.B. Laursen. 2008. Effect of a 5 min cold water immersion recovery on exercise performance in the heat. *Br J Sports Med* 44(6):461–465.

Peiffer, J.J., C.R. Abbiss, G. Watson, K. Nosaka, and P.B. Laursen. 2009. Effect of cold-water immersion duration on body temperature and muscle function. *J Sports Sci* 27(10):987–993.

———. 2010a. Effect of a 5-min cold-water immersion recovery on exercise performance in the heat. *Br J Sports Med* 44:461–465.

———. 2010b. Effect of cold water immersion on repeated 1 km cycling performance in the heat. *J Sci Med Sport* 13(1):112–116.

Périard, J.D., C. Caillaud, and M.W. Thompson. 2011. Central and peripheral fatigue during passive and exercise-induced hyperthermia. *Med Sci Sports Exerc* 43:1657–1665.

Périard, J.D., M.N. Cramer, P.G. Chapman, C. Caillaud, and M.W. Thompson. 2011. Cardiovascular strain impairs prolonged self-paced exercise in the heat. *Exp Physiol* 96:134–144.

Perrey, S. 2008. Compression garments: Evidence for their physiological effects. Paper presented at ISEA Conference, Biarritz, France.

Perrey, S., A. Bringard, S. Racinais, K. Puchaux, and N. Belluye. 2008. Graduated compression stockings and delayed onset muscle soreness. In *The engineering of sport, Vol. 1*, ed. M. Estivalet and P. Brisson, 547–554. Paris: Springer.

Peters Futre, E.M., T.D. Noakes, R.I. Raine, and S.E. Terblanche. 1987. Muscle glycogen repletion during active postexercise recovery. *Am J Physiol* 253:E305–E311.

Pettersson, U., B. Stalnacke, G. Ahlenius, K. Henriksson-Larsen, and R. Lorentzon. 1999. Low bone mass density at multiple skeletal sites, including the appendicular skeleton in amenorrheic runners. *Calcif Tissue Int* 64:117–125.

Phelain, J.F., E. Reinke, M.A. Harris, and C.L. Melby. 1997. Postexercise energy expenditure and substrate oxidation in young women resulting from exercise bouts of different intensity. *J Am Coll Nutr* 16:140–146.

Phillips, F., C.N. Chen, A.H. Crisp, J. Koval, B. McGuinness, R.S. Kalucy, E.C. Kalucy, and J.H. Lacey. 1975. Isocaloric diet changes and electroencephalographic sleep. *Lancet* 2:723–725.

Phillips, S.M., S.A. Atkinson, M.A. Tarnopolsky, and J.D. MacDougall. 1993. Gender differences in leucine kinetics and nitrogen balance in endurance athletes. *J Appl Physiol* 75:2134–2141.

Pichot, V., T. Busso, F. Roche, M. Garet, F. Costes, D. Duverney, J.R. Lacour, and J.C. Barthelemy. 2002. Autonomic adaptations to intensive and overload training periods: A laboratory study. *Med Sci Sports Exerc* 34(10):1660–1666.

Pichot, V., F. Roche, J.M. Gaspoz, F. Enjolras, A. Antoniadis, P. Minini, F. Costes, T. Busso, J.R. Lacour, and J.C. Barthelemy. 2000. Relation between heart rate variability and training load in middle-distance runners. *Med Sci Sports Exerc* 32(10):1729–1736.

Pitkanen, H.T., T. Nykamen, J. Kmutinen, K. Lahti, O. Keinamen, M. Alen, P.V. Komi, and A.A. Mero. 2003. Free amino acid pool and muscle protein balance after resistance exercise. *Med Sci Sports Exerc* 35:784–792.

Plisk, S.S., and M.H. Stone. 2003. Periodization strategies. *Strength Cond J* 25(6):19.

Plyley, M.J., R.J. Shephard, G.M. Davis, and R.C. Goode. 1987. Sleep deprivation and cardiorespiratory function. Influence of intermittent submaximal exercise. *Eur J Appl Physiol Occup Physiol* 56:338–344.

Pockelé, J., E. Cumps, F. Piacentini, and R. Meeusen. 2004. Online training diary for the early detection of overreaching & overtraining. Paper presented at the First International Conference on IT and Sport, Cologne, Germany.

Poortmans, J.R. 1993. Protein metabolism. In *Principles of exercise biochemistry*, ed. J.R. Poortmans, 186–229. Basel, Switzerland: Karger.

Portero, P., F. Canon, and F. Duforez. 1996. Massage et récupération: Approche électromyographique et biomécanique. Entretiens de Bichat—Journées de médecine physique et de rééducation, 114–119.

Portero, P., and J.M. Vernet. 2001. Effets de la technique LPG sur la récupération de la fonction musculaire après exercice physique intense. Ann Kinésithér 28:145–151.

Portier, H., J.C. Chatard, E. Filaire, M.F. Jamet-Devienne, A. Robert, and C.Y. Guézennec. 2008. Effects of branched-chain amino acids supplementation on physiological and psychological performance during an offshore sailing race. Eur J Appl Physiol 104:787–794.

Pournot, H., F. Bieuzen, R. Duffield, P.M. Lepretre, C. Cozzolino, and C. Hausswirth. 2010. Short term effects of various water immersions on recovery from exhaustive intermittent exercise. Eur J Appl Physiol 111(7):1287–1295.

Preisler, B., A. Falkenbach, B. Klüber, and D. Hofmann. 1990. The effect of the Finnish dry sauna on bronchial asthma in childhood. Pneumologie 44(10):1185–1187.

Prentice, W.E. 1982. An electromyographic analysis of the effectiveness of heat or cold and stretching for inducing relaxation in injured muscle. J Orthop Sports Phys Ther 3(3):133–140.

Price, T.B., G. Perseghin, and A. Duleka. 1996. NMR studies of muscle glycogen synthesis in insulin-resistant offspring of parents with non-insulin-dependent diabetes mellitus immediately after glycogen-depleting exercise. Proc Natl Acad Sci 93:5329–5334.

Proske, U., and D.L. Morgan. 1999. Do cross-bridges contribute to the tension during stretch of passive muscle? J Muscle Res Cell Motil 20:433–442.

Putkonen, P.T.S., and E. Elomaa. 1976. Sauna and physiological sleep: Increased slow-wave sleep after heat exposure. In Sauna studies: Papers read at the VI International Sauna Congress in Helsinki on August 15-17, 1974, ed. H. Teir, Y. Collan, and P. Valtakari, 270–279. Helsinki, Finland: Vammalan Kirjapaino.

Putman, C.T., N.L. Jones, L.C. Lands, T.M. Bragg, M.G. Hollidge-Horvat, and G.J. Heigenhauser. 1995. Skeletal muscle pyruvate dehydrogenase activity during maximal exercise in humans. Am J Physiol 269:E458–E468.

Pyne, D.B., I. Mujika, and T. Reilly. 2009. Peaking for optimal performance: Research limitations and future directions. J Sports Sci 27(3):195–202.

Quod, M.J., D.T. Martin, and P.B. Laursen. 2006. Cooling athletes before competition in the heat: Comparison of techniques and practical considerations. Sports Med 36:671–682.

Raedeke, T.D., and A.L. Smith. 2001. Development and preliminary validation of an athlete burnout measure. J Sport Exercise Psy 23:281–306.

———. 2004. Coping resources and athlete burnout: An examination of stress mediated and moderation hypotheses. J Sport Exercise Psy 26:525–541.

Raglin, J.S. 2001. Psychological factors in sport performance: The mental health model revisited. Sports Med 31:875–890.

Raglin, J., and W. Morgan. 1994. Development of a scale for use in monitoring training-induced distress in athletes. Int J Sports Med 15:84–88.

Raglin, J., W. Morgan, and P. O'Connor. 1991. Changes in mood state during training in female and male college swimmers. Int J Sports Med 12:585–589.

Rehrer, N.J. 2001. Fluid and electrolyte balance in ultraendurance sport. Sports Med 31:701–715.

Rehrer, N.J., M.C. Van Kemenade, T.A. Meester, F. Brouns, and W.H.M. Saris. 1992. Gastrointestinal complaints in relation to dietary intakes in triathletes. Int J Sports Nut 2:48–59.

Reilly, T., and M. Piercy. 1994. The effect of partial sleep deprivation on weight-lifting performance. Ergonomics 37:107–115.

Rennie, M.J., R.H.T. Edwards, M. Davies, S. Krywawych, D. Halliday, J.C. Waterlow, and D.J. Millward. 1981. Effect of exercise on protein turnover in man. Clin Sci 61:627–639.

Rhea, M.R., and B.L. Alderman. 2004. A meta-analysis of periodized versus nonperiodized strength and power training programs. Res Q Exerc Sport 75(4):413–422.

Richalet, J.P., M. Marchal, C. Lamberto, J.L. Le Trong, A.M. Antezana, and E. Cauchy. 1992. Alteration of aerobic and anaerobic performance after 3 weeks at 6542 m (Mt Sajama). (Abstract). Int J Sports Med 13:87.

Richardson, A., P. Watt, and N. Maxwell. 2009. Hydration and the physiological responses to acute normobaric hypoxia. Wild Environ Med 20:212–220.

Riché, D. 1998. Guide nutritionnel des sports d'endurance (2nd ed.). Paris: Vigot.

Richendollar, M.L., L.A. Darby, and T.M. Brown. 2006. Ice bag application, active warm-up, and 3 measures of maximal functional performance. J Athl Train 41:364–370.

Richmond, L., B. Dawson, D.R. Hillman, and P.R. Eastwood. 2004. The effect of interstate travel on sleep patterns of elite Australian Rules footballers. J Sci Med Sport 7:186–196.

Richter, E.A., K.J. Mikines, H. Galbo, and B. Kiens. 1989. Effects of exercise on insulin action in human skeletal muscle. J Appl Physiol 66:876–885.

Riddell, M.C., S.L. Partington, N. Stupka, D. Armstrong, C. Rennie, and M.A. Tarnopolsky. 2003. Substrate utilization during exercise performed with and without glucose ingestion in female and male endurance trained athletes. Int J Sport Nutr Exerc Metab 13:407–421.

Riedesel, M.L., D.Y. Allen, G.T. Peake, and K. Al Qattan. 1987. Hyperhydration with glycerol solutions. J Appl Physiol 63:2262–2268.

Rietjens, G., H. Kuipers, J. Adam, W. Saris, E. Van Breda, D. Van Hamont, and H. Keizer. 2005. Physiological, biochemical and psychological markers of strenuous training-induced fatigue. Int J Sports Med 26:16–26.

Rinard, J., P.M. Clarkson, L.L. Smith, and M. Grossman. 2000. Response of males and females to high-force eccentric exercise. *J Sports Sci* 18:229–236.

Robach, P., D. Biou, J.P. Herry, D. Deberne, M. Letournel, J. Vaysse, and J.P. Richalet. 1997. Recovery processes after repeated supramaximal exercise at the altitude of 4,350 m. *J Appl Physiol* 82(6):1897–1904.

Robazza, C., M. Pellizzari, M. Bertollo, and Y.L. Hanin. 2008. Functional impact of emotions on athletic performance: Comparing the IZOF model and the directional perception approach. *J Sports Sci* 26:1033–1047.

Roberts, A.C., G.E. Butterfield, A. Cymerman, J.T. Reeves, E.E. Wolfel, and G.A. Brooks. 1996. Acclimatization to 4,300-m altitude decreases reliance on fat as a substrate. *J Appl Physiol* 81(4):1762–1768.

Roberts, A.D., R. Billeter, and H. Howald. 1982. Anaerobic muscle enzyme changes after interval training. *Int J Sports Med* 3:18–21.

Roberts, G.C. 2001. *Advances in motivation in sport and exercise*. Champaign, IL: Human Kinetics.

Robertson, A., J.M. Watt, and S.D. Galloway. 2004. Effects of leg massage on recovery from high intensity cycling exercise. *Br J Sports Med* 38:173–176.

Robey, E., B. Dawson, C. Goodman, and J. Beilby. 2009. Effect of postexercise recovery procedures following strenuous stair-climb running. *Res Sports Med* 17(4):245–259.

Rodriguez, N.R., N.M. DiMarco, and S. Langley. 2009. Position of the American Dietetic Association, Dietitians of Canada, and the American College of Sports Medicine: Nutrition and athletic performance. *J Am Diet Assoc* 109:509–527.

Roehrs, T., and T. Roth. 2008. Caffeine: Sleep and daytime sleepiness. *Sleep Med Rev* 12:153–162.

Roehrs, T., J. Yoon, and T. Roth. 1991. Nocturnal and next day effect of ethanol and basal level of sleepiness. *Hum Psychopharmacol* 6:307–311.

Roepstorff, C., C.H. Steffensen, M. Madsen, B. Stallknecht, I.L. Kanstrup, E.A. Richter, and B. Kiens. 2002. Gender differences in substrate utilization during submaximal exercise in endurance-trained subjects. *Am J Physiol Endocrinol Metab* 282:E435–E447.

Roky, R., F. Chapotot, F. Hakkou, M.T. Benchekroun, and A. Buguet. 2001. Sleep during Ramadan intermittent fasting. *J Sleep Res* 10:319–327.

Romine, I.J., A.M. Bush, and C.R. Geist. 1999. Lavender aromatherapy in recovery from exercise. *Percept Mot Skills* 88:756–758.

Romo, M. 1976. Heart-attacks and the sauna. *Lancet* 9;2(7989):809.

Ronglan, L.T., T. Raastad, and A. Børgesen. 2006. Neuromuscular fatigue and recovery in elite female handball players. *Scand J Med Sci Sports* 16:267–273.

Rønsen, O., E. Børsheim, R. Bahr, B. Klarlund Pedersen, E. Haug, J. Kjeldsen-Kragh, and A.T. Høstmark. 2004. Immunoendocrine and metabolic responses to long distance ski racing in world-class male and female cross-country skiers. *Scand J Med Sci Sports* 14(1):39–48.

Roose, J., W.R. de Vries, S.L. Schmikli, F.J. Backx, and L.J. van Doornen. 2009. Evaluation and opportunities in overtraining approaches. *Res Q Exerc Sport* 80:756–764.

Roques, C.F. 2003. Agents physiques antalgiques. Données cliniques actuelles. *Annales de Réadaptation et de Médecine Physique* 46:565–577.

Rosekind, M.R. 2008. Sleep medicine: Past lessons, present challenges, and future opportunities. *Sleep Med Rev* 12:249–251.

Ross, A., and M. Leveritt. 2001. Long-term metabolic and skeletal muscle adaptations to short-sprint training: Implications for sprint training and tapering. *Sports Med* 31:1063–1082.

Ross, E.Z., N. Middleton, R. Shave, K. George, and A. Nowicky. 2007. Corticomotor excitability contributes to neuromuscular fatigue following marathon running in man. *Exp Physiol* 92:417–426.

Rousseaux, A. 1990. *Retrouver et conserver sa santé par le sauna*. Paris: Rousseaux.

Rowbottom, D., D. Keast, C. Goodman, and A. Morton. 1995. The haematological, biochemical and immunological profile of athletes suffering from the overtraining syndrome. *Eur J Appl Physiol* 70(6):502–509.

Rowell, L.B., G.L. Brengelmann, and J.A. Murray. 1969. Cardiovascular responses to sustained high skin temperature in resting man. *J Appl Physiol* 27:673–680.

Rowsell, G.J., A.J. Coutts, P. Reaburn, and S. Hill-Haas. 2009. Effects of cold-water immersion on physical performance between successive matches in high-performance junior male soccer players. *J Sports Sci* 27(6):565–573.

———. 2010. Effect of post-match cold-water immersion on subsequent match running performance in junior soccer players during tournament play. *J Sports Sci* 29(1):1–6.

Roy, B.D., K. Luttmer, M.J. Bosman, and M.A. Tarnopolsky. 2002. The influence of post-exercise macronutrient intake on energy balance and protein metabolism in active females participating in endurance training. *Int J Sport Nutr Exerc Metab* 12:172–188.

Rubini, E.C., A.L. Costa, and P.S. Gomes. 2007. The effects of stretching on strength performance. *Sports Med* 37:213–224.

Ruby, B.C., and R.A. Robergs. 1994. Gender differences in substrate utilisation during exercise. *Sports Med* 17:393–410.

Ruiz, D.H., J.W. Myrer, E. Durrant, and G.W. Fellingham. 1993. Cryotherapy and sequential exercise bouts following cryotherapy on concentric and eccentric strength in the quadriceps. *J Athl Train* 28:320–323.

Russ, D.W., I.R. Lanza, D. Rothman, and J.A. Kent-Braun. 2005. Sex differences in glycolysis during brief, intense isometric contractions. *Muscle Nerve* 32:647–655.

Rutkove, S.B. 2001. Effects of temperature on neuromuscular electrophysiology. *Muscle Nerve* 24(7):867–882.

Ryan, R.M., and E.L. Deci. 2001. On happiness and human potentials: A review of research on hedonic and eudaimonic well-being. *Annu Rev Psychol* 52:141–166.

Ryan, R.M., and C.M. Frederick. 1997. On energy, personality and health: Subjective vitality as a dynamic reflection of well-being. *J Pers* 65:529–565.

Rymaszewska, J., D. Ramsey, and S. Chladzinska-Keijna. 2008. Whole-body cryotherapy as adjunct treatment of depressive and anxiety disorders. *Arch Immunol Ther Exp* 56:63–68.

Rymaszewska, J., A. Tulczynski, Z. Zagrobelny, A. Kiejna, and T. Hadrys. 2003. Influence of whole-body cryotherapy on depressive symptoms: Preliminary study. *Acta Neuropsychiatrica* 15:122–128.

Saat, M., Y. Tochihara, N. Hashiguchi, R.G. Sirisinghe, M. Fujita, and C.M. Chou. 2005. Effects of exercise in the heat on thermoregulation of Japanese and Malaysian males. *J Physiol Anthropol Appl Human Sci* 24:267–275.

Saboisky, J., F.E. Marino, D. Kay, and J. Cannon. 2003. Exercise heat stress does not reduce central activation to non-exercised human skeletal muscle. *Exp Physiol* 88:783–790.

Sahlin, K., A. Alvestrand, R. Brandt, and E. Hultman. 1978. Intracellular pH and bicarbonate concentration in human muscle during recovery from exercise. *J Appl Physiol* 45(3):474–480.

Sandberg, M.L., M.K. Sandberg, and J. Dahl. 2007. Blood flow changes in the trapezius muscle and overlying skin following transcutaneous electrical nerve stimulation. *Phys Ther* 87:1047–1055.

Sargent, C., S. Halson, and G.D. Roach. 2013. Sleep or swim? Early-morning training severely restricts the amount of sleep obtained by elite swimmers. *Eur J Sport Sci.*

Sartori, S., and R. Poirrier. 1996. Syndrome affectif saisonnier et photothérapie: Concepts théoriques et applications cliniques. *Encéphale* 22:7–16.

Satoh, T., and Y. Sugawara. 2003. Effects on humans elicited by inhaling the fragrance of essential oils: Sensory test, multi-channel thermometric study and forehead surface potential wave measurement on basil and peppermint. *Annal Sci* 19:139–146.

Sawka, M.N., R. Francesconi, A. Young, and K. Pandolf. 1984. Influence of hydration level and body fluids on exercise performance in the heat. *J Am Med Assoc* 252:1165–1169.

Sawka, M.N., R.R. Gonzalez, and K.B. Pandolf. 1984. Effects of sleep deprivation on thermoregulation during exercise. *Am J Physiol* 246:R72–R77.

Sawka, M.N., and K. Pandolf. 1990. Effects of body water loss on physiological function and exercise performance. In *Perspectives in exercise science and sports medicine, Vol 3. Fluid homeostasis during exercise*, ed. C. Gisolfi and D.R. Lamb, 1–38. Carmel, IN: Benchmark Press.

Sawka, M.N., C. Wenger, and K. Pandolf. 1996. Thermoregulatory responses to acute exercise: Heat stress and heat acclimatation. In *Handbook of physiology: Environmental physiology*, ed. M.J. Fregly and C.M. Blatteis, 157–186. New York: Oxford University Press.

Sawyer, P.C., T.L. Uhl, C.G. Mattacola, D.L. Johnson, and J.W. Yates. 2003. Effects of moist heat on hamstring flexibility and muscle temperature. *J Strength Cond Res* 17(2):285–290.

Sayers, S.P., and P.M. Clarkson. 2001. Force recovery after eccentric exercise in males and females. *Eur J Appl Physiol* 84:122–126.

Scanlan, A.T., B.J. Dascombe, P.R. Reaburn, and M. Osborne. 2008. The effects of wearing lower-body compression garments during endurance cycling. *Int J Sports Physiol Perform* 3:424–438.

Schaal, K., M. Tafflet, H. Nassif, V. Thibault, C. Pichard, M. Alcotte, T. Guillet, N. El Helou, G. Berthelot, S. Simon, and J.F. Toussaint. 2011. Psychopathology within high level sport: Gender-based differences and sport-specific patterns. *PLoS ONE* 6(5):e19007.

Schaser, K.D., J.F. Stover, I. Melcher, A. Lauffer, N.P. Haas, H.J. Bail, U. Stockle, G. Puhl, and T.W. Mittlmeier. 2006. Local cooling restores microcirculatory hemodynamics after closed soft-tissue trauma in rats. *J Trauma* 61:642–649.

Schmidt-Nielsen, K. 1995. *Animal physiology: Adaptation and environment*. Cambridge, UK: Cambridge University Press.

Schniepp, J., T.S. Campbell, K.I. Powell, and D.M. Pincivero. 2002. The effects of cold-water immersion on power output and heart rate in elite cyclists. *J Strength Cond Res* 16(4):561–566.

Schoene, R.B. 1997. Control of breathing at high altitude. *Respiration* 64:407–415.

Scott, J.P., L.R. McNaughton, and R.C. Polman. 2006. Effects of sleep deprivation and exercise on cognitive, motor performance and mood. *Physiol Behav* 87:396–408.

Seggar, J.F., D.M. Pedersen, N.R. Hawkes, and C. McGown. 1997. A measure of stress for athletic performance. *Percept Mot Skills* 84:227–236.

Seiler, S., O. Haugen, and E. Kuffel. 2007. Autonomic recovery after exercise in trained athletes: Intensity and duration effects. *Med Sci Sports Exerc* 39:1366–1373.

Seligman, M.E.P., and M. Csikszentmihalyi. 2000. Positive psychology: An introduction. *Am Psychol* 55:5–14.

Sellwood, K.L., P. Brukner, D. Williams, A. Nicol, and R. Hinman. 2007. Ice-water immersion and delayed-onset muscle soreness: A randomised controlled trial. *Br J Sports Med* 41(6):392–397.

Selye, H. 1936. A syndrome produced by diverse nocuous agents. *Nature* 138:32.

Sen, C.K. 1995. Oxidants and antioxidants in exercise. *J Appl Physiol* 79:675–686.

Seo, J.Y. 2009. [The effects of aromatherapy on stress and stress responses in adolescents]. *J Korean Acad Nurs* 39:357–365.

Sewright, K.A., M.J. Hubal, A. Kearns, M.T. Holbrook, and P.M. Clarkson. 2008. Sex differences in response to maximal eccentric exercise. *Med Sci Sports Exerc* 40:242–251.

Shephard, R.J. 2000. Exercise and training in women, Part I: Influence of gender on exercise and training responses. *Can J Appl Physiol* 25:19–34.

Sherman, W.M., L.E. Armstrong, T.M. Murray, F.C. Hagerman, D.L. Costill, R.C. Staron, and J.L. Ivy. 1984. Effect of a 42.2-km footrace and subsequent rest or exercise on muscular strength and work capacity. *J Appl Physiol* 57:1668–1673.

Sherman, W.M., and D.R. Lamb. 1988. Nutrition and prolonged exercise. In *Perspectives in exercise science and sports medicine, Vol. 1: Prolonged exercise,* ed. D.R. Lamb and R. Murray, 213–280. Indianapolis: Benchmark Press.

Sherry, J.E., K.M. Oehrlein, K.S. Hegge, and B.J. Morgan. 2001. Effect of burst-mode transcutaneous electrical nerve stimulation on peripheral vascular resistance. *Phys Ther* 81:1183–1191.

Shilo, L., H. Sabbah, R. Hadari, S. Kovatz, U. Weinberg, S. Dolev, Y. Dagan, and L. Shenkman. 2002. The effects of coffee consumption on sleep and melatonin secretion. *Sleep Med* 3:271–273.

Shintani, F., T. Nakaki, S. Kanba, K. Sato, G. Yagi, M. Shiozawa, S. Aiso, R. Kato, and M. Asai. 1995. Involvement of interleukin-1 in immobilisation stress-induced increase in plasma adrenocorticotropic hormones and in release of hypothalamic monoamines in rat. *J Neurosci* 15:1961–1970.

Shirreffs, S.M., M.N. Sawka, and M. Stone. 2006. Water and electrolyte needs for football training and match play. *J Sports Sci* 24:699–707.

Shirreffs, S.M., A.J. Taylor, J.B. Leiper, and R.J. Maughan. 1996. Post exercise rehydration in man: Effects of volume consumed and drink sodium content. *Med Sci Sports Exerc* 28:1260–1271.

Shoemaker, J.K., P.M. Tiidus, and R. Mader. 1997. Failure of manual massage to alter limb blood flow: Measures by Doppler ultrasound. *Med Sci Sports Exerc* 29:610–614.

Shoenfeld, Y., E. Sohar, A. Ohry, and Y. Shapiro. 1976. Heat stress: Comparison of short exposure to severe dry and wet heat in saunas. *Arch Phys Med Rehabil* 57(3):126–129.

Shrier, I. 2004. Does stretching improve performance? A systematic and critical review of the literature. *Clin J Sport Med* 14:267–273.

Shumate, J.B., M.H. Brooke, J.E. Carroll, and J.E. Davis. 1979. Increased serum creatine kinase after exercise: A sex-linked phenomenon. *Neurology* 29:902–904.

Siegel, R., J. Maté, M.B. Brearley, G. Watson, K. Nosaka, and P.B. Laursen. 2010. Ice slurry ingestion increases core temperature capacity and running time in the heat. *Med Sci Sports Exerc* 42:717.

Siems, W., and R. Brenke. 1992. Changes in the glutathione system of erythrocytes due to enhanced formation of oxygen free radicals during short-term whole body cold stimulus. *Arct Med Res* 51:3–9.

Silva, J.M. 1990. An analysis of the training stress syndrome in competitive athletics. *J Appl Sport Psychol* 2:5–20.

Simmons, S.E., T. Mündel, and D.A. Jones. 2008. The effects of passive heating and head cooling on perception of exercise in the heat. *Eur J Appl Physiol* 104:281–288.

Skein, M., R. Duffield, J. Edge, M.J. Short, and T. Mundel. 2011. Intermittent-sprint performance and muscle glycogen following 30 h sleep deprivation. *Med Sci Sports Exerc* 43(7):1301–1311.

Skurvydas, A., S. Kamandulis, A. Stanislovaitis, V. Streckis, G. Mamkus, and A. Drazdauskas. 2008. Leg immersion in warm water, stretch-shortening exercise, and exercise-induced muscle damage. *J Athl Train* 43(6):592–599.

Sleivert, G.G., J.D. Cotter, W.S. Roberts, and M.A. Febbraio. 2001. The influence of whole-body vs. torso pre-cooling on physiological strain and performance of high-intensity exercise in the heat. *Comp Biochem Physiol A Mol Integr Physiol* 128:657–666.

Sluka, K.A., and D. Walsh. 2003. Transcutaneous electrical nerve stimulation: Basic science mechanisms and clinical effectiveness. *J Pain* 4:109–121.

Smith, D.J. 2003. A framework for understanding the training process leading to elite performance. *Sports Med* 33(15):1103–1126.

Smith, L.L., M.N. Keating, D. Holbert, D.J. Spratt, M.R. McCammon, S.S. Smith, and R.G. Israel. 1994. The effects of athletic massage on delayed onset muscle soreness, creatine kinase, and neutrophil count: A preliminary report. *J Orthop Sports Phys Ther* 19:93–99.

Smith, M., L. Robinson, J. Saisan, and R. Segal. 2011. How to sleep better: Tips for getting a good night's sleep. www.helpguide.org/life/sleep_tips.htm.

Smith, R.E. 1986. Toward a cognitive-affective model of athletic burnout. *J Sport Psychol* 8:36–50.

Smolander, J., J. Leppäluoto, T. Westerlund, J. Oksa, B. Dugué, M. Mikkelson, and A. Ruokonen. 2009. Effects of repeated whole-body cold exposures on serum concentrations of growth hormone, thyrotropin, prolactin and thyroid hormones in healthy women. *Cryobiology* 58:275–278.

Snellen, J., and D. Mitchell. 1972. Calorimetric analysis of the effect of drinking saline solution in whole-body sweating. II. Response to different volumes, salinities and temperature. *Pflügers Arch* 331:134–144.

Snow, R.J., M.J. McKenna, S.E. Selig, J. Kemp, C.G. Stathis, and S. Zhao. 1998. Effect of creatine supplementation on sprint exercise performance and muscle metabolism. *J Appl Physiol* 84:1667–1673.

Snyder, A.C. 1998. Overtraining and glycogen depletion hypothesis. *Med Sci Sports Exerc* 30:1146–1150.

Snyder, A., A. Jeukendrup, M. Hesselink, H. Kuipers, and C. Foster. 1993. A physiological/psychological indicator of overreaching during intensive training. *Int J Sports Med* 14:29–32.

Sparling, P.B. 1995. Expected environmental conditions for the 1996 summer Olympic Games in Atlanta. *Clin J Sport Med* 5:220–222.

———. 1997. Editorial: Environmental conditions during the 1996 Olympic Games—A brief follow-up report. *Clin J Sports Med* 7:159–161.

Speechly, D.P., S.R. Taylor, and G.G. Rogers. 1996. Differences in ultra-endurance exercise in performance-matched male and female runners. *Med Sci Sports Exerc* 28:359–365.

Spencer, M., D. Bishop, B. Dawson, and C. Goodman. 2005. Physiological and metabolic responses of repeated-sprint activities specific to field-based team sports. *Sports Med* 35:1025–1044.

Spencer, M., D. Bishop, B. Dawson, C. Goodman, and R. Duffield. 2006. Metabolism and performance in repeated cycle sprints: Active versus passive recovery. *Med Sci Sports Exerc* 38:1492–1499.

Spencer, M., B. Dawson, C. Goodman, B. Dascombe, and D. Bishop. 2008. Performance and metabolism in repeated sprint exercise: Effect of recovery intensity. *Eur J Appl Physiol* 103:545–552.

Stacey, D.L., M.J. Gibala, K.A. Martin Ginis, and B.W. Timmons. 2010. Effects of recovery method on performance, immune changes, and psychological outcomes. *J Orthop Sports Phys Ther* 40(10):656–665.

Stahl, M.L., W.C. Orr, and C. Bollinger. 1983. Postprandial sleepiness: Objective documentation via polysomnography. *Sleep* 6:29–35.

Stamford, B.A., A. Weltman, R. Moffatt, and S. Sady. 1981. Exercise recovery above and below anaerobic threshold following maximal work. *J Appl Physiol* 51:840–844.

Stampi, C., J. Mullington, M. Rivers, J.P. Campos, and R. Broughton. 1990. Ultrashort sleep schedules: Sleep architecture and recuperative value of 80-, 50- and 20-min naps. In *Sleep'90*, ed. J.A. Horne, 71–74. Bochum, Germany: Pontenagel Press.

Standage, M., D.C. Treasure, J.L. Duda, and K.A. Prusak. 2003. Validity, reliability, and invariance of the situational motivation scale (SIMS) across diverse physical activity contexts. *J Sport Exer Psychol* 25:19–43.

Steffensen, C.H., C. Roepstorff, M. Madsen, and B. Kiens. 2002. Myocellular triacylglycerol breakdown in females but not in males during exercise. *Am J Physiol Endocrinol Metab* 282:E634–E642.

Stein, M.D., and P.D. Friedmann. 2005. Disturbed sleep and its relationship to alcohol use. *Subst Abus* 26:1–13.

Steinacker, J.M., and M. Lehmann. 2002. Clinical findings and mechanisms of stress and recovery in athletes. In *Enhancing recovery: Preventing underperformance in athletes,* ed. M. Kellmann, 103–118. Champaign, IL: Human Kinetics.

Steinacker, J.M., W. Lormes, Y. Liu, A. Opitz-Gress, B. Baller, K. Günther, U. Gastmann, K.G. Petersen, M. Lehmann, and D. Altenburg. 2000. Training of junior rowers before world championships. Effects on performance, mood state and selected hormonal and metabolic responses. *J Phys Fitness Sports Med* 40:327–335.

Steinacker, J.M., W. Lormes, S. Reissnecker, and Y. Liu. 2004. New aspects of the hormone and cytokine response to training. *Eur J Appl Physiol* 91:382–393.

Stone, M.H., H.S. O'Bryant, B.K. Schilling, R.L. Johnson, K.C. Pierce, G. Haff, and M. Stone. 1999. Periodization: Effects of manipulating volume and intensity. Part 1. *Strength Cond J* 21(2):56–62.

Stupka, N., S. Lowther, K. Chorneyko, J.M. Bourgeois, C. Hogben, and M.A. Tarnopolsky. 2000. Gender differences in muscle inflammation after eccentric exercise. *J Appl Physiol* 89:2325–2332.

Stupka, N., and P.M. Tiidus. 2001. Effects of ovariectomy and estrogen on ischemia-reperfusion injury in hindlimbs of female rats. *J Appl Physiol* 91:1828–1835.

Suhonen, O. 1983. Sudden coronary death in middle age and characteristics of its victims in Finland. A prospective population study. *Acta Med Scand* 214(3):207–214.

Sullivan, S.J., L.R. Williams, D.E. Seaborne, and M. Morelli. 1991. Effects of massage on alpha motoneuron excitability. *Phys Ther* 71:555–560.

Sutton, J.R., N.L. Jones, and C.J. Toews. 1981. Effect of pH on muscle glycolysis during exercise. *Clin Sci* 61:331–338.

Suzuki, K., J. Peake, K. Nosaka, M. Okutsu, C.R. Abbiss, R. Surriano, D. Bishop, M.J. Quod, H. Lee, D.T. Martin, and P.B. Laursen. 2006. Changes in markers of muscle damage, inflammation and HSP70 after an Ironman triathlon race. *Eur J Appl Physiol* 98:525–534.

Suzuki, M., T. Umeda, S. Nakaji, T. Shimoyama, T. Mashiko, and K. Sugawara. 2004. Effect of incorporating low intensity exercise into the recovery period after a rugby match. *Br J Sports Med* 38:436–440.

Swank, A., and R.J. Robertson. 1989. Effect of induced alkalosis on perception of exertion during intermittent exercise. *J Appl Physiol* 67:1862–1867.

Swart, J., and C. Jennings. 2004. Use of blood lactate concentration as a marker of training status. *S Afr J Sports Med* 16(3):3–7.

Sweet, T.W., C. Foster, M.R. McGuigan, and G. Brice. 2004. Quantitation of resistance training using the session rating of perceived exertion method. *J Strength Cond Res* 18(4):796–802.

Swenson, C., L. Sward, and J. Karlsson. 1996. Cryotherapy in sports medicine. *Scand J Med Sci Sports* 6:193–200.

Symons, J.D., T. VanHelder, and W.S. Myles. 1988. Physical performance and physiological responses following

60 hours of sleep deprivation. *Med Sci Sports Exerc* 20:374–380.

Taggart, P., P. Parkinson, and M. Carruthers. 1972. Cardiac responses to thermal, physical, and emotional stress. *Br Med J* 3:71–76.

Takahashi, M., A. Nakata, T. Haratani, Y. Ogawa, and H. Arito. 2004. Post-lunch nap as a worksite intervention to promote alertness on the job. *Ergonomics* 47:1003–1013.

Takahashi, T., and Y. Miyamoto. 1998. Influence of light physical activity on cardiac responses during recovery from exercise in humans. *Eur J Appl Physiol Occup Physiol* 77:305–311.

Takase, B., T. Akima, K. Satomura, F. Ohsuzu, T. Mastui, M. Ishihara, and A. Kurita. 2004. Effects of chronic sleep deprivation on autonomic activity by examining heart rate variability, plasma catecholamine, and intracellular magnesium levels. *Biomed Pharmacother* 58(1):S35–S39.

Taoutaou, Z., P. Granier, B. Mercier, J. Mercier, S. Ahmaidi, and C. Prefaut. 1996. Lactate kinetics during passive and partially active recovery in endurance and sprint athletes. *Eur J Appl Physiol Occup Physiol* 73:465–470.

Tarnopolsky, L.J., J.D. MacDougall, S.A. Atkinson, M.A. Tarnopolsky, and J.R. Sutton. 1990. Gender differences in substrate for endurance exercise. *J Appl Physiol* 68:302–308.

Tarnopolsky, M.A. 2000. Gender differences in metabolism, nutrition and supplements. *J Sci Med Sport* 3:287–298.

———. 2008. Sex differences in exercise metabolism and the role of 17-beta estradiol. *Med Sci Sports Exerc* 40:648–654.

Tarnopolsky, M.A., M. Bosman, J.R. MacDonald, D. Vandeputte, J. Martin, and B.D. Roy. 1997. Postexercise protein-carbohydrate supplements increase muscle glycogen in men and women. *J Appl Physiol* 83:1877–1883.

Tarnopolsky, M.A., M. Gibala, A.E. Jeukendrup, and S.M. Philips. 2005. Nutritional needs of elite endurance athletes. Part I: Carbohydrate and fluid requirements. *Eur J Sport Sci* 33:117–144.

Tarnopolsky, M.A., J.D. MacDougall, and S.A. Atkinson. 1988. Influence of protein intake and training status on nitrogen balance and lean body mass. *J Appl Physiol* 64:187–193.

Tarnopolsky, M.A., C. Zawada, L.B. Richmond, S. Carter, J. Shearer, T. Graham, and S.M. Phillips. 2001. Gender differences in carbohydrate loading are related to energy intake. *J Appl Physiol* 91:225–230.

Tatár, P., M. Vigas, J. Jurcovicová, R. Kvetnanský, and V. Strec. 1986. Increased glucagon secretion during hyperthermia in a sauna. *Eur J Appl Physiol Occup Physiol* 55(3):315–317.

Taylor, D.C., J.D. Dalton Jr., A.V. Seaber, and W.E. Garrett Jr. 1990. Viscoelastic properties of muscle-tendon units. The biomechanical effects of stretching. *Am J Sports Med* 18:300–309.

Taylor, S.R., G.G. Rogers, and H.S. Driver. 1997. Effects of training volume on sleep, psychological, and selected physiological profiles of elite female swimmers. *Med Sci Sports Exerc* 29:688–693.

Terrados, S.N. 1992. Altitude training and muscular metabolism. *Int Sports Med* 13:206–209.

Terry, L. 1985. *Stretching and muscle soreness. An investigation into the effects of hold-relax stretch on delayed post-exercise soreness in the quadriceps muscle: A pilot study (Thesis).* Adelaide, Australia: South Australian Institute of Technology.

Terry, L. 1987. Stretching and muscle soreness: An investigation into the effects of a hold-relax stretch on delayed post-exercise soreness in the quadriceps muscle: A pilot study. *Aust J Physiother* 33(1):69.

Tessitore, A., R. Meeusen, C. Cortis, and L. Capranica. 2007. Effects of different recovery interventions on anaerobic performances following preseason soccer training. *J Strength Cond Res* 21(3):745–750.

Tessitore, A., R. Meeusen, R. Pagano, C. Benvenuti, M. Tiberi, and L. Capranica. 2008. Effectiveness of active versus passive recovery strategies after futsal games. *J Strength Cond Res* 22:1402–1412.

Thevenet, D., M. Tardieu-Berger, S. Berthoin, and J. Prioux. 2007. Influence of recovery mode (passive vs. active) on time spent at maximal oxygen uptake during an intermittent session in young and endurance-trained athletes. *Eur J Appl Physiol* 99:133–142.

Thiriet, P., D. Gozal, D. Wouassi, T. Oumarou, H. Gelas, and J.R. Lacour. 1993. The effect of various recovery modalities on subsequent performance, in consecutive supramaximal exercise. *J Sports Med Phys Fitness* 33:118–129.

Thomas, L., I. Mujika, and T. Busso. 2008. A model study of optimal training reduction during pre-event taper in elite swimmers. *J Sport Sci* 26:643–652.

———. 2009. Computer simulations assessing the potential performance benefit of a final increase in training during pre-event taper. *J Strength Cond Res* 23:1729–1736.

Thomas, M.M., S.S. Cheung, G.C. Elder, and G.G. Sleivert. 2006. Voluntary muscle activation is impaired by core temperature rather than local muscle temperature. *J Appl Physiol* 100:1361–1369.

Thompson, E.R. 2007. Development and validation of an internationally reliable short-form of the positive and negative affect schedule (PANAS). *J Cross Cult Psychol* 38:227–242.

Thorell, A., M.F. Hirshman, J. Nygren, L. Jorfeldt, J.F.P. Wojtaszewski, and S.D. Dufresne. 1999. Exercise and insulin cause GLUT-4 translocation in human skeletal muscle. *Am J Physiol* 277:E733–E741.

Tiidus, P.M., K.A. Dawson, L. Dawson, A. Roefs, and E. Bombardier. 2004. Massage does not influence muscle soreness or strength recovery following a half-marathon. *Med Sci Sport Exer* 36:S15–S16.

Tiidus, P.M., M. Deller, E. Bombardier, M. Gul, and X.L. Liu. 2005. Estrogen supplementation failed to attenuate biochemical indices of neutrophil infiltration or damage in rat skeletal muscles following ischemia. *Biol Res* 38:213–223.

Tiidus, P.M., D. Holden, E. Bombardier, S. Zajchowski, D. Enns, and A. Belcastro. 2001. Estrogen effect on post-exercise skeletal muscle neutrophil infiltration and cal-pain activity. *Can J Physiol Pharmacol* 79:400–406.

Tipton, C.M. 1997. Sports medicine: A century of progress. *J Nutr* 127(5):S878–S885.

Tipton, K.D., E. Borsheim, A.P. Sanford, S.E. Wolf, and R.R. Wolfe. 2003. Acute response of net muscle protein balance reflects 24-h balance after exercise and amino acid ingestion. *Am J Physiol Endocrinol Metab* 284:E76–E89.

Tipton, K.D., B.B. Rasmussen, S.L. Miller, S.E. Wolf, S.K. Owens-Stovall, and B.E. Petrini. 2001. Timing of amino acid carbohydrate ingestion alters response of muscle to resistance exercise. *Am J Physiol Endocrinol Metab* 281:E177–E206.

Tochikubo, O., S. Ri, and N. Kura. 2006. Effects of pulse synchronized massage with air cuffs on peripheral blood flow and autonomic nervous system. *Circ J* 70:1159–1163.

Todd, G., J.E. Butler, J.L. Taylor, and S.C. Gandevia. 2005. Hyperthermia: A failure of the motor cortex and the muscle. *J Physiol* 1:621–631.

Tong, K.C., S.K. Lo, and G.L. Cheing. 2007. Alternating frequencies of transcutaneous electric nerve stimulation: Does it produce greater analgesic effects on mechanical and thermal pain thresholds? *Arch Phys Med Rehabil* 88:1344–1349.

Toyokawa, H., Y. Matsui, J. Uhara, H. Tsuchiya, S. Teshima, H. Nakanishi, A.H. Kwon, Y. Azuma, T. Nagaoka, T. Ogawa, and Y. Kamiyama. 2003. Promotive effects of far-infrared ray on full-thickness skin wound healing in rats. *Exp Biol Med* 228(6):724–729.

Tremblay, F., L. Estephan, M. Legendre, and S. Sulpher. 2001. Influence of local cooling on proprioceptive acuity in the quadriceps muscle. *J Athl Train* 36:119–123.

Trenell, M.I., N.S. Marshall, and N.L. Rogers. 2007. Sleep and metabolic control: Waking to a problem? *Clin Exp Pharmacol Physiol* 34:1–9.

Turner, A. 2011. The science and practice of periodization: A brief review. *Strength Cond J* 33(1):12.

Turner, R.W. 1980. Fats and heart disease: Points for controversy. *Nurs Times* 76(50):2189–2190.

Uckert, S., and W. Joch. 2007. Effects of warm-up and precooling on endurance performance in the heat. *Br J Sports Med* 41:380–384.

Udagawa, Y., and H. Nagasawa. 2000. Effects of far-infrared ray on reproduction, growth, behaviour and some physiological parameters in mice. *In Vivo* 14(2):321–326.

Uhde, T.W., B.M. Cortese, and A. Vedeniapin. 2009. Anxiety and sleep problems: Emerging concepts and theoretical treatment implications. *Curr Psychiatry Rep* 11:269–276.

Urhausen, A., H. Gabriel, and W. Kindermann. 1995. Blood hormones as markers of training stress and overtraining. *Sports Med* 20:251–276.

———. 1998. Impaired pituitary hormonal response to exhaustive exercise in overtrained endurance athletes. *Med Sci Sports Exer* 30:407–414.

Urhausen, A., H. Gabriel, B. Weiler, and W. Kindermann. 1998. Ergometric and psychological findings during overtraining: A long-term follow-up study in endurance athletes. *Int J Sports Med* 19:114–120.

Urhausen, A., and W. Kindermann. 2002. Diagnosis of overtraining: What tools do we have? *Sports Med* 32:95–102.

Uusitalo, A.L.T. 2001. Overtraining. Making a difficult diagnosis and implementing targeted treatment. *Physician Sportsmed* 29:35–50.

Uusitalo, A.L., A.J. Uusitalo, and H.K. Rusko. 2000. Heart rate and blood pressure variability during heavy training and overtraining in the female athlete. *Int J Sports Med* 21:45–53.

Uusitalo, A.L.T., M. Valkonen-Korhonen, P. Helenius, E. Vanninen, K. Bergstrom, and J. Kuikka. 2004. Abnormal serotonin reuptake in an overtrained insomnic and depressed team athlete. *Int J Sports Med* 25:150–153.

Vaile, J.M., N.D. Gill, and A.J. Blazevich. 2007. The effect of contrast water therapy on symptoms of delayed onset muscle soreness. *J Strength Cond Res* 21(3):697–702.

Vaile, J., S. Halson, N. Gill, and B. Dawson. 2008a. Effect of cold water immersion on repeat cycling performance and thermoregulation in the heat. *J Sports Sci* 26:431–440.

———. 2008b. Effect of hydrotherapy on recovery from fatigue. *Int J Sports Med* 29(7):539–544.

———. 2008c. Effect of hydrotherapy on the signs and symptoms of delayed onset muscle soreness. *Eur J Appl Physiol* 102(4):447–455.

Vaile, J., C. O'Hagan, B. Stefanovic, M. Walker, N. Gill, and C.D. Askew. 2011. Effect of cold water immersion on repeated cycling performance and limb blood flow. *Br J Sports Med* 45:825–829.

Vanderthommen, M., K. Soltani, D. Maquet, J.M. Crielaard, and J.L. Croisier. 2007. Does neuromuscular electrical stimulation influence muscle recovery after maximal isokinetic exercise? *Isokinet Exerc Sci* 15:143–149.

Van Dongen, H.P., and D.F. Dinges. 2005. Sleep, circadian rhythms, and psychomotor vigilance. *Clin Sports Med* 24:237–249, vii–viii.

van Hall, G., J.A.L. Calbet, H. Søndergaard, and B. Saltin. 2001. The re-establishment of the normal blood lactate response to exercise in humans after prolonged acclimatization to altitude. *J Physiol* 536:963–975.

van Hall, G., S.M. Shirreffs, and J.A.L. Calbert. 2000. Muscle glycogen resynthesis during recovery from cycle exercise: No effect of additional protein ingestion. *J Appl Physiol* 88:1631–1636.

VanHelder, T., and M.W. Radomski. 1989. Sleep deprivation and the effect on exercise performance. *Sports Med* 7:235–247.

van Loon, L.J.C., W.H.M. Saris, M. Kruijshoop, and A.J.M. Wagenmakers. 2000. Maximizing post-exercise muscle glycogen synthesis: Carbohydrate supplementation and the application of amino acid or protein hydrolysate mixture. *Am J Clin Nutr* 72:106–111.

Verducci, F.M. 2000. Interval cryotherapy decreases fatigue during repeated weight lifting. *J Athl Train* 35:422–426.

Versey, N., S. Halson, and B. Dawson. 2010. Effect of contrast water therapy duration on recovery of cycling performance: A dose-response study. *Eur J Appl Physiol* 111(1):37–46.

Vgontzas, A.N., D.A. Papanicolaou, E.O. Bixler, A. Kales, K. Tyson, and G.P. Chrousos. 1997. Elevation of plasma cytokines in disorders of excessive daytime sleepiness: Role of sleep disturbance and obesity. *J Clin Endocrinol Metab* 82:1313–1316.

Vogel, J.A., L.H. Hartley, J.C. Cruz, and R.P. Hogan. 1974. Cardiac output during exercise in sea level residents at sea level and high altitude. *J Appl Physiol* 36:169–172.

Vogt, M., A. Puntschart, H. Howald, B. Mueller, C. Mannhart, L. Gfeller-Tuescher, P. Mullis, and H. Hoppeler. 2003. Effects of dietary fat on muscle substrates, metabolism, and performance in athletes. *Med Sci Sports Exerc* 35:952–960.

Volek, J.S., C.E. Forsythe, and W.J. Kraemer. 2006. Nutritional aspects of women strength athletes. *Br J Sports Med* 40:742–748.

Volek, J.S., and M.J. Sharman. 2005. Diet and hormonal responses: Potential impact on body composition. In *The endocrine system in sports and exercise*, ed. W.J. Kraemer and A.D. Rogol, 426–443. Oxford, UK: Blackwell.

Vollestad, N.K., and O.M. Sejersted. 1988. Biochemical correlates of fatigue. *Eur J Appl Physiol* 25:960–965.

Von Duvillard, S.P., W.A. Braun, M. Markofski, R. Beneke, and R. Leithauser. 2004. Fluids and hydration in prolonged endurance performance. *Nutrition* 20:651–656.

Vuori, H., H. Urponen, and T. Peltonen. 1978. The prevalence of chronic disease in children in Finland. *Public Health* 92(6):272–278.

Wallas, C.H., J.R. Warren, and M.M. Kowalski. 1979. Energy metabolism and Na+, K+ redistribution in human erythrocytes treated with lipopolysaccharide endotoxin. *Proc Soc Exp Biol Med* 161(3):255–259.

Walsh, N.P., M. Gleeson, D.B. Pyne, D.C. Nieman, F.S. Dhabhar, R.J. Shephard, S.J. Oliver, S. Bermon, and A. Kajeniene. 2011. Position statement. Part two: Maintaining immune health. *Exerc Immunol Rev* 17:64–103.

Walsh, N.P., M. Gleeson, R.J. Shephard, J.A. Woods, N.C. Bishop, M. Fleshner, C. Green, B.K. Pedersen, L. Hoffman-Goetz, C.J. Rogers, H. Northoff, A. Abbasi, and P. Simon. 2011. Position statement. Part one: Immune function and exercise. *Exerc Immunol Rev* 17:6–63.

Walsh, R.M., T.D. Noakes, J.A. Hawley, and S.C. Dennis. 1994. Impaired high-intensity cycling performance time at low levels of dehydration. *Int J Sports Med* 15:392–398.

Walter, G., K. Vandenborne, K.K. McCully, and L.S. Leigh. 1997. Noninvasive measurement of phosphocreatine recovery kinetics in single human muscles. *Am J Physiol* 272:C525–C534.

Walters, P.H. 2002. Sleep, the athlete, and performance. *Strength Cond J* 24:17–24.

Ward-Smith, A.J., and P.F. Radford. 2000. Investigation of the kinetics of anaerobic metabolism by analysis of the performance of elite sprinters. *J Biomech* 33:997–1004.

Wassinger, C.A., J.B. Myers, J.M. Gatti, K.M. Conley, and S.M. Lephart. 2007. Proprioception and throwing accuracy in the dominant shoulder after cryotherapy. *J Athl Train* 42:84–89.

Waterhouse, J., G. Atkinson, B. Edwards, and T. Reilly. 2007. The role of a short post-lunch nap in improving cognitive, motor, and sprint performance in participants with partial sleep deprivation. *J Sports Sci* 25:1557–1566.

Waterhouse, J., B. Drust, D. Weinert, B. Edwards, W. Gregson, G. Atkinson, S. Kao, S. Aizawa, and T. Reilly. 2005. The circadian rhythm of core temperature: Origin and some implications for exercise performance. *Chronobiol Int* 22:207–225.

Waterhouse, J., S. Folkard, H. Van Dongen, D. Minors, D. Owens, G. Kerkhof, D. Weinert, A. Nevill, I. Macdonald, N. Sytnik, and P. Tucker. 2001. Temperature profiles, and the effect of sleep on them, in relation to morningness-eveningness in healthy female subjects. *Chronobiol Int* 18:227–247.

Watson, D., L.A. Clark, and A. Tellegen. 1988. Development and validation of brief measures of positive and negative affect: The PANAS scales. *J Pers Soc Psychol* 54:1063–1070.

Watts, P.B., M. Daggett, P. Gallagher, and B. Wilkins. 2000. Metabolic response during sport rock climbing and the effects of active versus passive recovery. *Int J Sports Med* 21:185–190.

Waylonis, G.W. 1967. The physiologic effects of ice massage. *Arch Phys Med Rehabil* 48:37–42.

Weber, S.T., and E. Heuberger. 2008. The impact of natural odors on affective states in humans. *Chem Senses* 33:441–447.

Webster, J., E.J. Holland, G. Sleivert, R.M. Laing, and B.E. Niven. 2005. A light-weight cooling vest enhances performance of athletes in the heat. *Ergonomics* 48:821–837.

Weerapong, P., P.A. Hume, and G.S. Kolt. 2005. The mechanisms of massage and effects on performance, muscle recovery and injury prevention. *Sports Med* 35:235–256.

Weinberg, R., and A. Jackson. 1988. The relationship of massage and exercise to mood enhancement. *Sport Psychol* 2:202–211.

Weitzmann, M.N., and R. Pacifici. 2006. Estrogen deficiency and bone loss: An inflammatory tale. *J Clin Invest* 116:1186–1194.

Wells, A.S., N.W. Read, K. Uvnas-Moberg, and P. Alster. 1997. Influences of fat and carbohydrate on postprandial sleepiness, mood, and hormones. *Physiol Behav* 61:679–686.

Wendt, D., L.J. van Loon, and W.D. Lichtenbelt. 2007. Thermoregulation during exercise in the heat: Strategies for maintaining health and performance. *Sports Med* 37:669–682.

Weppler, C.H., and S.P. Magnusson. 2010. Increasing muscle extensibility: A matter of increasing length or modifying sensation? *Phys Ther* 90(3):438–449.

Wessel, J., and A. Wan. 1994. Effect of stretching on the intensity of delayed onset muscle soreness. *Clin J Sport Med* 4:83–87.

White, S.C., C.A. Berry, and R.R. Hessberg. 1972. Effects of weightlessness on astronauts—A summary. *Adv Space Res* 10:47–55.

Whitehead, N.P., J.E. Gregory, D.L. Morgan, and U. Proske. 2001. Passive mechanical properties of the medial gastrocnemius muscle of the cat. *J Physiol* 536(3):893–903.

Whitelaw, G.P., O.J. Oladipo, B.P. Shah, K.A. DeMuth, J. Coffman, and D. Segal. 2001. Evaluation of intermittent pneumatic compression devices. *Orthopedics* 24:257–261.

Wiemann, K., and M. Kamphövner. 1995. Verhindert statisches Dehnen das Auftreten von Muskelkater nach exentrisxhem Training? *Deutsche Zeitschrift für Sportmedizin* 46:411–421.

Wigernaes, I., A.T. Hostmark, P. Kierulf, and S.B. Stromme. 2000. Active recovery reduces the decrease in circulating white blood cells after exercise. *Int J Sports Med* 21:608–612.

Wigernaes, I., A.T. Hostmark, S.B. Stromme, P. Kierulf, and K. Birkeland. 2001. Active recovery and post-exercise white blood cell count, free fatty acids, and hormones in endurance athletes. *Eur J Appl Physiol* 84:358–366.

Wilcock, I.M., J.B. Cronin, and W.A. Hing. 2006. Physiological response to water immersion: A method for sport recovery? *Sports Med* 36(9):747–765.

Willems, M.E.T., T. Hale, and C.S. Wilkinson. 2009. Effects of manual massage on muscle-specific soreness and single leg jump performance after downhill treadmill walking. *Medicina Sportiva* 13:91–96.

Williams, C. 2004. Carbohydrate intake and recovery from exercise. *Science & Sports* 19:239–244.

Williams, M.B., P.B. Raven, D.L. Fogt, and J.L. Ivy. 2003. Effects of recovery beverages on glycogen restoration and endurance exercise performance. *J Strength Cond* 17:12–19.

Wilson, G.J., A.J. Murphy, and J.F. Pryor. 1994. Musculotendinous stiffness: Its relationship to eccentric, isometric, and concentric performance. *J Appl Physiol* 76:2714–2719.

Wojtaszewski, J.P.F., P. Nielson, B. Kiens, and E.A. Richter. 2001. Regulation of glycogen synthase kinase-3 in human skeletal muscle: Effects of food intake and bicycle exercise. *Diabetes* 50:265–269.

Wolever, T.M.S., Z. Cohen, L.U. Thompson, M.J. Thorne, M.J.A. Jenkins, E.J. Propikchuk, and D.J.A. Jenkins. 1986. Ideal loss of available carbohydrate in man: Comparison of a breath hydrogen method with direct measurement using a human ileostomy model. *Am J Gastro* 81:115–122.

Wong, T.S., and F.W. Booth. 1990. Protein metabolism in rat gastrocnemius muscle after stimulated chronic concentric exercise. *J Appl Physiol* 69:1707–1717.

Wurtman, R.J., J.J. Wurtman, M.M. Regan, J.M. McDermott, R.H. Tsay, and J.J. Breu. 2003. Effects of normal meals rich in carbohydrates or proteins on plasma tryptophan and tyrosine ratios. *Am J Clin Nutr* 77:128–132.

Wüst, R.C., C.I. Morse, A. de Haan, D.A. Jones, and H. Degens. 2008. Sex differences in contractile properties and fatigue resistance of human skeletal muscle. *Exp Physiol* 93:843–850.

Wyndham, C.H. 1974. 1973 Yant Memorial Lecture: Research in the human sciences in the gold mining industry. *Am Ind Hyg Assoc J* 35(3):113–136.

Yackzan, L., C. Adams, and K.T. Francis. 1984. The effects of ice massage on delayed muscle soreness. *Am J Sports Med* 12:159–165.

Yamane, M., H. Teruya, M. Nakano, R. Ogai, N. Ohnishi, and M. Kosaka. 2006. Post-exercise leg and forearm flexor muscle cooling in humans attenuates endurance and resistance training effects on muscle performance and on circulatory adaptation. *Eur J Appl Physiol* 96(5):572–580.

Yamauchi, T. 1989. Whole-body cryotherapy is a method of extreme cold: 175 °C treatment initially used for rheumatoid arthritis. *Z Phys Med Baln Med Klin* 15:311.

Yanagisawa, O., Y. Miyanaga, H. Shiraki, H. Shimojo, N. Mukai, M. Niitsu, and Y. Itai. 2003. The effects of various therapeutic measures on shoulder strength and muscle soreness after baseball pitching. *J Sports Med Phys Fitness* 43:189–201.

Yanagisawa, O., M. Niitsu, H. Takahashi, K. Goto, and Y. Itai. 2003. Evaluations of cooling exercised muscle with MR imaging and 31P MR spectroscopy. *Med Sci Sports Exerc* 35:1517–1523.

Yokota, M., G.P. Bathalon, and L.G. Berglund. 2008. Assessment of male anthropometric trends and the effects on simulated heat stress responses. *Eur J Physiol* 104:297–302.

Yoshida, T. 2002. The rate of phosphocreatine hydrolysis and resynthesis in exercising muscle in humans using 31P-MRS. *J Physiol Anthropol Appl Human Sci* 21:247–255.

Yoshida, T., H. Watari, and K. Tagawa. 1996. Effects of active and passive recoveries on splitting of the inorganic phosphate peak determined by 31P-nuclear magnetic resonance spectroscopy. *NMR Biomed* 9:13–19.

Yu, J.G., D.O. Fürst, and L.E. Thornell. 2003. The mode of myofibril remodelling in human skeletal muscle affected by DOMS induced by eccentric contractions. *Histochem Cell Biol* 119:383–393.

Yu, S.Y., J.H. Chiu, S.D. Yang, Y.C. Hsu, W.Y. Lui, and C.W. Wu. 2006. Biological effect of far-infrared therapy on increasing skin microcirculation in rats. *Photodermatol Photoimmunol Photomed* 22(2):78–86.

Zagrobelny, Z. 2003. *Local and whole body cryotherapy.* Wroclaw, Poland: Wydaw Mictwo Medyczne Urban and Partner.

Zainuddin, Z., M. Newton, P. Sacco, and K. Nosaka. 2005. Effects of massage on delayed-onset muscle soreness, swelling, and recovery of muscle function. *J Athl Train* 40:174–180.

Zawadzki, K.M., B.B. Yaspelkis, and L.L. Ivy. 1992. Carbohydrate-protein complex increases the rate of muscle glycogen storage after exercise. *J Appl Physiol* 72:1854–1859.

Zehnder, M., M. Ith, R. Kreis, W. Saris, U. Boutellier, and C. Boesch. 2005. Gender-specific usage of intramyocellular lipids and glycogen during exercise. *Med Sci Sports Exerc* 37:1517–1524.

Zelikovski, A., C.L. Kaye, G. Fink, S.A. Spitzer, and Y. Shapiro. 1993. The effects of the modified intermittent sequential pneumatic device (MISPD) on exercise performance following an exhaustive exercise bout. *Br J Sports Med* 27:255–259.

Zemke, J.E., J.C. Andersen, W.K. Guion, J. McMillan, and A.B. Joyner. 1998. Intramuscular temperature responses in the human leg to two forms of cryotherapy: Ice massage and ice bag. *J Orthop Sports Phys Ther* 27:301–307.

Zhong, X., H.J. Hilton, G.J. Gates, S. Jelic, Y. Stern, M.N. Bartels, R.E. Demeersman, and R.C. Basner. 2005. Increased sympathetic and decreased parasympathetic cardiovascular modulation in normal humans with acute sleep deprivation. *J Appl Physiol* 98:2024–2032.

Index

NOTE: Page numbers followed by an italicized *f* or *t* indicate a figure or table will be found on that page, respectively. Italicized *ff* or *tt* indicate multiple figures or tables will be found on that page, respectively.

About the Editors

Christophe Hausswirth, PhD, has been the senior physiologist at l'Institut National du Sport, de l'Expertise et de la Performance (National Institute of Sport, Expertise and Performance, or INSEP) since 1995 and is an associate professor and leader in recovery and nutrition guidelines in the research department. In 1996 he earned his PhD in biomechanics and physiology of human movement at the University of Orsay, France. In 2000 he earned his diploma in supervising research dealing with the energy cost of locomotion in long-duration sport events. With the recent creation of the new Sport, Expertise and Performance (SEP) laboratory at INSEP, he is responsible for the research focusing on multi-disciplinary approaches to optimize high-level performance. He also serves on the scientific board of the *International Journal of Sports Physiology and Performance*.

Hausswirth has performed extensive research on the physiological aspects of endurance sport performance by manipulating cadence in several sports. He has published more than 70 articles in peer-reviewed journals, 3 books, and 10 book chapters. He is responsible for running a mission providing clinical counseling and education of athletes, research, student supervision and teaching, development of educational resources, and organization of food service via fact sheets. His research interests include fluid needs for optimal performance; carbohydrate metabolism and performance of exercise in BMX cycling; pacing strategies in triathlon; recovery strategies in synchronized swimming, soccer, and handball players; precooling strategies for exercise in temperate and hot conditions; and postexercise recovery.

Iñigo Mujika, PhD, is an associate professor in the department of physiology, faculty of medicine and odontology, at the University of the Basque Country in Leioa, Spain. As a researcher, sport science practitioner, and coach, Mujika is widely considered one of the most respected experts on tapering and peaking for optimal performance.

Since 1992 Mujika has been devoted to the research of applied sport physiology. He has published 3 books, more than 80 peer-reviewed scientific articles, and 28 book chapters. He has also presented nearly 200 lectures on sport physiology and training at conferences and seminars worldwide.

As a sport physiologist, Mujika works closely with elite athletes and coaches in a variety of individual and team sports. From 2003 to 2004, Mujika was senior physiologist at the Australian Institute of Sport. In 2005, he worked as physiologist and trainer of the professional road bicycle racing team Euskaltel Euskadi. Between 2006 and 2008 Mujika was the head of the department of research and development for the Spanish professional football team Athletic Club Bilbao. In the lead-up to the London 2012 Olympics he was the physiologist of the Spanish swimming team. He is also a coach of world-class triathletes, including Olympians Ainhoa Murua (Athens 2004, Beijing 2008, and London 2012) and Eneko Llanos (Athens 2004).

Mujika serves as associate editor of the *International Journal of Sports Physiology and Performance*. In 1995 he earned a doctoral degree in biology of muscular exercise from the University of Saint-Etienne in France, and in 1999 he earned a second doctorate in physical activity and sport sciences along with an Extraordinary Doctorate Award from the University of the Basque Country in Spain. In 2002 and 2007 he received the National Award for Sport Medicine Research from the University of Oviedo in Spain. He has also received two awards for his work with triathletes: Best Coach of Female Athlete (2006) from the Spanish Triathlon Federation and the High Performance Basque Sport Award (2007) from the Basque Sport Foundation.

Fluent in four languages (Basque, English, French, and Spanish), Mujika has lived in California, France, South Africa, and Australia. He currently resides in the Basque Country, Spain. Mujika enjoys surfing, cycling, swimming, strength training, and hiking, as well as cinema and traveling.